FUNDAMENTAL PRINCIPLES OF SYSTEMS ANALYSIS AND DECISION-MAKING

FUNDAMENTAL PRINCIPLES OF SYSTEMS ANALYSIS AND DECISION-MAKING

Paul J. Ossenbruggen
University of New Hampshire

JOHN WILEY & SONS, INC.
New York Chichester Brisbane Toronto Singapore

ACQUISITIONS EDITOR	Cliff Robichaud
MARKETING MANAGER	Susan Elbe
PRODUCTION EDITOR	Richard Blander
DESIGNER	Maddy Lesure
MANUFACTURING MANAGER	Andrea Price
ILLUSTRATION COORDINATOR	Edward Starr
COVER DESIGN	David Levy

This book was set in Times Roman by Digital Production and the Fine Print Design Company and printed and bound by Malloy Lithographing. The cover was printed by Phoenix Color Corp.

Recognizing the importance of preserving what has been written, it is a policy of John Wiley & Sons, Inc. to have books of enduring value published in the United States printed on acid-free paper, and we exert our best efforts to that end.

The paper in this book was manufactured by a mill whose forest management programs include sustained yield harvesting of its. timberlands. Sustained yield harvesting principles ensure that the number of trees cut each year does not exceed the amount of new growth.

Library of Congress Cataloging in Publication Data:

Ossenbruggen, Paul J. (Paul John)

Fundamental principles of systems analysis and decision-making / by Paul J. Ossenbruggen.

p. cm.

Includes index.

ISBN 0-471-52156-6 (paper)

1. Decision-making. 2. System analysis. I. Title.

QA279.4.087 1993

302.3'011--dc20 93-35998

CIP

Printed in the United States of America

10 9 8 7 6 5 4 3 2

Preface

This book deals with fundamental issues of infrastructure that modern industrialized societies have to address and solve. This is accomplished by using systems analysis and modeling techniques for decision-making. Systems analysis is more than an exercise in mathematics; it offers a logical framework for organizing studies dealing with complex issues that face our society.

The student is required to learn five basic concepts:

- The Systems Approach: How to address complex problems in an orderly and thorough manner.
- Optimization: How to formulate and analyze mathematical models for finding the best solution subject to constraints, such as physical, economic, legal, and other societal restrictions.
- Economic Evaluation: How to effectively measure and compare the performance of alternatives offered as solutions to a problem.
- Variability and Uncertainty: How to analyze processes involving chance and uncertainty.
- Model Calibration: How to extract information from data.

This book is intended for engineering, mathematics, and science students interested in technological, social, and economic issues. A prerequisite course in freshmen calculus is essential as well as an elementary course in probability/statistics, and physics or chemistry is desirable. Depending on the students' background, the material can be covered in one or two semesters. I have taught the material given in this book in a two-semester sequence—at the second-semester junior level, and at the first-semester senior years.

Exercises

Exercises are chosen to reinforce fundamental concepts and to enhance practical skills. There are two types of questions:

1. ***Questions for Discussion*** are designed to stimulate thought about a particular topic and reinforce the idea that real world problems and answers are not "right" or "wrong" but are judged by rules of adequacy. Some problems deal with personal opinion or controversial issues. Sometimes, students are asked to take a position on an issue and support it with factual evidence.

Since knowledge of the vocabulary of a subject is so important to understanding and communicating about it, every section of this book first asks for the definitions of terms. Terms can be defined by word definitions but, in some instances, it might be clearer to use mathematical definitions, sketches, or diagrams. When possible, the student is urged to give an example to illustrate the meaning of the term.

2. ***Questions for Analysis*** are designed to reinforce understanding of the material and enhance problem-solving skills. These problems generally require more than plugging numbers into a formula. The student might be asked to formulate a problem as a mathematical model and then to solve it. In many cases, the real challenge is in the formulation.

An instructor's manual, which includes detailed solutions to problems and the author's opinions and comments to open-ended questions, is available to instructors through the publisher.

Computer Programs

The computer plays an important role in the quantitative analysis and the decision-making processes. It is particularly helpful in performing sensitivity analyses or answering "what if" questions and solving other problems where tables and graphic output can help interpret the results. Assignments that are most easily solved with the aid of a computer are indicated with an asterisk (*).

Students are expected to have a working knowledge of how to use a spreadsheet program. No other computer programming skill is necessary. Students can use a spreadsheet program of their choice. For Chapters 2 and 3, it is necessary for students to have access to the simplex method for solving linear programming problems. The program called "simplex," written for the Macintosh, is available from the author and publisher. Other programs written for personal and mainframe computers can be used.

Case Study

The methods of analysis presented in this book deal with planning, design, and management problems, and so have broad application. In order to truly appreciate the methods being introduced, we must know how the system works and understand the scope of problems and the consequences associated with an alternative that is offered as a solution. Technology alone cannot solve a societal problem. With systems modeling and analysis, various alternatives are investigated and critically evaluated, and solutions are sought. Systems analysis and modeling are not offered as a panacea, but as a way to achieve more informed decision-making.

To stimulate interest and motivate understanding, a case study of collection, treatment, and disposal of municipal solid waste (MSW) is used in parts of the book. It is a real-world, multidisciplinary problem involving economic, technical, social, and political issues. It must be emphasized that the MSW case study has been chosen to bridge the gap between mathematical modeling and real-world application. Of more than 250 examples and exercises in this book, fewer than 20 percent of the problems deal with the MSW case study. The same approaches used to address it have direct application to other fields. Let us consider the breadth and depth of this case study.

The MSW case study examines a variety of issues including transportation, construction, economic, and other important problems. For example, incineration, recycling, and waste source reduction and disposing of MSW in landfills are offered as solutions to the problem (Chapter 1). As communities seek a solution to this problem in an environmentally safe and cost effective manner, they are finding that no one method of disposal is proving to be completely acceptable. As the cost for disposal has become increasingly expensive and new demands for environmental quality are required, solid waste management has become a major social and political issue in many communities. A comprehensive plan should take into consideration the monetary and social benefits and costs of each alternative. Clearly, one must have more than superficial understanding of the problem to appreciate the consequences of a solution. Data and other pertinent information have been collected and is used in examples and exercises to give deeper insight and meaning to the issues at hand.*

This same philosophy used for addressing resource allocation and capital investment problems (Chapters 1 through 4) is carried forward to discuss issues dealing with capital investment, design, and planning problems under conditions of uncertainty (Chapters 5 through 8). Since a comprehensive MSW management plan should consider the long-term public health and environmental impacts associated with incineration and landfilling, analysis and discussion of these systems are used to answer questions of technological risk associated with incineration, to the design of a landfill, and to the operation of incinerators. Of course, problems from other fields are used to illustrate the principles and applications of systems analysis.

I thank the following people for their assistance: Paul C. Ossenbruggen, for the development of the computer programs; John Bonneville, Marie Gaudard, Árpád Horváth, Scott Nerny, Karoly Palovits, and David Tolli, for their support and assistance, particularly in evaluating the manuscript and its pedagogical integrity; Nickolas Buckeridge, Christopher Dwinal, Rita McCarroll, Jonathan Hilt, and other members of the junior level classes of 1991, 1992 and 1993, for their patience in using this material during its development.

October 1993 Paul J. Ossenbruggen

*A user-friendly computer program, SWAP (Solids Waste Analysis Program), has been developed to aid in the analysis of the MSW management problems (Section 3.3). This program, written for the Macintosh, uses computer graphics to establish a network diagram, formulates a linear programming model to minimize cost of disposal, and uses the simplex method to determine the least cost solution. It is available from the author and publisher. While desirable, it is not an imperative that this program be used with this book.

Contents

Notation

The following units are generally given in parentheses within the chapter text:

T = time

E = energy

L = length

M = mass

= unitless number

anw = annual net worth as a uniform payment series ($/T where T in year),

a = ash content of a material (% of dry mass)

c = unit cost ($/#)

d = distance (L)

e = efficiency (% or $0 \leq e \leq 1$)

h = unit heat content of a material (E/M of dry mass)

f = annual inflation rate (% or $0 \leq f \leq 1$)

i = discount as an annual interest rate (% or $0 \leq i \leq 1$)

i_b = bank loan interest rate (% or $0 \leq i \leq 1$)

i_f = risk free, rate of return interest rate on investment (% or $0 \leq i \leq 1$)

i' = internal rate of return interest rate (% or $0 \leq i \leq 1$)

$k, k(t)$ = total cost or amortized debt in year t ($/T)

m = moisture content of a material (%)

n = project life (T in years)

nw = net worth ($),

pw = present worth of discounted dollars where t = year 0 ($)

npw = net present worth expressed in year zero $t = 0$ dollars ($)

p = price ($/#)

q = consumer demand (#)

r, r' = real and corrected real interest rate where $0 \leq r \leq 1$

s = shipping fee ($/M-L)

t = time (T)

u = tipping fee ($/M)

v = fraction of people who volunteer to recycle, (%)

w = total worth such as a benefit, revenue, or savings ($)

x = control variable (M), or number of items produced (#)

Vectors and Matrix Notation

$$\mathbf{b} = \begin{bmatrix} b_1 \\ b_2 \\ \cdot \\ \cdot \\ \cdot \\ b_m \end{bmatrix} \text{ where } j=0,1,2,\ldots,m$$

$$\mathbf{x} = \begin{bmatrix} x_1 \\ x_2 \\ \cdot \\ \cdot \\ \cdot \\ x_n \end{bmatrix} \text{ where } j = 0, 1, 2, \ldots, n$$

$A^T, \mathbf{x}^T$ = transpose of **x** and A

$$\mathbf{A} = m \times n \text{ matrix} = \begin{bmatrix} a_{11} & a_{12} & \cdots & a_{1n} \\ a_{21} & a_{22} & \cdots & a_{2n} \\ \cdot & \cdot & \cdots & \cdot \\ a_{m1} & a_{m2} & \cdots & a_{mn} \end{bmatrix}$$

For example, **A x = b**.

For Optimization Models

$\boldsymbol{x}^o$ = critical point, $f'(\boldsymbol{x}^o) = 0$

$\boldsymbol{x}$ = vector of control variables

$\boldsymbol{x}^*$ = location of optimum solution

j = subscript index, 0, 1, 2, . . . ,

$f(\mathbf{x})$ = performance function = function of the control variables of **x**

b_j = right-hand side of the constraint function, for example, $f_j(\mathbf{x}) \leq b_j$

y = value of $f(\boldsymbol{x})$

y^* = optimum value of $f(\boldsymbol{x}) = f(\boldsymbol{x}^*)$

For Probability Models

$a(t)$ = availability at time t (% or $0 \leq a(\mathrm{t}) \leq 1$)

$f(y)$ = probability function of Y, when Y is a discrete variable or a probability density function of Y, when Y is a continuous variable

$F(y)$ = $P(Y \leq y)$ = cumulative probability function of Y

h = hazard or damage cost as a consequence of system failure ($)

L_Y, U_Y = lower and upper bound confidence or prediction endpoints for Y

$r(t)$ = system reliability at time t (% or $0 \le r(t) \le 1$)

$R(t)$ = risk as an expected annual cost = $h\theta(t)$

R = risk as an expected total cost = $h\theta$

S = failure event

S' = complement of S or success event or working state.

Y = random variable

y = observed outcome of Y

Z = sample space, for example, $Z = \{S, S'\}$ where S = event and S' = complementary event with $P(S) + P(S') = 1$ and $0 \le P(S), P(S') \le 1$

θ = lifetime risk probability = $n\theta(t)$ where $\theta(t)$ is generally assumed to be constant for all t and n = project life or human lifespan (% or $0 \le \theta \le 1$)

$\theta(t)$ = annual risk probability in year t (% or $0 \le \theta(t) \le 1$)

$\theta(d)$ = risk probability for intake dosage d (% or $0 \le \theta(d) \le 1$)

$\hat{\theta}, \hat{\theta}(t), \hat{\theta}(d)$ = estimate of θ, $\theta(t)$ and $\theta(d)$

μ = $E(Y)$ = mean or expectation of Y

σ^2 = $E[(Y - \mu)^2]$ = variance of Y

σ = $\sqrt{\sigma^2}$ = standard deviation of Y

Chapter 1

The Systems Approach

Why does this magnificent applied science which saves work and makes life easier bring us so little happiness? The simple answer runs: we have not yet learned how to make sensible use of it.

Albert Einstein (1879–1955)

A *system* is an organized, integrated unit that serves a common purpose. Typically, it is formed from many, often diverse components. This definition implies that the components act in unison and in accordance with a coherent plan for smooth and orderly operation.

Systems analysis is a systematic methodology for defining problems and evaluating solutions. It requires that a clearly defined project objective be established, that alternative solutions be proposed and carefully analyzed in light of this objective, and that the best solution be selected. It is a general, logical method of analysis that is used to investigate physical, social, and political systems.

This book introduces approaches that lead to the creation of successful systems and discusses issues that surround their implementation, particularly those issues that have an impact on society.

1.1 AN OPTIMIZATION PROCESS

Systems analysis uses mathematics to determine how a set of interconnected components behave in response to a given set of inputs. It is a goal-oriented discipline, so it is imperative that a clearly stated project objective be established. In this book, considerable attention is given to the *problem definition*, the establishment of an objective that properly reflects the needs and desires of the client. The *client* is a generic term used to describe the individual or individuals who pay for and/or receive benefits from the system.

Systems analysis can be used for *decision making*. After the project objective is established, the benefits and cost of each alternative solution are identified, analyzed, and weighed; then a final choice is made by the client. Since the analysis is structured in a mathematical framework, a suitable measure of effectiveness or merit must be defined. A *measure of effectiveness* or *merit* is a standard or benchmark that serves as a basis for judging how well a solution achieves the project objective. All alternatives are evaluated using the same criteria and are given an equal chance of being selected; therefore, the process is considered to be inclusive, rational, and fair.

Typically, systems analysis is applied to processes that are complex, interactive, and expensive to build, maintain, and operate. Our task in this book is to demonstrate how to address and solve decision-making problems. Hence, we illustrate

- How information is gathered and used
- How mathematical models are derived and analyzed
- How solutions are obtained

Planning, design, and management are all facilitated by system analysis. Planning involves the establishment of goals, policies, and strategies for social and economic units. Design, on the other hand, deals with the arrangement of parts to make an engineered object work in a desired manner, whereas management deals with the control and operation of a system. Depending on the problem definition and boundaries placed on it, a problem can be considered one of planning, design, or management. The differences between them can be subtle, and the distinction is not that important. Whatever the application, the approach is the same; it requires innovation and creativity.

Problems associated with systems analysis can be broadly classified as those of (1) resource allocation, (2) capital investment, (3) probability, and (4) statistics. Regardless of the classification, the aim is the same—to allocate resources in a manner that best satisfies a defined goal.

Money is needed to purchase, operate, and maintain system components. In the context of systems analysis, money, labor, and material are considered to be primary *resources*. Our task is to determine the best way to use them to achieve the stated objective. We are free to use resources within certain restrictions. In a problem treated as one of a resource allocation, we introduce these restrictions as constraint equations, which are derived from laws of nature, from component performance and resource limitations, and from regulations imposed by society. The combination of resources, an alternative solution, that best meets the stated objective and satisfies all constraint equations is called an *optimum solution*.

The methods introduced in this book are general and can be applied to a broad range of problems, including structural design, water resources planning, air quality management, and other important economic concerns.

The Systems Approach

Problem analysis involves the following stages:

- Problem definition (PD)
- Generation of alternatives (GA)
- Model formulation (MF)
- Analysis and alternative selection (A&AS)

The overall strategy, called the *systems approach*, endeavors to simplify a problem by reducing it into these stages. The overall process is depicted in Figure 1.

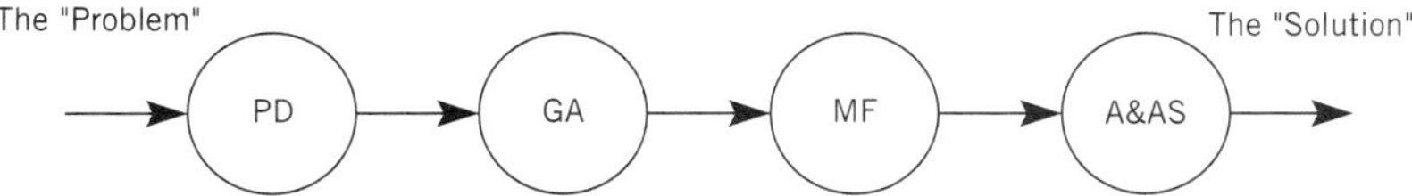

Figure 1 The systems approach as a flow of information.

Problem Definition

In the problem definition stage, the most important stage of the systems approach, we seek to identify the client's single most important goal. In this book, optimization models are limited to mathematical models of resource allocation problems with a single objective function. This may appear to be a major limitation, but in most situations, it is not as severe as we might first imagine.

Closely coupled with identifying the objective function is defining a measure of effectiveness or merit. These two factors have a major impact on the analysis and solution; therefore, they must be chosen with care.

The following steps are recommended for establishing the project objective and its measure of effectiveness:

- Determine your client.
- Establish the needs of the client.
- Identify the client's single most important objective.
- Choose a measure of effectiveness.
- Discuss the project objective with the client.
- Insure that the client clearly understands and agrees with it.

If the client is a group, then the last two steps should be presented to key representatives of the group.

When these seemingly simple steps are overlooked, considerable effort, time, and money can be wasted and the plan or design can be rejected because the recommended solution does not meet the client's need.

Generation of Alternatives

Alternative solutions that can satisfy the project objective should be identified. As much information about each alternative as possible should be obtained at this point. Gathering information may require searching the literature, obtaining technical and cost data on equipment, operation, and maintenance, and finding other pertinent information. Through this information, mathematical models can be formulated that accurately simulate the performance of each alternative.

It is also recommended that the availability of resources and anything that may impose a constraint on the system be established. In this book, constraint equations are typically derived from mass balance considerations, limitations on labor, materials, and budget. Codes, laws, and other regulations may impose additional constraints on the system.

Model Formulation

A primary task is to develop a mathematical model that reflects reality. This stage usually involves two steps:

- Model decoupling
- Model integration

Model decoupling is a simplifying step whereby system components are modeled and analyzed as subsystems. Simplifying in this manner often leads to a better understanding of how the whole system works. In the model integration step, the interaction that was lost by decoupling is regained.

The primary purpose of this two-step procedure is to simplify the overall model while maintaining reality. A delicate balance exists between model detail and the ability to effectively and efficiently analyze the model. Modeling detail may offer better reality at increased computational expense. Under certain circumstances, a simple model may prove to be more valuable than a complex one.

Analysis and Alternative Selection

The aim here is to find an optimum solution, the alternative that best achieves the client's objective and satisfies all constraint conditions. The analysis and alternative selection stages must be conducted in a fair and unbiased manner. In other words, all alternatives are given an equal chance of being selected, and the final choice is based on its merits. Analytical techniques are introduced in Chapter 2; in this chapter, the qualitative aspects of systems analysis are stressed.

Feedback and Sensitivity Analyses

An initial analysis may lead not to a final solution but to more questions. Nevertheless it is useful because it leads to an improved understanding of the problem and ultimately, a better solution. The systems approach is flexible, allowing the model to be modified and stages to be repeated. This kind of analysis is called a *feedback analysis*.

Through feedback analysis, the problem is often redefined, additional resources are found, models reformulated, and analyses repeated. Figure 2, for example, shows a feedback loop indicating a generation of new alternatives. It indicates that the mathematical model is reformulated and that a new solution is obtained.

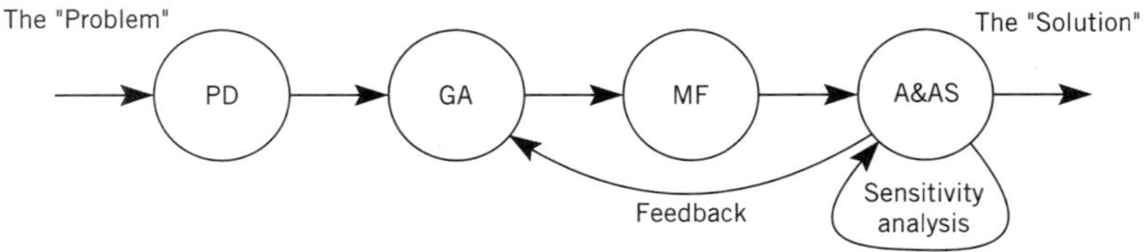

Figure 2 The concepts of feedback and sensitivity analyses as a flow of information.

Systems analysis models are ordinarily used to investigate "what if" problems. Essentially, model parameters are varied, and the response is studied to obtain a better overall understanding of the system. The importance of this kind of study, called *sensitivity analysis*, cannot be overemphasized.

The "Problem" and the "Solution"

The complete systems analysis process, including feedback and sensitivity analyses, implies that a "solution" to the "problem" has been obtained. A systems analyst may recommend an optimum solution as the best solution, but there is no assurance that it will be selected by the client. Other, more subjective factors often play a significant and decisive role in a selection. In many real-world applications, particularly when the client is a group representing diverse opinions and desires, no single "solution" may satisfy everyone. In fact, under some circumstances fundamental disagreements arise within the group and no consensus can be reached on a problem definition. When a group has no alternative solution of unanimous choice or it cannot reach a compromise on one, the problem is called an *open-ended problem*.

Both subjective evaluation of qualitative factors and an objective analysis as presented here are used in the decision-making process. A qualitative analysis most often proves to be critical in reaching a decision. However, after a comprehensive analysis has been performed and thoroughly discussed, people can still prefer the status quo. In systems analysis, we call this a no-active or do-nothing alternative. A do-nothing alternative can often be the best solution.

An Overview

The following describes the kind of information that is gathered and the thinking process used to address a "problem" using the systems approach. In this section, we focus on gathering information, and we qualitatively put it to use, or the PD stage. In Section 1.2, network diagrams are used as an aid in the GA and MF stages—in other words, selecting and arranging system components to achieve the project objective. The details of the A&AS stage are reserved for Chapters 2 and 3.

Systems analysis does not always require an analytical solution. It is well suited to qualitative analysis. Even though our eventual aim in this book is to cast problems in an analytical framework, here in this chapter we take a basically qualitative view of the systems approach.

Remember that the systems analyst does not work in isolation. The recommendations of the study are offered to a client. Since many projects significantly impact society in various ways, the acceptance or rejection of a recommended solution may hinge on social and political pressure, and not simply on an objective analysis as advocated here. It is extremely important that the problem definition stage be taken very seriously and that the client's needs be identified and incorporated into the analysis at the very beginning.

The primary purpose of the following example on municipal solid waste (MSW) management is to demonstrate how the systems approach is applied to a practical, real-world problem. It is an interdisciplinary problem involving transportation, environmental science and engineering, and economics. The issues, problems, and data are based on practical experience and on information taken from journal articles, reports, textbooks, and news stories. We will learn how to apply proven analytical techniques and methods, and, as importantly, to challenge opinions, philosophies, and the use of data in model formulation. The MSW management issue possesses many facets; it was chosen to provoke thought and creativity. It impacts society in many, often conflicting ways, and we should consider the enormity of the problem.

Industrialized countries have become "throwaway" societies. The average American generates over 4 pounds of trash per day, which is at least twice the amount produced by other industrialized countries. The problem is exacerbated by an increase in individual waste generation and population growth. The total amount of MSW has increased from 180 million tons in 1988 to 196 million tons in 1990, a 7.5% increase. Owing to increased awareness of public health risks and the desire to protect the environment, the practices of dumping household and commercial wastes in unsecured landfills and open-air burning are restricted or outlawed in the United States. As a result, 5000 to 10,000 unlined landfills have been or will be closed, resulting in a serious shortage of suitable disposal capacity throughout the country.

Improved and new disposal practices and technologies are being offered as replacements. Recycling, composting, waste reduction at the source, and various methods of incineration and energy recovery are some of the options. Selecting the best option requires a knowledge of each technology and a knowledge of the codes, laws, and regulations that limit and control their use. Budgetary considerations are also an integral part of the decision-making process. Indeed, the final solution must satisfy a broad range of demands.

As communities seek to solve this problem in an environmentally safe and cost effective manner, they are finding that no one method of disposal is proving to be completely acceptable. With disposal becoming increasingly expensive and new demands for environmental quality required, solid waste management has become a major social and political issue in many communities.

EXAMPLE 1.1 *Problem Definition for an MSW Management Plan*

Suppose a local government has hired us to assist it in developing a comprehensive MSW management plan. At the present rate of disposal, the community landfill will be filled and closed in a few years.

Our ultimate task is to conduct an analysis and to prepare a report including a recommendation for action. The plan will be implemented only after the final proposed plan is approved by the executive branch of government, presented to the citizens at public hearings, and funded by executive and legislative action.

The following guidelines summarize the policy, laws, and regulations of the community:

- MSW should be regarded as a natural resource, such as a forest or a wetland.
- MSW must be handled, processed, and disposed of in a manner that protects the environment. Environmental protection and public health standards must be met.

- No one method of waste disposal should be emphasized.
- The most economic methods and proven technologies will be given priority.

Prepare a more detailed description of the problem by outlining issues and information that needs to be gathered for a systems analysis of an MSW management plan.

SOLUTION

The first stage of the systems approach is to define the problem. A key issue is to determine *who* is the client. Is it the local government officials that hired us, the executive and legislative branches of government, or the citizens of the community? Since, in democratic societies, voters elect government officials to represent their points of view, the community is the client. Public hearings and the legislative process give citizens ample opportunity to learn about the project and to express their support or concerns.

Obviously, the government agency that hired us, pays our fee, and receives our report will have great influence on the direction of the study and so it makes good common sense to please this agency. This statement implies that there is a potential for conflict of opinion between the government officials and the public it serves, and the public may not be served. Ideally, from ethical, legal, and practical grounds, there should be none. The government agency should have the same responsibility to the public as we have, as well as the same overall aim of improving community service.

Now that the *who* question has been answered, the following community-needs questions are addressed:

What are the community resources?

Where is the community located?

When will the plan be implemented?

How will it be implemented?

These questions are asked primarily to help us learn more about the community and its needs. Community resources include existing monetary allocation for MSW collection, processing, and disposal, anticipated revenues from the community, and subsidies from other government and private sources. Government agencies can provide this information, but it is important to talk to managers and workers to gain insights into day-to-day operations. They can supply important data on waste-generation rates, operation cost, and other data.

It is essential to inspect the existing landfill, visit the neighborhoods, and talk to the people. Population statistics and maps are also a good source of general information for the *what, where,* and *when* questions. For example, the population of the community has an important bearing on the kind of disposal facilities that might be considered. Some disposal options require that a certain amount of waste be processed for the facility to be operated efficiently. If the population of the community is too small, that community may have to work with other communities on a regional basis. Zoning regulations will have a bearing on where sites may be located. The *how* question is addressed through the choice of alternatives to be considered for analysis.

AN OPEN-ENDED QUESTION

Once sufficient data on the client's background and the overall project objective are obtained, an objective function—a mathematical statement of the project objective—is prepared. An *optimization model* consists of

- An objective function
- A set of constraint equations

Without going into detail, each equation of the constraint set imposes some physical, financial, legal, and regulatory limitation on the system. Let us repeat: our task is to find a solution, the optimum solution, that (1) meets the objective as stated in the model objective function, and at the same time (2) satisfies all system limitations as stated in the constraint set.

In the problem definition stage, our task is to establish a single objective function that satisfies the overall objective of the community living within the boundary of the solid waste district under study. For purposes of discussion, suppose there are two factions within the community, supporting the following objectives:

- *Cost conscientious:* Minimize the total cost of MSW collection, processing, and disposal.
- *Environmentally conscientious:* Legislative action that will require manufacturers to use recycled goods for packaging and a community recycling program.

The environmentally conscientious faction's objective is to reduce waste at the source and to reuse materials. This group supports, for example, deposits on beverage containers. States with this system claim that 80% of consumers return their beverage containers.

The two objectives are sufficiently different as to require the formulation of two mathematical models. We will begin by discussing the advantages of using a monetary measure of effectiveness and then follow by presenting a nonmonetary measure of effectiveness, a measure of merit, for the environmental group.

Experience has shown that a monetary measure of effectiveness is a good choice for many systems analysis problems. First, the client can easily understand a monetary measure. For example, the optimum solution, measured as total annual cost for the community, can be calculated as the annual user fee or property tax. Each individual household should be able to compare and evaluate the new fee or tax against the existing fee or tax for such service.

From the analytical view of mathematical modeling, a monetary measure is an excellent overall measure of effectiveness. System components have different functions that are measured on different scales. For instance, the amount of material that is recycled can be measured on a tons per year basis and the amount of energy recovery from incineration on a kWh per year basis. Through the use of selling prices, these incompatible measures can be converted to a dollar scale and added together to obtain an annual revenue. In a similar manner, the annual costs of equipment, materials, labor, and other items and services can be determined. The annual revenue can be subtracted from the total annual costs, to obtain the annual net cost for solid waste collection, processing, and disposal.

In contrast, since the environmental group supports an MSW source reduction and recycling program, a different measure of effectiveness is needed. The following objective function is proposed:

- Minimize the total MSW disposed of in the landfill.

The environmentalists accept this function because they feel that a successful source reduction and recycling program will significantly reduce the amount of material currently being disposed of in the landfill. As part of the overall analysis, both factions want to compare optimum solutions from the two models against the historical cost records for MSW collection, processing, and disposal and the tonnage records for the landfill.

The environmental faction is taking an advocacy position. In systems analysis, selection of the best alternative will be based on weighing the merits of the competing alternatives. The major point for the analyst is to establish all pertinent information, incorporate it into the model, and let the analysis take care of finding the optimum solution.

Since we are using optimization models with different objective functions, we have implied that two different, possibly conflicting optimum solutions will be obtained. This may not be the case. The differences between the two models may be interpretive, and the same optimum solution may be obtained for each model. For example, the source reduction and recycling program may minimize the mass of material disposed of in the landfill at minimum cost. The analysis will reveal if this conjecture is true. (See Example 2.3 for further discussion.)

In complex situations, it is important to critically analyze model formulation and the solutions. Feedback and sensitivity analyses prove most useful. Systems analysis is more than a mathematical exercise in finding the optimum solution; it is a process of discovery and learning.

QUESTIONS FOR DISCUSSION

1.1.1 Define the following terms:

client	optimum solution
constraint equations	resources
feedback analysis	sensitivity analysis
measure of effectiveness or merit	system
open-ended problems	systems analysis
optimization model	systems approach

1.1.2 What is a major advantage of using a monetary measure of effectiveness? Give an example.

1.1.3 (a) Give an example of an open-ended problem. A local newspaper may help you to identify one.

(b) In this book, optimization models are used with different objective functions. Therefore, how would you analyze an open-ended question using the systems approach and this model characteristic?

1.1.4 (a) In your opinion, what is the most critical problem facing your home community? Is it crime, educational opportunity, environmental pollution, health-care delivery, decaying infrastructure (i.e., unsafe bridges, inadequate water treatment, etc.), traffic congestion, water supply, wastewater treatment, MSW processing, or some other problem?

(b) Compare your answer with the following tabulated survey results for University of New Hampshire students in 1992:

What is the most critical problem facing your home community?

	Rural	Town	City	Total
Crime	0	1	2	3
Education	2	7	0	9
Pollution	2	3	1	6
Health	2	1	0	3
Infrastructure	4	3	0	7
Traffic	2	1	1	4
Water	2	0	0	2
Wastewater	0	1	0	1
MSW	6	3	2	11
Other	3	1	0	4
Total	23	21	6	50

(c) Write an objective function for the critical problem that you identified in (a).

(d) Specify a measure of effectiveness for (c). If possible, specify a monetary and nonmonetary measure of effectiveness for (c). Give the units.

1.1.5 (a) What is the most equitable plan to charge for MSW trash pickup service?

- A property tax where each house is charged on a percentage of its assessed value of the property.
- A user fee based on the amount of trash generated.
- A graduated income tax based on the ability to pay. No user fee or tax is paid for MSW trash pickup services.

Explain your answer.

(b) Is this the least expensive plan to implement?

(c) Would the majority of people in your home community agree with your choice? Explain.

(d) Compare your answer for (a) and (c) with the following tabulated survey results for University of New Hampshire students in 1992:

What is the fairest plan to charge for MSW trash pickup service?

	Rural	Town	City	Total
Property tax	6	2	2	10
User fee	17	14	4	35
Income tax	0	5	0	5

Would the residents in your home community agree with your choice?

Agree	19	16	6	41
Disagree	4	3	0	7

(e) Does your plan encourage the reuse of MSW materials and waste reduction at the source? Why? If no, modify or select a payment plan that will achieve this goal. If yes, explain how it will achieve this goal.

(f) If you had to choose between (a) and (b), which plan would you select? Why?

1.1.6 Suppose a public hearing were held in your home community that addressed the disposal of MSW by

- Landfill disposal
- Incineration
- Recycling

(a) Which would you favor the most and the least, and why?

(b) Which would be the most favored and the least favored, and why?

1.1.7 The systems approach relies heavily on mathematical modeling. In your opinion, what is the greatest advantage of this approach and what is its greatest disadvantage?

1.2 A QUALITATIVE VIEW

In Section 1.1, the discussion focused on problem definition (PD), the first and most important stage of the systems analysis process. This section investigates the next two stages, namely, the generation of alternatives (GA) and model formulation (MF). Once again, we emphasize the qualitative aspects of systems analysis, particularly using systems analysis as an organizing tool and as a means of communicating ideas to the client.

Net Worth

The profit motive using net worth as a measure of effectiveness is a fundamental principle in a market economy for evaluating projects and decision making. This principle permits the formulation of least cost optimization models and the evaluation and selection of alternatives using capital investment analysis. Net worth or net profit is defined to be the difference between total revenue and total production cost, or

$$nw = w - k$$

where

nw = net worth or profit for a product (\$)
w = total revenue or sales (\$)
k = total production cost (\$)

Net worth can be written in terms of the unit selling price p (\$/#), unit production cost c (\$/#) and output or production level x. (The symbol # represents the number of items produced and sold.) Assuming that all items produced by a firm are sold at price p, the relationship is

$$\begin{aligned} nw &= px - cx = (p - c)x \\ \$ &= \left(\frac{\$}{\#}\right)\# - \left(\frac{\$}{\#}\right)\# = \left(\frac{\$}{\#} - \frac{\$}{\#}\right)\# \end{aligned}$$

The difference, $p - c$, is called the unit net profit. The term "unit," as in p and c, indicates selling price and production cost of one item, respectively.

When $nw \geq 0$, a firm is profitable; otherwise $nw < 0$, and it is operating at a loss. If the project objective is to maximize net profit, then an alternative solution satisfying the condition, $nw \geq 0$, is considered to be a feasible solution and worthy of consideration for selection. Otherwise, it is an infeasible solution and is rejected from further consideration.

Generation of Alternatives

In order to make intelligent and informative decisions, performance and cost data on each alternative must be obtained. This information will be used for model formulation, but just as important, it will give qualitative insights into the strengths and weaknesses of alternative solutions. Remember that not all decisions are made exclusively with qualitative data supplied from mathematical models. Qualitative and quantitative information is weighed, and then a decision is made.

Nevertheless, total project cost is an important consideration. The cost of an alternative solution, that is, resource and component costs, can be classified as capital investment, and long-term costs. A *capital investment* is an expenditure of money for plant construction and equipment purchase. This cost is usually a one-time, lump-sum expense with units of \$. Since capital costs for plant and equipment can be significant, ranging in the hundreds of thousands or millions of dollars, decision makers sometimes choose an alternative based on capital investment costs only. *Long-term costs* include operation, maintenance, labor, and other costs. They are incurred over the operating life of the project and are often calculated as annual costs with units of \$ per year, or (\$/T). Since long-term costs can be significant, they should neither be overlooked nor undervalued in the alternative selection process.

In order to combine capital and long-term costs, we must convert these costs to the same scale. In other words, convert \$ → \$/T or convert \$/T → \$. This topic will be discussed in Section 4.2, as a topic on the amortization of debt. For purposes of model formulation (see Chapters 2 and 3), unit prices p and unit costs c are given. It is assumed that a monetary scale conversion has been performed and that both capital and long-term costs are reflected in p and c.

Since a practical example is most suitable to convey the basic principles of GA and MF, we will once again discuss the MSW management problem introduced as Example 1.1. We reiterate that the systems approach is a wide application and is not limited to this case study. We will see that qualitative information gives us insights into the difficulty in providing an MSW management system that is satisfactory to everyone.

EXAMPLE 1.2 *Generation of Alternatives*

The following technologies are most often cited as viable alternatives to the MSW problem:

- Landfill
- Incineration with waste-to-energy recovery
- Recycling
- Source reduction

We will qualitatively evaluate the benefits and costs of each alternative in turn. We will also briefly comment on other technologies.

LANDFILL

A landfill is a final resting place for MSW. Regardless of the optimum alternative selected in this case study, a landfill must be provided as part of the final MSW management system.

As a result of problems with vermin, water, and air pollution, open dumping of MSW and the use of unlined landfills are no longer permitted. To protect groundwater, solid waste is typically placed and compacted into landfill refuse cells. The landfill is often constructed with leachate collection and leak detection devices as shown in the figure. These environmental and public health protection measures have added significantly to MSW disposal costs.

A landfill design is highly dependent on the topography, geology, and the surface and groundwater hydrology of the area. If the soil is pervious, then a multiple liner system or groundwater protection system may be required to protect groundwater against possible leakage. It is recommended that the capital construction, annual operating and labor, and possible cleanup costs be considered in the evaluation.

INCINERATION

Incineration is an effective means to reduce the MSW volume; it can reduce volume by as much as 90%. Depending on its composition, mass reduction can range from 60% to 75%.

A typical solid waste has a gross heat content (dry weight basis) of about 3350 kcal per kg as shown in the accompanying table. It is less than peat and dry wood, which has a heat content of about 4790 kcal per kg. Common fuel oils have heat values between 10,180 and 11,100 kcal per kg compared to 6600 kcal per kg for coke and bituminous coal and 7790 kcal per kg for anthracite coal.

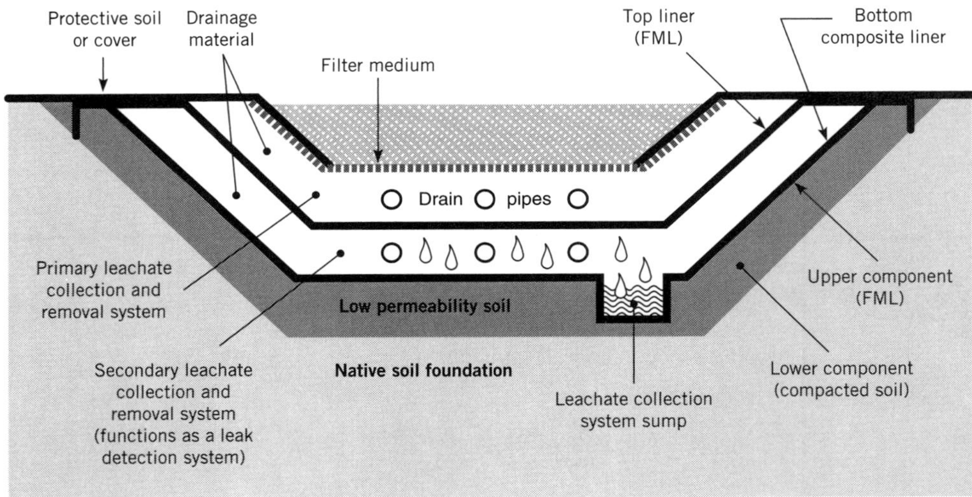

Source: Environmental Defense Fund, *Recycling and Incineration, Evaluating the Choices* (1990).

Composition of a Typical MSW in the United States
The composition of MSW varies by neighborhood and time of year

	As Collected, or Wet weight (kg)	Water Content (%)	Gross Heat Value (Dry) (kcal/kg)	Ash (Wet) (% of as collected mass)
Paper	0.46	10	4207	5.4
Food waste	0.23	72	4713	4.5
Dust and cinder	0.10	3	2106	70
Rag and textiles	0.05	10	4250	2.2
Plastics, mixed	0.03	2	7982	10
Metal	0.09	0	412	100
Glass	0.06	0	0	100
Mixture	1.00	21.7	3352	33[a]

Source: Niessen, W. R. (1978).
[a]Given as % of dry mixture. The bottom ash is 26% of the wet weight mixture.

In spite of its modest heat content rating, MSW is considered an important source of energy. In the United States, it is estimated that 180 million tons of MSW are collected each year. This amounts to 4 pounds or 1.8 kg per person per day, or 1500 pounds of material per person per year. The industrialized countries in Europe and Japan produce between 1.7 to 2.3 pounds per person per day. The United States has the potential to burn its MSW waste and produce energy amounting to 5 x 10^{11} kWh per year. This is equivalent to 57,000 thermal MW or 20 large central electrical generating facilities, assuming 40% efficiency in converting from thermal to electrical energy, using coal, oil, gas, or hydropower as an energy source.

Incinerators supplied with energy recovery equipment are called waste-to-energy plants. The heat produced from burning MSW is used to produce steam, which in turn is used to generate electricity or to produce community heating. Of the alternatives considered in the case study, waste-to-energy plants are generally the most expensive alternative to build and operate. Operators of these plants require special training; therefore, this factor should not be overlooked when compiling long-term costs.

A typical plant consists of special furnaces to withstand a highly corrosive atmosphere and equipment for shredding, sorting, flue gas cleaning, and steam or electrical power generation. A refused derived fuel (RDF) system separates noncombustible material from combustible waste prior to burning and then shreds the combustible waste to enhance its burning characteristics. A mass-burn system, on the other hand, burns MSW directly without separation.

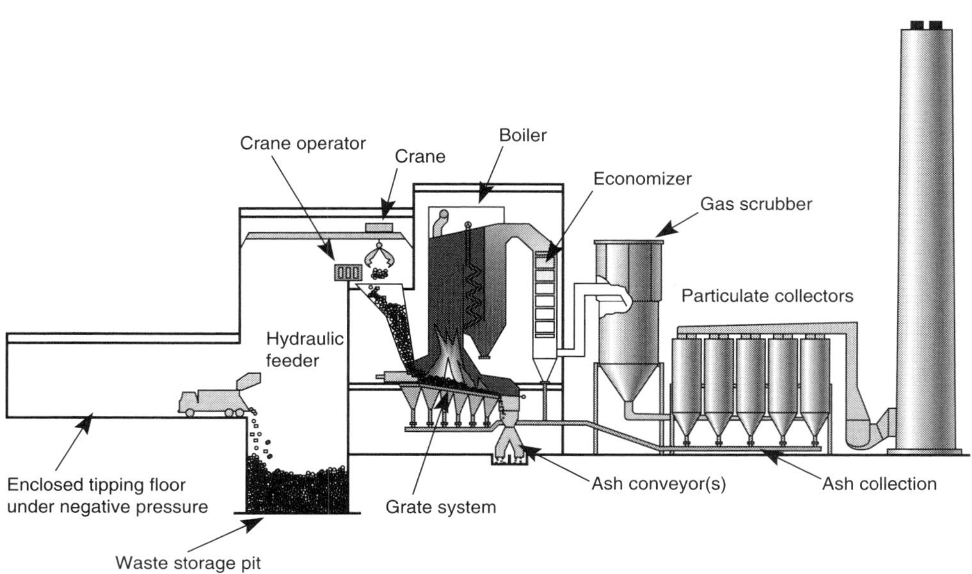

Mass-burn MSW Waste-to-Energy System. *Source:* Ogden Projects, Inc. (1993).

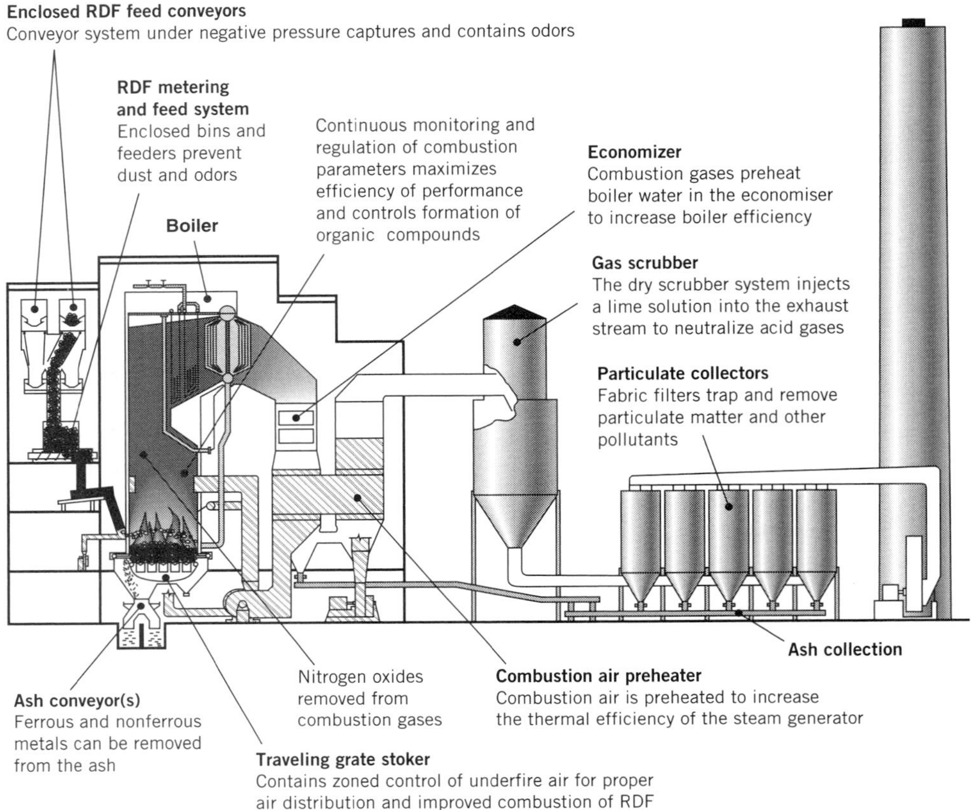

Refused Derived Fuel (RDF) Waste-to-Energy Plant. *Source:* Ogden Projects, Inc. (1993).

RECYCLING

Recycling requires society to alter its habits. As compared to landfill and combustion technologies, recycling is relatively inexpensive to implement. It requires that glass, metals, paper products, and other materials be separated. Some communities rely on volunteer programs that provide bins for each type of recycled material. Individuals must take their materials to special dropoff locations. Other communities use curbside collection systems in which recycled materials are placed in separate containers for pickup. A materials recycling center (MRC) equipped with a conveyor belt system avoids separation at the generating source, either the home or business. Here, bulk MSW is collected and shipped to an MRC where MSW is separated, packaged, and then shipped to an appropriate marketplace for recycled material.

Whereas recycling has a potential for profit making, its overall success is highly dependent on community cooperation. Some communities have made recycling mandatory, whereas others have found some success with voluntary efforts. Special education campaigns and other incentives have been used to encourage recycling efforts.

Another major difficulty associated with recycling is the highly variable demand for recycled materials, resulting in unstable prices. When a price falls, the recycled materials are often stored in anticipation of a return of higher prices. When prices are deflated for long time periods, the accumulation of these materials exceeds storage capacities, forcing operators to dispose of recyclable materials without selling them. Obviously, this defeats the purpose of recycling.

SOURCE REDUCTION

Source reduction deals primarily with the manufacture, packaging, and distribution of finished goods. In the long term, it may be the most desirable alternative of those considered here because it permits effective use of raw materials and it minimizes waste by-products during production.

Despite these advantages, industries have been reluctant to implement these source reduction methods in their plants. The reasons given center primarily on high capital investment costs, price and competition, and risk and uncertainty associated with consumer acceptance. Certainly, it is a complicated issue. Consider the issue of who bears the cost of MSW disposal and shopping convenience.

Convenience has been a major selling point for the sale of many items. For example, let's consider the implications of selling products in reusable containers. Since containers are to be returned, fundamental changes in consumer buying habits as well as in transportation and production practices are required. Suppliers will have to collect the empty containers and return them to the manufacturer for cleaning and refilling. Clearly, this scheme places extra burdens on the consumer, supplier, and manufacturer. Under the present system of using throwaway containers, suppliers and manufacturers avoid these disposal costs. These costs are indirectly borne by consumers, in the form not of higher prices but of taxes and user fees for MSW disposal.

Some observers believe that including the true disposal cost in the price of the item indicates the true worth of the item. If it is costly to dispose of an item, then this higher cost will be reflected in its price. The burden will be on the manufacturer to find a way to reduce the disposal cost. This kind of pricing is called a "green fee."

Even without changing the pricing structure as suggested, some feel that the savings gained from new waste-reduction manufacturing methods are more cost effective than continuing to use current methods. In order to encourage waste-reduction efforts, government incentives, such as tax cuts, have been advocated. Of the four alternatives presented in this example, source reduction is the most difficult to implement.

OTHER TECHNOLOGIES

MSW has other potential uses, including:

- Recovering methane gas from landfills; methane is a product of the natural biological degradation process.
- Producing agricultural products such as fertilizer and animal feed by composting MSW with sludge from wastewater treatment plants.
- Producing low- to high-grade fuels by the processes of gasification, pyrolysis, and liquefaction.

Most of these technologies are in the research and development stage; thus, they are not currently cost competitive with the other methods cited.

DISCUSSION

From this brief overview of alternatives, we can conclude that no alternative is completely satisfactory or unobjectionable. From a systems point of view, it may be cost effective to use a combination of alternatives. For example, recycling and waste-to-energy plants can be used together. Systems modeling enables us to address this issue as well as many others.

In any event, clearly the final decision will not be an easy one and will be based on economic, technical, environmental, public health, and social considerations. Not all these factors can be introduced into a resources allocation model for objective analysis; subjective evaluation is needed to deal with such a complex, open-ended question.

Model Formulation

Good communication with the client should be maintained throughout all the stages of the systems approach. The client is generally interested in the impact of the result and in obtaining an overall understanding of the model, not mathematical modeling details. We must remember that even the simplest mathematical model can be difficult for some to understand.

A network diagram is an effective communication tool that can be used for a variety of applications. In Figures 1 and 2, for example, network diagrams are used as conceptual diagrams describing the systems approach as a flow of information. Consider the network shown in Figure 3. Here, a town is considering three alternative solutions:

A. Disposing of MSW in a landfill.

B. Recycling paper and disposing of the remaining MSW in a landfill.

C. Incinerating MSW and disposing of the ash in a landfill.

The system components are a recycling facility, a marketplace for paper, an incinerator, and a landfill. Our task is to optimize the use of these facilities to satisfy the project objective. Whether the objective is to minimize total cost of MSW collection, processing, and disposal or to minimize MSW mass disposal at the landfill, the network diagram is the same. The task is also the same: to determine the optimum use of these facilities to achieve the desired result.

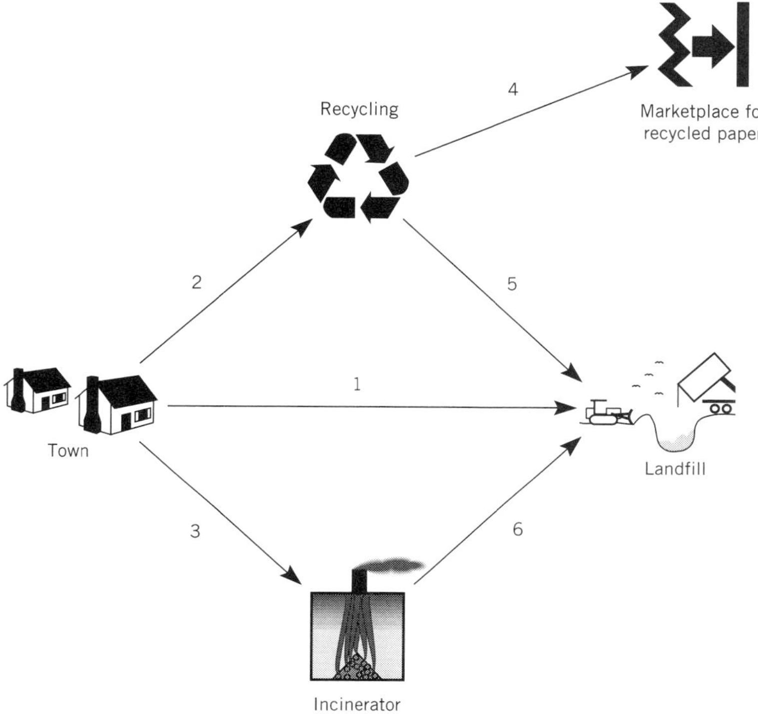

Figure 3 An MSW management model as a network.

The essential elements of a network are nodes, links, and paths. *Nodes* are used generally to describe the location of some activity. In Figure 3, the nodes represent the waste source, transfer and final disposal locations. The town is a *source* node or waste-generation site; the landfill and market place for recycled paper are *sink* nodes and final disposal sites; and recycling center and incinerator are *intermediate* nodes and places for waste processing and transfer.

Links are directed arcs that show the direction of shipment of material between nodes. A *path* is a set of connected links between source, intermediate, and sink nodes. Our task as systems analysts is to find optimum paths for the mass of recyclable paper and nonrecyclable materials shipping over these links that satisfy the project objective. Since MSW is a mixture, an optimum solution will include two paths from the source to sink nodes, as shown in the following table.

Alternative	Recyclable Paper	Nonrecyclable Materials
A	Path 1	Path 1
B	Path 2 $\rightarrow$ 4	Path 2 $\rightarrow$ 5
C	Path 3 $\rightarrow$ 6	Path 3 $\rightarrow$ 6

EXAMPLE 1.3 *Model Formulation*

Consider the following alternative solutions:

- Landfill
- Recycling
- Waste-to-energy

We will draw a network diagram to consider shipping recycled paper and plastic to one marketplace, metals to another marketplace, and garbage and ash to respective landfills.

SOLUTION

The network diagram shows six distinctly different alternatives available to the town.

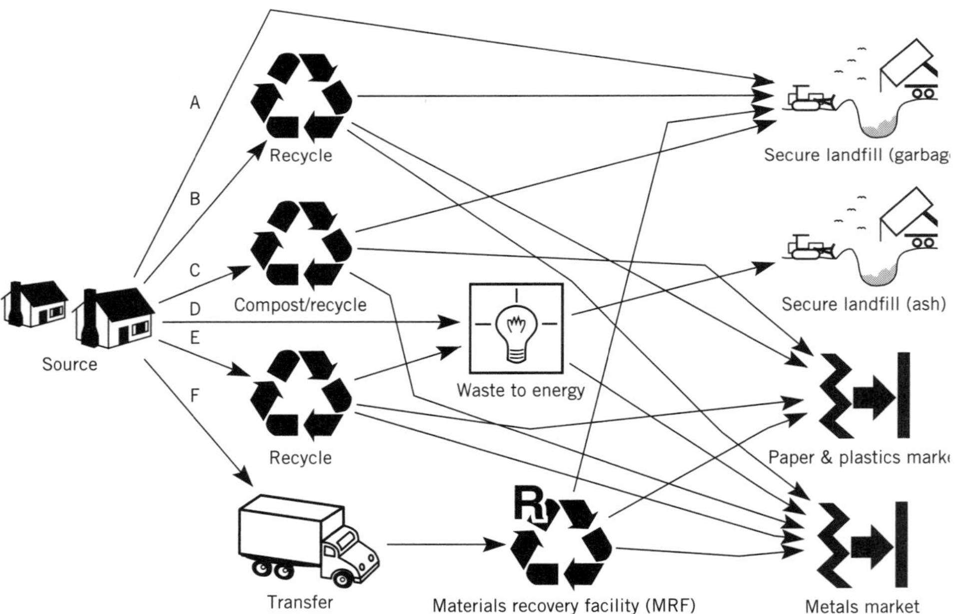

The alternative solutions are:

A. Dispose of MSW in a landfill (garbage) without processing.

B.* Recycle paper, plastics, and metals and dispose of remaining MSW in a landfill (garbage).

C.* Recycle paper, plastics, and metals; combine MSW with sludge from a domestic wastewater treatment plant to produce compost; and dispose of the unusable MSW portion in a landfill (garbage). The compost material can be considered a resource and be sold.

D. Use MSW as an energy source and dispose of its ash in a landfill.

E.* Recycle paper, plastics, and metals, use remaining MSW as an energy source, and dispose of its ash in a landfill.

F. Ship unseparated MSW to a transfer facility and then to an MRF for processing.

In the parlance of solid waste management, the charge to dispose of MSW at a facility is a *tipping fee*. It is a unit cost, that is, usually charged on a $ per ton basis, which reflects the total cost to operate a facility, including construction, operating, maintenance, and labor costs. As the network diagram makes evident, shipping costs and income from recycling and energy recovery will also be decisive in selecting the best alternative for this MSW management problem.

Summary

The systems approach has been presented in this chapter in the broadest sense. Emphasis here is on the first three stages: problem definition (PD), generation of alternatives (GA), and model formulation (MF). It is not accidental that the conceptual framework is presented, that network diagrams are used for communication purposes, and that the mathematical details of analysis and alternative selection (A&AS) are omitted. By prematurely entering into the analytical phases of GA, MF and A&AS, one easily loses sight of the primary aim of the project. The key message of this chapter is to concentrate on the problem definition stage, assuring that the primary goals are being addressed and adequately measured. Once this stage is satisfactorily completed, the generation of alternatives and model formulation stages will logically follow and fall into place, as demonstrated by the MSW case study examples.

QUESTIONS FOR DISCUSSION

1.2.1 Define the following terms:

capital investment
green fee
long-term cost
network diagram, including its principal parts
net worth or net profit

---

* MSW may be separated at the source (town), taken to a dropoff location for pickup and shipped to the marketplace. Another option is to ship directly to an MRC without separation where it is processed and sold. The critical point of discussing these alternatives is to demonstrate various possibilities; there are other options that are not shown.

profit motive
tipping fee

1.2.2 (a) Prepare a table listing, in your opinion, the major advantages and disadvantages of each of the following solid waste disposal options:
landfill
recycling
source waste reduction
waste-to-energy

(b) What are other potential uses of MSW?

1.2.3 (a) Draw a network diagram where towns A and B can ship its MSW to a materials recovery facility where paper can be separated and sold, and where MSW can be incinerated, with the option of removing metals from the ash and selling them. The diagram will include nodes for towns A and B, a materials recovery facility, an incinerator, and marketplaces for paper and metals.

(b) Redraw the network diagram from (a) to allow the residents of the towns to separate metal beverage containers and paper from their waste and sell them directly at the marketplace for metals and paper.

1.2.4 Draw a network diagram where residents have two recycling alternatives:

A. *A 4-Bag System:* Curbside pickup where residents separate and place nonrecyclable materials, paper, glass, and metals in different bags. The bags are picked up for them and shipped and sold at marketplaces for these items. The nonrecyclable materials are picked up and disposed of in a landfill.

B. *A 2-Bag System:* Curbside pickup where residents separate and place nonrecyclable and recyclable materials in two different bags. The bags with recyclable materials are picked up for them and shipped to a materials recovery facility for processing. Paper, glass, and metals are sold at marketplaces for these items. The nonrecyclable materials are picked up and disposed of in a landfill.

1.2.5 (a) Suppose a public opinion survey were held in your home community. Each person is asked to select one item from the following list for the most critical MSW issue facing the community:

i. Extending the usable life of the landfill.
ii. Protecting public health.
iii. Reducing taxes and utility fees for MSW removal, processing, and disposal.
iv. No problem.
v. Other (please describe).

In your opinion, what one issue would be considered the most critical? Why?

(b) Which one of the following alternatives would be most favored or is currently being used in your community?

- Landfill
- Combustion
- Recycling
- Source reduction

(c) In your opinion, which is the best technical solution?

(d) In your opinion, which is the most politically viable solution?

(e) In your opinion, which is the best overall solution to the MSW removal, processing, and disposal problem?

(f) Compare your answers with the following tabulated survey results for a class of 50 engineering students in 1992, in which they were asked their opinions about questions (a) – (e):

Question (a)	Rural	Town	City	Total
Life of landfill	8	9	4	21
Public health	4	5	1	10
Reducing taxes	5	4	1	10
No problem	3	1	0	4
Other	3	2	0	5

Question (b)*	Rural	Town	City	Total
Landfill	3	5	0	8
Combustion	0	2	0	2
Recycling	15	9	3	27
Source Reduction	2	3	1	6

Question (c)*	Rural	Town	City	Total
Landfill	1	0	0	1
Combustion	11	10	1	22
Recycling	2	3	3	8
Source Reduction	6	4	0	10

Question (d)*	Rural	Town	City	Total
Landfill	2	2	0	4
Combustion	0	0	0	0
Recycling	13	12	5	30
Source Reduction	5	4	1	10

Question (e)*	Rural	Town	City	Total
Landfill	1	0	0	1
Combustion	2	4	0	6
Recycling	8	5	3	16
Source Reduction	9	8	1	18

*No opinion responses are not included.

(g) Describe your home community. How relevant are these results to it?

(h) Discuss these results in terms of the Problem Definition phase of the systems approach—that is, in terms of defining the "problem" and determining the "solution."

REFERENCES

Beckwith, M., and M. Hager (1989). "Buried Alive," *Newsweek, The International Magazine,* New York (Nov. 27).

Buekens, A., and P. K. Patrick (1974). "Incineration," *Energy from Solid Waste*, Noyes Data Corporation, Park Ridge, NJ, 79–150.

Combustion Engineering (1981). *Fossil Power Systems*, Windsor, CT.

Connor, D. M. (1988), "Breaking Through the 'NIMBY' Syndrome," *Civil Engineering*, New York (December).

Crawford, J. F., and P. G. Smith, (1985). *Landfill Technology*, Butterworths, London, England.

Environmental Defense Fund (1990). *Recycling and Incineration, Evaluating the Choices*, R.A. Denison and J. Ruston (eds.), Island Press, Washington, DC.

Kaiser, E. (1978). *Combustion and Incineration Processes*, Dekker, New York.

Neal, H. A., and J. R. Schubel,(1987). *Solid Waste Management and the Environment*, Prentice Hall, Englewood Cliffs, NJ.

Niessen, W. R. (1978). *Combustion and Incineration Processes*, Dekker, New York.

Rathje, W., and C. Murphy (1992). *Rubbish, The Archaeology of Garbage*, HarperCollins, New York.

Thomas, H. A. (1963). "The Animal Farm: A Mathematical Model for the Discussion of Social Standards for Control of the Environment," *The Quarterly Journal of Economics*, 250–256.

U.S. Conference of Mayors (1990). *Incineration of Municipal Solid Waste, Scientific and Technical Evaluation of the State-of-the-Art*, Report of the Expert Panel, Washington, DC.

Chapter 2

The Principles of Linear Programming

A really valuable factor is intuition.

Albert Einstein (1879–1955)

In Chapter 1, the systems approach was introduced as a four-step process of problem definition (PD), generation of alternatives (GA), model formulation (MF), and analysis and alternative selection (A&AS). Assuming that the PD and GA stages are complete, our task in this chapter is to demonstrate how to translate this qualitative information into a mathematical model, which we call an optimization model, and then solve for its optimum solution.

In the MF stage, an optimization model is formulated with an objective function y and a set of constraint equations, both of which are written as functions of the resource variables, $x_1, x_2, \ldots, x_n$ or $\mathbf{x}$, the resource vector. In the A&AS stage, the optimum solution, y^* and $\mathbf{x}^*$, is found. The overall process is depicted in the flow chart.

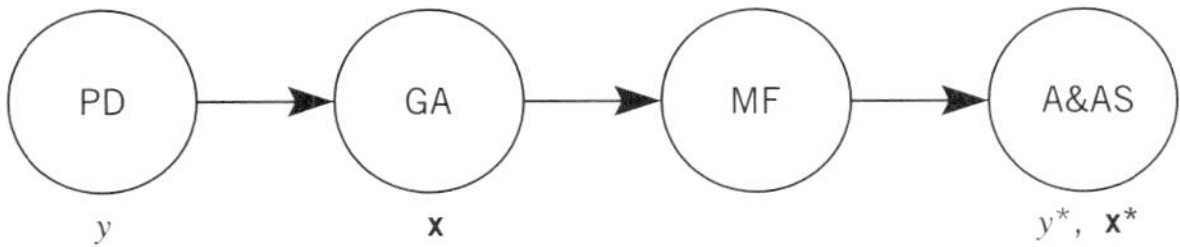

One of our primary aims in this chapter is to learn the fundamental principles of optimization. We begin by investigating the mathematical form of the most generalized optimization models. In order to simplify the discussion, the principles of optimization will be demonstrated with the use of one-dimensional and two-dimensional models where the resource vector is limited to $\mathbf{x} = [x]$ and $\mathbf{x} = [x_1, x_2]$, respectively. In addition, we limit the discussion to linear models and will find y^* and $\mathbf{x}^* = [x^*]$ and $\mathbf{x}^* = [x_1^*, x_2^*]$ by graphical means and the method of substitution. Higher dimensional models of **x**, which are solved with the aid of electronic computer and special algorithms, are discussed in Section 2.3. Nevertheless, the same basic principles used for finding y^* and $\mathbf{x}^*$ in the one- and two-dimensional models by graphical means are the ones used for higher order models.

2.1 ONE- AND TWO-DIMENSIONAL OPTIMIZATION MODELS

An optimization model is a mathematical statement of the problem. It consists of a single objective or merit function and a set of constraint equations.

$$\begin{array}{ll}
\text{objective:} & \text{minimize } y = f_0(\mathbf{x}) \\
\text{constraint 1:} & f_1(\mathbf{x}) \le b_1 \\
\text{constraint 2:} & f_2(\mathbf{x}) \le b_2 \\
\vdots & \vdots \\
\text{constraint } m\text{:} & f_m(\mathbf{x}) \le b_m
\end{array}$$

where $\mathbf{x} = [x_1, x_2, \ldots, x_n]^T$ = the transpose of the vector **x**.

The goal is to find the vector **x**, such that y is minimized and all constraint equations are satisfied.

The variables of the vector **x** or x_j where $j = 1, 2, \ldots, n$, are called *control* or *decision variables* because they are adjusted to minimize the objective function within the confines of the constraint set. The constants b_i are called the resource constraint parameters,where $i = 1, 2, \ldots, m$. Each function, $f(\mathbf{x})$, is called a *performance function* because it describes the functional relationship between some measure of effectiveness and the resource vector **x**. A *feasible solution* is a vector of **x** that satisfies all conditions specified in the constraint set. An *optimum solution*, $\mathbf{x}^*$, is a feasible solution that has the added property of minimizing y.

The optimization model in its most general form can consist of a set of nonlinear performance functions. Even with the aid of a powerful computer, finding a solution to a constrained nonlinear model can be most challenging. For some problems, there is no guarantee that a solution is optimum, or even that a feasible solution can be found. In this book, attention is focused on simpler models: (1) linear programming models and (2) nonlinear unconstrained optimization models. A broad range of practical problems fall within these two categories. Nonlinear optimization models are used primarily for estimating parameters of empirical models and are discussed in Chapter 8.

The LP Model

When all performances of a constrained optimization model are linear functions, the model is called a linear programming or LP model. It has the following structure:

$$\text{minimize } y = \sum_{j=1}^{n} c_j x_j$$

$$\sum_{j=1}^{n} a_{ij} x_j \le b_i, \quad i = 1, \ldots, p$$

$$\sum_{j=1}^{n} a_{ij} x_j \geq b_i, \quad i = p+1, \ldots, q$$

$$\sum_{j=1}^{n} a_{ij} x_j = b_i, \quad i = q+1, \ldots, m$$

$$x_j \geq 0, \quad j = 1, \ldots, n$$
$$b_i \geq 0, \quad i = 1, \ldots, m$$

The model consists of n control variables in $\mathbf{x}$ and m constraint equations depending on $\mathbf{b} = [b_1, b_2, \ldots, b_m]$. All control variables in $\mathbf{x}$ and constraint parameters in $\mathbf{b}$ must be nonnegative numbers. The control variables may be integer or continuous variables. The coefficients a_{ij} and c_j are constants. No sign restrictions are placed on a_{ij} and c_j. The coefficients c_j are often referred to as unit costs, y as the total cost, and b_i as resource constraint parameters. When a nonmonetary measure of effectiveness is employed, the objective function y is often called a *merit function*.

Now that the mathematical definitions and other formalities are given, let us turn our attention to making practical use of them.

Model Formulation

The model is more than a mathematical simulation of the system under investigation. It is a mathematical statement reflecting the client's goals stated in terms of a project objective function and all restrictions that are placed on the system. The following guidelines are used to formulate a model:

1. Define the control variables.
2. Establish the objective function.
3. Establish constraints placed on the system.

The approach is sufficiently general to apply to the formulation of both linear and nonlinear models. The final form of the optimization model is dictated by the problem statement, the system under investigation, and other practical matters. Model formulation is not an exercise of finding a problem that fits a particular linear or nonlinear programming format.

Once the model is formulated, then a method of analysis is chosen. However, since linear models are generally well behaved—that is, an optimum solution can be calculated—it may be desirable to introduce simplifying assumptions and to transform a nonlinear model into a linear form. Of course, simplifying assumptions must be justified. If this is not possible, then a nonlinear method of analysis must be used.

The Feasible Region

The constraint set of an optimization model serves two major functions.

1. It simulates the system or process under investigation.
2. It is a statement of restrictions reflecting policy, resource, budgetary, and other limiting factors.

The region specified by the constraint set is called a *feasible region*. The optimum solution $\mathbf{x}^*$ will lie within or on the boundary of the feasible region. For linear models, $\mathbf{x}^*$ must occur at an *extreme point*, a point formed by the intersection of boundary lines. These concepts can be most easily seen by solving a one-dimensional optimization model.

Consider the example:

$$\text{minimize } y = x$$
$$\text{subject to the constraint,}$$
$$1 \leq x \leq 3$$

The objective function is drawn, and the boundaries of the constraint equation are indicated as shown in Figure 1. In the one-dimensional case, the boundary points of $x = 1$ and $x = 3$ are the extreme points. Since the objective function is a linear function, the search for minimum y is restricted to the two extreme points. The optimum solution is determined by inspection. It is $x^* = 1$ and $y^* = 1$. When $x = 3$, the maximum value of y is obtained.

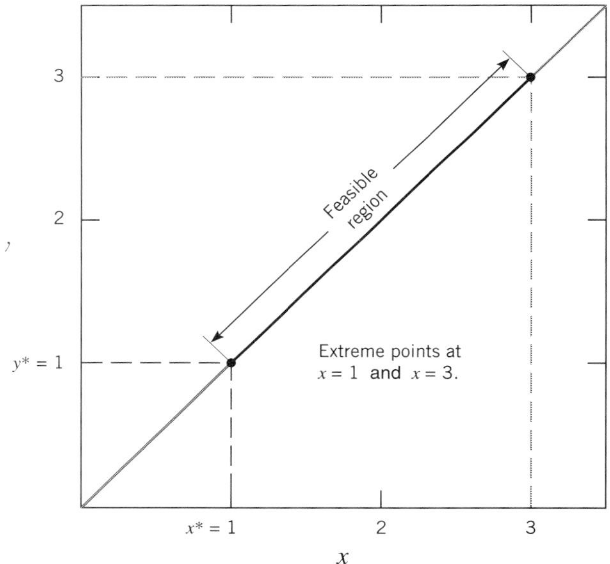

Figure 1 A one-dimensional optimization model.

The LP optimization procedure for higher dimension models uses this simple, but elegant, search procedure. Note that the interior of the feasible region, $1 < x < 3$, will never contain x^*; therefore, there is never any need to consider an interior point of the feasible region for x^*, only extreme points.

Mass Balance

For many problems, especially problems that are described as networks, the mass conservation law is used to develop a set of equations that simulates the system. The conservation law requires that a mass balance relationship be satisfied. That is, the mass entering a system must equal the mass leaving it,

$$\textit{Mass in} = \textit{Mass out}$$

For networks, selected nodes are isolated, and the mass balance relationship is applied at each node. One or more constraint equations will be derived for each isolated node. A mass balance constraint will be a strict equality constraint. For example,

$$100 = x_1 + x_2$$

Resource, budgetary, and other societal limitations, on the other hand, tend to be formulated as side constraints and inequality constraints. For example:

$$x_1 \leq 3 \text{ is a side constraint,}$$

and

$$x_1 + x_2 \leq 5 \text{ is an inequality constraint}$$

As a way of verifying these relationships, unit checks should be made. The units on each side of the equality or inequality function of the objective function and constraint equation must be the same. The symbols integer or count (#), total cost ($), percentage (%), mass (M), length (L), time (T), and energy (E) are used for this purpose. Percentages and fractions are unitless.

Rather than strict rules, there are only guidelines for model formulation. In every case, control variables are carefully defined, and network diagrams are used whenever possible to describe the system.

The Method of Substitution

The method of substitution can be used with LP models that have at least one strict equality constraint equation. When this condition is present, a model can be reduced from a higher to a lower dimension. Reducing a two-dimensional model to a one-dimensional model increases the chance of obtaining an optimum solution by inspection. The method will be demonstrated by example.

EXAMPLE 2.1 *A 1-D MSW Network Model*

A town is considering operating a materials recovery facility (MRC) where paper can be separated and sold for p_4 = \$25 per ton. The town generates 10,000 tons of MSW per year. Fifty percent of the solid waste is recyclable paper. The tipping fees are:

	k	u_k, Tipping Fees ($ per ton)
MRC	2	10
Landfill	3	50

All shipping fees are assumed to be equal to zero.
Given this information, would you recommend operating an MRC facility?

SOLUTION

These steps will be followed:

1. Establish the system, that is, network diagram.
2. Define control variables of **x**.
3. Formulate mathematical model.
4. Solve for **x*** and y^*.

Since all MSW is to be collected and shipped to the MRC, the network diagram for this plan is

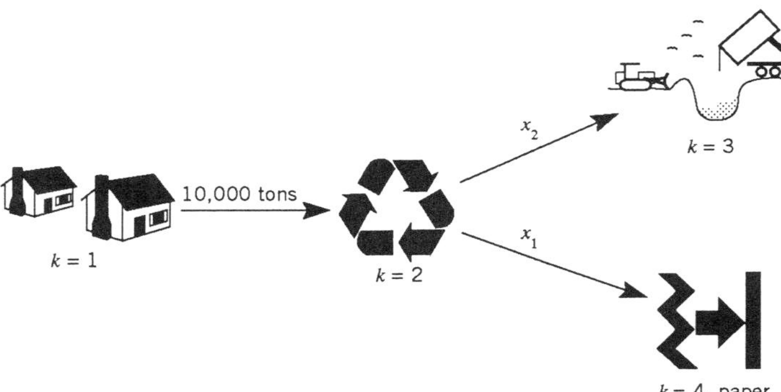

Note that recycled paper is sold only if it reduces the total cost of operating the MRC.
The control variables are defined to be

x_1 = amount of recyclable paper sold at the marketplace (M)
x_2 = amount of bulk mass that is shipped to the landfill (M)

The total annual cost for operating this system y is the sum of the tipping fees at the recycling center and landfill minus the profit received from selling paper. The second line shows that a unit balance has been met.

$$\begin{array}{ccccc} y = u_2(10{,}000) & - & p_4 x_1 & + & u_3 x_2, \text{ or} \\ (\$) = \ (\$/\text{M})(\text{M}) & & (\$/\text{M})(\text{M}) & & (\$/\text{M})(\text{M}) \end{array}$$

$$y = \$10\ (10{,}000) - 25x_1 + 50x_2$$

Since a mass balance must exist at the MRC, the following equality constraint must be satisfied.

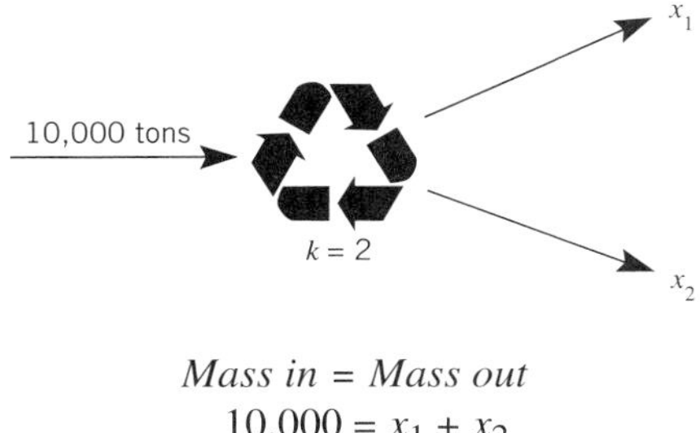

$$\textit{Mass in} = \textit{Mass out}$$
$$10{,}000 = x_1 + x_2$$
$$(\mathrm{M}) = (\mathrm{M})\ (\mathrm{M})$$

Since 50 percent of the MSW is paper, the following inequality constraint must be satisfied,

$$x_1 \le 5000$$
$$(\mathrm{M}) \le (\mathrm{M})$$

This information is summarized in an LP model;

$$\text{minimize } (y - 100{,}000) = \text{minimize } y' = -25\,x_1 + 50\,x_2$$
$$x_1 + x_2 = 10{,}000$$
$$x_1 \le 5000$$
$$x_1 \ge 0$$
$$x_2 \ge 0$$

Since $x_1 + x_2 = 10{,}000$, this 2-D model may be reduced to a 1-D model. We use a method of substitution. We will solve for x_2 and simplify the problem to be a function of x_1 only.

$$x_2 = 10{,}000 - x_1$$

Substitute this equation into every equation containing x_2, including the side constraint equation, $x_2 \ge 0$. In this example, x_2 appears in the objective function and side constraint equation only. Thus,

$$\text{minimize } y' = -25x_1 + 50(10{,}000 - x_1) = \$500{,}000 - 75x_1,$$

and

$$x_2 = 10{,}000 - x_1 \ge 0, \text{ or } x_1 \le 10{,}000$$

After gathering and simplifying, the 1-D LP optimization model is written as

$$\text{minimize } y' = \$500{,}000 - 75x_1$$
$$0 \le x_1 \le 5000$$

The minimum cost solution is found by inspection. It occurs at the extreme point, $x_1^* = 5000$ tons with $x_2^* = 5000$ tons, or

$$\mathbf{x}^* = (5000, 5000),\ y'^* = \$125{,}000 \text{ and } y^* = \$225{,}000$$

The optimum solution shows that a cost saving is received by recycling paper. The MRC is recommended.

DISCUSSION

The optimum solution is an all or nothing assignment. In other words, as much recyclable paper that can be assigned to the control variable x_1 is assigned to it. In this case, all available recyclable paper is recycled, $x_1^* = 5000$ tons. All or nothing assignments are typical assignments in LP optimization problems.

Graphical Solution of 2-D LP Models

Optimization models with two control variables are most easily interpreted and solved with a graph. Since models are linear, problems may be solved with the aid of a pencil and a straight-edge or ruler.

The following approach is recommended:

Establish the Feasible Region

1. Temporarily ignore the nonequality constraints and plot each constraint equation on the graph assuming it to be an equality constraint. Label each constraint equation for easy identification.
2. Determine the feasible region for each constraint equation. Use an arrow to show the direction of feasibility. For equality constraints, only those values that fall on the line are feasible; thus, no arrow is needed.
3. Determine the intersection of the feasible regions for the individual constraint equations and indicate this intersection with shading or some other identification. If no feasible region exists, stop.

Find the Optimum Solution x*

4. Choose a value of y_0 and plot the corresponding contour on the graph. The choice of initial guess of y_0 is arbitrary, but it is desirable for it to intersect the feasible region. It makes the next step easier.
5. The slope of y_0 is critical for finding $\mathbf{x}^*$. The contour line of y^* must be parallel to y_0. Choose another y, call it y_1, and plot the corresponding contour on the graph. Since relative magnitudes of y_0 and y_1 are known, the slope of the objective function is known.
6. Draw the contour line parallel to y_0 and y_1 that intersects the boundary of the feasible region and will minimize y. This point is the location of $\mathbf{x}^*$. Calculate the optimum value y^*.

An alternative to steps 4 through 6 is to superimpose a contour plot of the objective function on the two-dimensional feasible region plot; then $\mathbf{x}^*$ is found by inspection. The procedure will be demonstrated by example.

EXAMPLE 2.2 *Formulating a 2-D MSW Network Model*

A privately operated recycling recovery facility can purchase MSW from two towns, separate newsprint and computer paper, and sell it. Household and commercial business surveys show that the composition and supply of MSW are:

	k	Computer Paper %	Newprint %	Available MSW Supply (tons per year)
Town 1	1	5	20	10,000
Town 2	2	15	30	20,000

The company will offer the towns a price of $p = \$35$ per ton of MSW. It will sign contracts with each town to ensure that it will receive the optimum amount of material to maximize its profit. The company seeks to determine how much MSW it should purchase from each town.

To remain in business, analyses indicate that the company must collect at least 1500 tons of computer paper per year. When the supply of newsprint becomes large, the price that the company receives will fall. To gain maximum market price, the company will sell no more than 6000 tons of newsprint per year.

How much computer paper and newsprint should the company purchase from each town?

SOLUTION

The network diagram shows the flow of material.

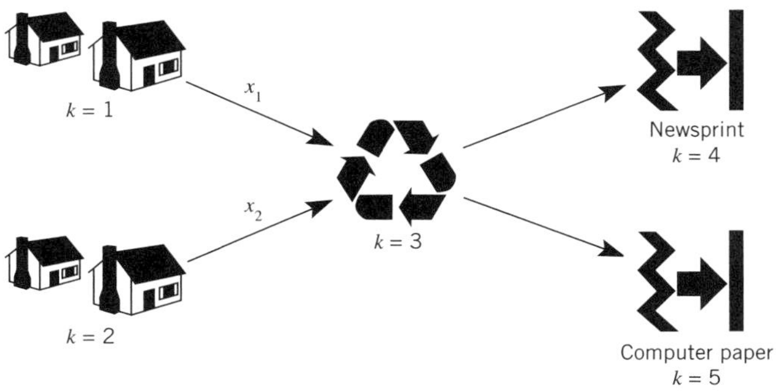

Control variables are defined as

x_1, x_2 = the annual amount of bulk mass material (M) purchased from towns 1 and 2, respectively.

The company will maximize its profit by minimizing its contracting costs. The minimum cost objective function is

$$y = \$35\,x_1 + \$35\,x_2$$
$$(\$) = (\$/\text{M})(\text{M}) + (\$/\text{M})(\text{M})$$

The second line shows that a unit balance is satisfied.

Mass balance relationships are used to determine the total mass of computer paper and newsprint the facility can receive. The mass balance for computer paper is determined by isolating the recycling node. The amount of computer paper from Towns 1 and 2 are $0.05\,x_1$ and $0.15\,x_2$, respectively.

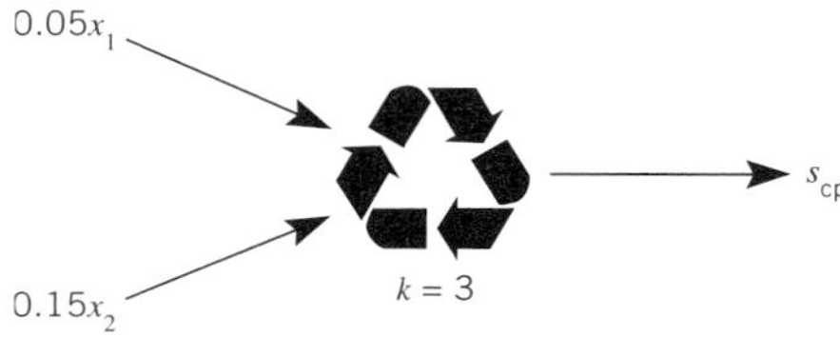

$$\textit{Mass in} = \textit{Mass out}$$
$$0.05x_1 + 0.15x_2 = s_{cp} \quad \text{(Computer paper)}$$
$$(\%)(\text{M}) + (\%)(\text{M}) = (\text{M})$$

where s_{cp} = supply of computer paper

A mass balance for newsprint can be established in a similar manner.

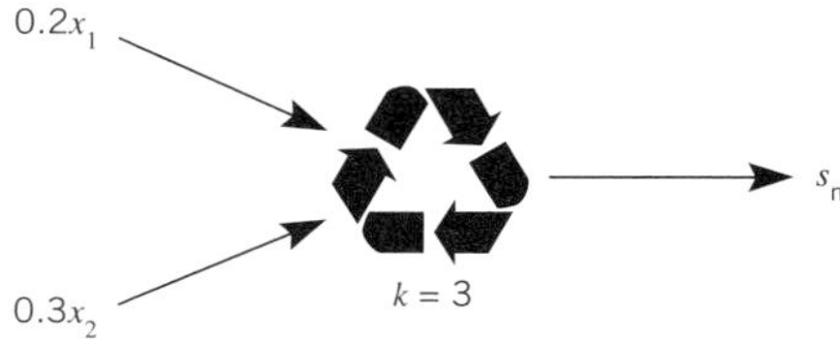

$$\textit{Mass in} = \textit{Mass out}$$
$$0.2x_1 + 0.3x_2 = s_n \quad \text{(Newsprint)}$$
$$(\%)(\text{M}) + (\%)(\text{M}) = (\text{M})$$

where s_n = supply of newsprint

These two equations simulate the separation process. In order to complete this description, the resource of computer paper and newsprint must be introduced into the constraint set. Since the company is under no obligation to collect all the MSW from the towns, the town supplies are introduced as side constraints.

$$x_1 \le 10{,}000 \quad \text{(Town 1)}$$
$$(\text{M}) \le (\text{M})$$

and

$$x_2 \leq 20{,}000 \quad \text{(Town 2)}$$
$$\text{(M)} \leq \text{(M)}$$

In order for the company to remain in business and gain maximum profit, it must supply at least 1500 tons of computer paper per year and no more than 6,000 tons of newsprint. By isolating nodes $k = 4$ and $k = 5$ above, these constraints, written in terms of s_{cp} and s_n, are side constraints:

$$s_{cp} \geq 1500 \quad \text{(Computer paper)}$$
$$\text{(M)} \geq \text{(M)}$$

$$s_n \leq 20{,}000 \quad \text{(Newsprint)}$$
$$\text{(M)} \leq \text{(M)}$$

Since s_{cp} = 15 tons and s_n = 6000 tons, the constants s_{cp} and s_n are eliminated from the model. The complete optimization model is written in terms of two control variables x_1 and x_2:

$$\text{minimize } y = \$35x_1 + \$35x_2$$
$$\text{subject to the constraint set,}$$
$$\begin{array}{ll} 0.05x_1 + 0.15x_2 \geq 1500 & \text{(Computer paper)} \\ 0.2x_1 + 0.3x_2 \leq 6000 & \text{(Newsprint)} \\ 0 \leq x_1 \leq 10{,}000 & \text{(Town 1)} \\ 0 \leq x_2 \leq 20{,}000 & \text{(Town 2)} \end{array}$$

In this case, there is no equality constraint equation to simplify this model from a 2-D to 1-D model as in Example 2.1. The optimum solution $\mathbf{x}^*$ may be obtained by graphical means. The steps for establishing a feasible region are as follows:

1. The feasible region is constructed by first temporarily changing the inequality signs in each constraint to equal signs. For example, the town constraint is temporarily written as

$$0.05\,x_1 + 0.15\,x_2 = 1500$$

This line is plotted on the $x_1 - x_2$ plane. This procedure is repeated for the newsprint and supply equations.

$$0.2x_1 + 0.3x_2 = 6000$$
$$x_1 = 10{,}000$$
$$x_2 = 6000$$

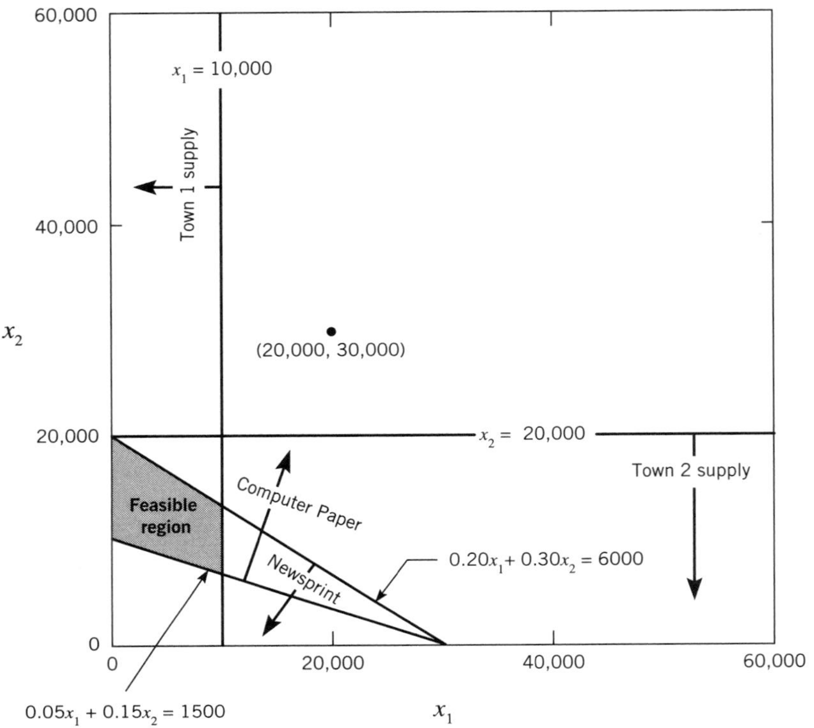

2. The direction of feasibility for each equation is determined. An arbitrary point, x_1 = 20,000 and x_2 = 30,000, is chosen as a test point. It satisfies the inequality condition, for constraint labelled computer paper.

$$0.05x_1 + 0.15x_2 = 0.05\ (20{,}000) + 0.15(30{,}000) > 1500$$

Thus, it lies in the feasible region. All points that lie on this side of the line, $0.05\ x_1$ + $0.15\ x_2$, are feasible solutions. An arrow indicates the direction of feasibility.

The same point is used to determine the feasible region for the other three constraints. An arrow showing the feasible region is drawn for each one.

3. The intersection of the individual feasible regions for each constraint specifies the feasible region. Since the objective function is linear, the optimum solution must lie on the boundary of the feasible region, or more specifically at an extreme point. The extreme points, (0, 10,000), (0, 20,000), (10,000, 13,333), and (10,000, 6667), are defined by the intersection of the boundary lines as shown in the following figure.

The procedure for finding the optimum solution is:

4. Assume y_0 = \$1,750,000 and plot the contour,

$$\$35x_1 + \$35x_2 = \$1{,}750{,}000$$

5. Assume $y_1 \leq y_0$. Let y_1 = \$1,050,000 and plot the contour,

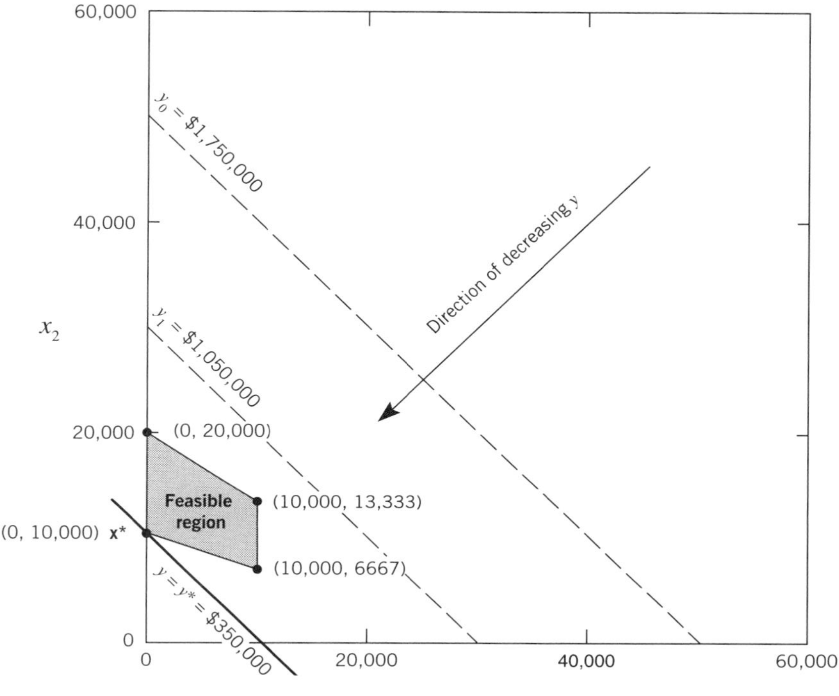

$$\$35x_1 + \$35x_2 = \$1{,}050{,}000.$$

The direction of decreasing y is denoted on the graph.

6. The location of optimum solution **x*** must be parallel to y_0 and y_1 and must be constrained within the feasible region. The point satisfying both requirements occurs at the boundary or extreme point of the feasible as illustrated. Inspection shows that it is

$$\mathbf{x}^* = (0, 10{,}000) \text{ and } y^* = \$350{,}000$$

The optimum solution is for the company to sign a contract with town 2 exclusively for 10,000 tons of MSW per year.

EXAMPLE 2.3 ***A Minimum Cost System Versus a Minimum Mass Disposal System***

A town has two choices for disposal of its solid waste. It can dispose of it in the landfill directly, or burn it, recover the waste energy, and dispose of the ash in the landfill. There are two organizations in the town. One group advocates cost efficiency in government-run operations, whereas the other group is concerned with the closure of the existing landfill and subsequent difficulties with replacing the closed landfill. The town officials are faced with two demands:

- To minimize the total cost to the town for collecting, shipping, and operating its landfill and incineration facilities.
- To minimize the amount of material deposited in the landfill.

Clearly, town officials would like to satisfy the demands of both groups.

The town produces a minimum of 100 tons of waste per year. In order to extend the life of the landfill, dumping restrictions have been imposed. It will accept no more than 80 tons of untreated waste per year and no more than 15 tons of incinerator ash per year.

The heat produced by the incinerator can be used for community district heating. The plant, operating at 75% thermal efficiency, produces 2000 kWh of usable energy per ton of MSW. Energy is sold at $0.05 per kWh. There is a 75% reduction in mass for every ton of solid waste that is incinerated.

The following costs are expressed as landfill and incinerator tipping fees.

Combined Shipping and Tipping Fee	k	cost (unit per ton)
Landfill: untreated waste	2	$50
bottom ash		$80
Incinerator	3	$30

The tipping fee for bottom ash reflects the cost of protecting groundwater against the leaching of toxic chemicals contained in this material. Metals, for example, become water soluble and more easily transported after combustion.

Formulate and solve optimization models to investigate the demands of each advocacy group. Analyze the solutions to determine if the two groups have compatible or opposing objectives.

SOLUTION

The network diagram for both groups is:

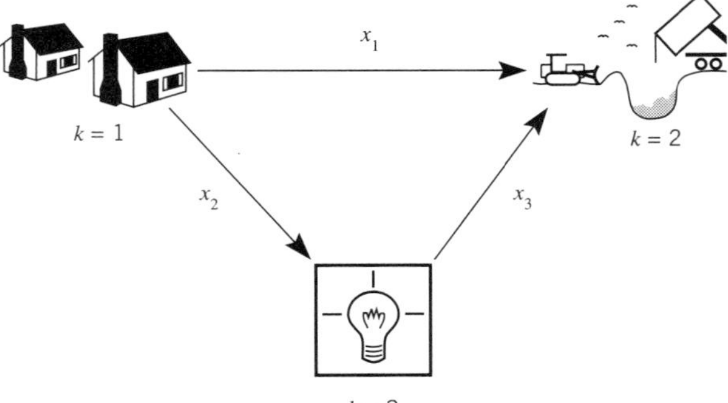

The control variables and objective function variables are defined as

x_1 = mass of MSW shipped from $k = 1$ to 2 in tons per year (M)
x_2 = mass of MSW shipped from $k = 1$ to 3 in tons per year (M)
x_3 = mass of bottom ash shipped from $k = 3$ to 2 in tons per year (M)
y = annual disposal costs ($)
z = annual mass disposed of in the landfill (M)

The same network diagram and control variables are used for the minimum cost and minimum MSW mass disposal problems.

MINIMUM COST MODEL

The annual cost to the town, y, is the sum of user costs minus profits from power generation, or

$$\begin{array}{c} y = 50x_1 + 30x_2 + 80x_3 - (2000)(0.05)x_2 \\ (\$) = (\$)(\text{M}) \quad (\$)(\text{M}) \quad (\$)(\text{M}) \quad (\text{E/M})(\$/\text{E})(\text{M}) \end{array}$$

The second line shows a unit balance on both sides of the performance function. Since the ash content is $x_3 = 0.25x_2$, x_3 can be eliminated from the model by substitution. The objective function reduces to:

$$\text{minimize } y = 50x_1 - 50x_2$$

A constraint set is obtained by evaluating the mass balance at the town and by considering the restrictions placed on landfill dumping.

The mass balance and constraint equations for the town are:

$$\begin{array}{c} \textit{Mass in} = \textit{Mass out} \\ 100 = x_1 + x_2 \quad \text{(Town)} \\ (\text{M}) = (\text{M}) + (\text{M}) \end{array}$$

Since the problem states that a minimum of 100 tons of MSW is produced by the town, the equality constraint equation is rewritten as an inequality constraint,

$$x_1 + x_2 \geq 100 \quad \text{(Town)}$$

The total amount of untreated waste being shipped from the town must be less than 80 tons per year, or

$$\begin{array}{c} x_1 \leq 80 \quad \text{(Landfill 1)} \\ (\text{M}) \leq (\text{M}) \end{array}$$

The total amount of ash being shipped from the incinerator must be less than 15 tons per year,

$$x_3 = 0.25x_2 \leq 15 \quad \text{(Landfill 2)},$$

or

$$x_2 \le 60 \quad \text{(Landfill 2)}$$
$$(\text{M}) \le (\text{M})$$

The minimum cost model is

$$\text{minimize } y = 50x_1 - 50x_2$$
subject to the constraints,
$$x_1 + x_2 \ge 100 \qquad \text{(Town)}$$
$$0 \le x_1 \le 80 \qquad \text{(Landfill 1)}$$
$$0 \le x_2 \le 60 \qquad \text{(Landfill 2)}$$

The following two-dimensional plot shows the feasible region. The point $x = (100, 100)$ is a test point used to help establish the feasible region.

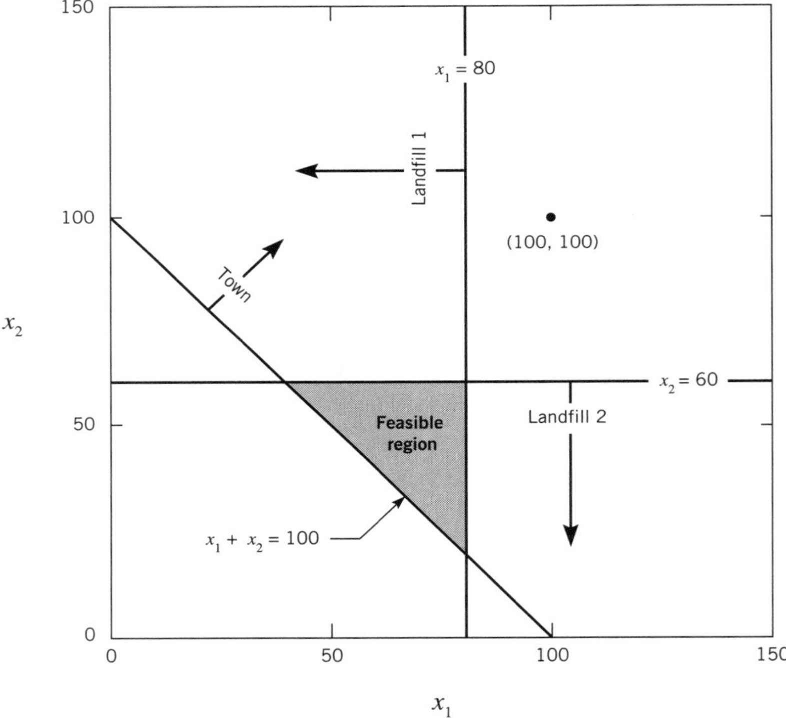

The contour plot of the objective function,

$$\text{minimize } y = 50x_1 - 50x_2,$$

is superimposed on the $x_1 - x_2$ plane. The following figure shows the contour lines for y_0 = \$0 and y_1 = \$5000. The direction of decreasing cost is shown by an arrow. The optimum solution $\mathbf{x}^*$ must occur at an extreme point. Inspection shows that $\mathbf{x}^*$ lies at the intersection of the town and landfill 2 constraints. The optimum solution is

$$\mathbf{x}^* = (40, 60),$$

and

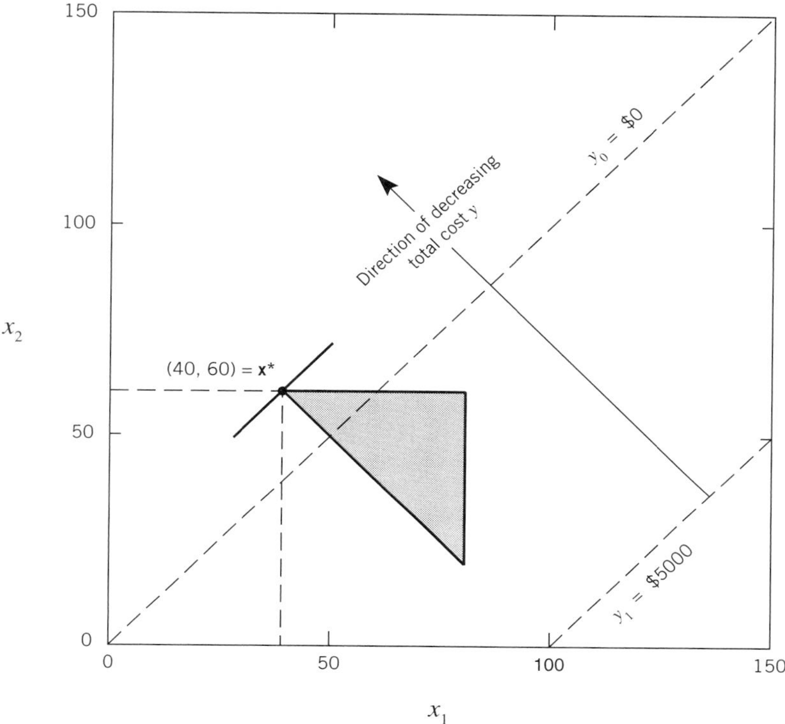

$$y^* = 50(40) - 50(60) = -\$1000 \text{ per year}$$

The town makes a net profit of \$1000 per year by shipping 40 tons to the landfill and 60 tons to the incinerator. The amount of MSW disposed of in the landfill is

$$z^* = 40 + 15 = 55 \text{ tons per year}$$

THE MINIMUM MSW MASS MODEL

We can make use of the minimum cost model in formulating this model.

Here, the objective is to minimize the disposal of landfill waste. The objective or merit function is

$$\begin{aligned} \text{minimize } z &= x_1 + x_3 \\ (\text{M}) &= (\text{M}) + (\text{M}) \end{aligned}$$

But since $x_3 = 0.25x_2$, it is expressed as

$$\text{minimize } z = x_1 + 0.25x_2$$

The supply of MSW from the town is the same as in the minimum cost solution:

$$x_1 + x_2 \geq 100 \text{ (Town)}$$

Since the minimum cost solution had an optimum solution or net profit of \$1000 per year, we will require an annual profit of at least \$1000 per year. This is entered into the formulation as budgetary constraint, and in terms of the minimum cost, optimum $y^* = -\$1000$. For above, the total cost is $y = 50x_1 - 50x_2$. Since $y \leq y^*$, the budget constraint is

$$\begin{array}{c} 50x_1 - 50x_2 \quad \leq -1000 \text{ (Budget)} \\ (\$/M)(M) + (\$/M)(M) \leq (\$) \end{array}$$

or

$$-50x_1 + 50x_2 \geq 1000$$

The landfill constraints are the same as in the minimum cost model:

$$\begin{array}{c} 0 \leq x_1 \leq 80 \text{ (Landfill 1)} \\ 0 \leq x_2 \leq 60 \text{ (Landfill 2)} \end{array}$$

The final model is

$$\begin{array}{rl} & \text{minimize } z = x_1 + 0.25x_2 \\ & \text{subject to the constraints,} \\ -50x_1 + 50x_2 \geq \$1000 & \text{(Budget)} \\ x_1 + x_2 = 100 & \text{(Town)} \\ 0 \leq x_1 \leq 80 & \text{(Landfill 1)} \\ 0 \leq x_2 \leq 60 & \text{(Landfill 2)} \end{array}$$

A two-dimensional plot shows the feasible region as point **a**.

Since the feasible region is the point **a**, no search is necessary. The optimum solution is the same as for the optimum minimum cost solution:

$$\begin{aligned} \mathbf{x}^* &= (40, 60) \\ y^* &= -\$1000 \\ z^* &= 55 \text{ tons per year} \end{aligned}$$

DISCUSSION

Even though the feasible regions for the minimum cost and minimum mass models differ, the same essential information is included in both models. This exercise demonstrates that the goals for minimum cost and minimum landfill mass disposal have compatible objectives.

With regard to model formulation, it makes no theoretical difference whether the models are derived with a minimum cost or a minimum landfill mass disposal objective function. There is an important caveat. The same descriptive information must be included in both

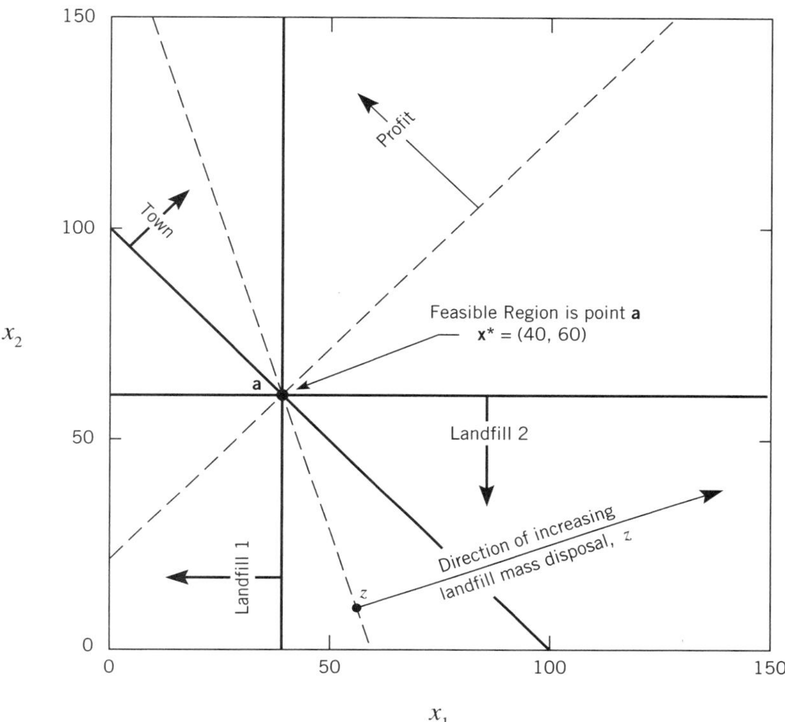

models. In essence, the only difference in this model formulates how the costs and profits are treated. In one model, it is treated as an objective function, and in the other, it is treated as a budget constraint. Of course, selecting the budget constraint in the minimum mass model equal to y^* is a critical assumption. All other information is identically the same in both models.

The important point is to determine how the system operates, establish an objective, formulate the model, and let the mathematics find $\mathbf{x}^*$.

QUESTIONS FOR DISCUSSION

2.1.1 Define the following terms:

decision variables	feasible solution
extreme point	mass conservation law
feasible region	merit function

2.1.2 Some people believe that the most serious drawback of the optimization model is the restriction that the objective function must be a single function. In practice, it is possible to use a weighing function w so that dissimilar measures can be combined to obtain unit balance.

For example,

$$y = f_0(\mathbf{x}) + wg_0(\mathbf{x})$$

where

$$y = \text{total cost (\$)}$$
$$f_0(\mathbf{x}) = \text{operation cost (\$)}$$
$$g_0(\mathbf{x}) = \text{number of lives lost (\#)}$$

(a) What are the units for w and what does w represent (i.e., define w)?

(b) While this definition of y is viable mathematically, discuss the difficulties of assigning a value to w.

QUESTIONS FOR ANALYSIS

2.1.3 The flow of wastewater x_1 and x_2 (L^3/T), which is sent to a mixing tank, can be controlled. The maximum flows are b_1 and b_2 (L^3/T) for towns 1 and 2, respectively. The concentration of a contaminant from these waters is a_1 and a_2 (M/L^3). The system is assumed to operate under steady-flow conditions, and no chemical or biological reactions are assumed to take place. The maximum flow that the mixing tank can accept is b_3 (L^3/T), and the maximum contaminant concentration in the effluent line is limited to b_4 (M/L^3). Let c_1 ($\$/L^3$) and c_2 ($\$/L^3$) be unit pumping costs for links 1 and 2.

(a) Derive an LP model to minimize total pumping cost y in terms of the control variables x_1 and x_2.

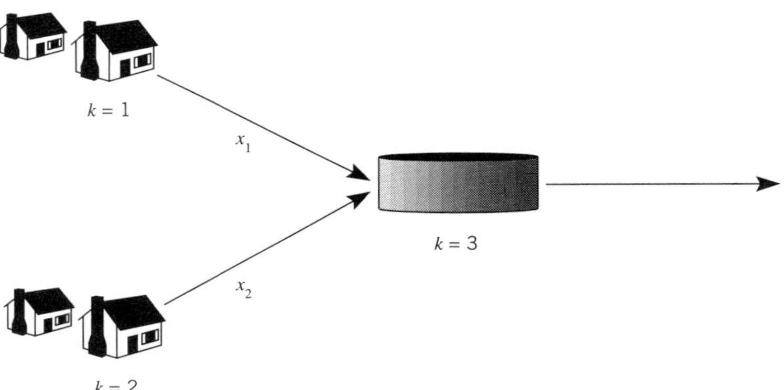

(b) Can the LP model from (a) be solved by the method of substitution? Explain your answer.

2.1.4 (a) Derive a performance function, where

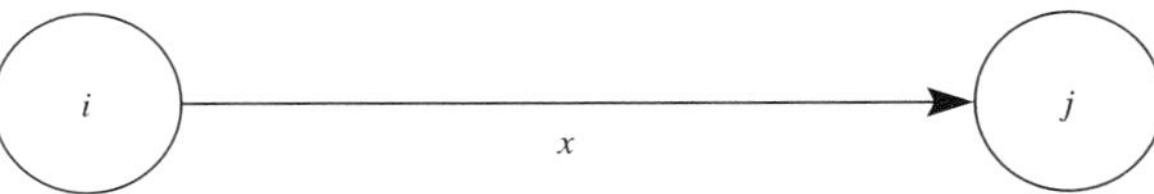

k or $k(x)$ = total cost of shipping mass x between the two nodes plus paying tipping fee charges for mass x

Let

x = mass of MSW shipped (M) from node i to j
u = tipping or processing fee (\$/M)
s = shipping fee (\$/M-L)
d = distance (L)

(b) Suppose sink node j is a waste-to-energy plant. Derive the performance function for $w = w(x)$ = gross revenue from the sale of recovered energy from mass x.

Let

p = selling price of the recovered energy (\$/E)
h = unit heat content of MSW (E/M)
e = recovered or usable energy as a percentage of total energy produced (%)

(c) The net profit or worth $nw = nw(x)$ is the difference between the gross revenue minus total cost. Use the results from (a) and (b) to derive a net worth function for the waste-to-energy plant.

(d) What is the breakeven price of recovered energy? (The selling price for recovered energy equals the total cost of shipping and processing the MSW.)

2.1.5 Let

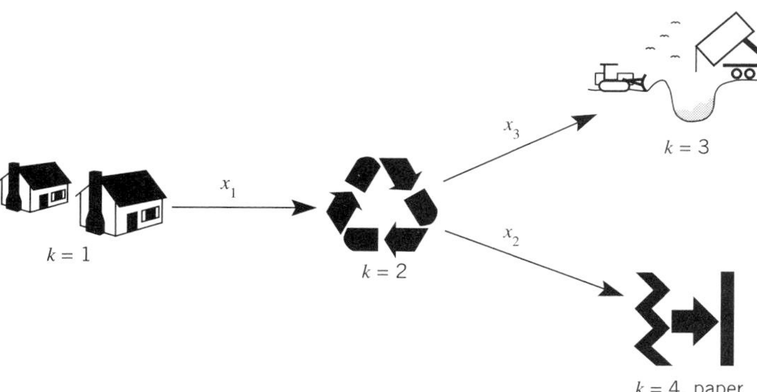

u_2, u_3 = tipping fee at the recycling center and landfill (\$/M)
s_1, s_2, s_3 = unit shipping cost over links 1, 2, and 3 (\$/M-L)
p = selling price of the recycled paper (\$/M),
d_1, d_2, d_3 = travel distance over links 1, 2, and 3 (L)
x_1, x_2, x_3 = mass of MSW shipped (M) over links 1, 2 and 3

(a) The net worth $nw = w - k$

where

w = gross revenue from recycled paper,

and

k = total cost of shipping and processing MSW

Determine the total net worth for the network, assuming that x_1, x_2, and x_3 are control variables for each respective link. Clearly define x_1, x_2, and x_3.

(b) Assume that q is the total bulk mass per year generated by the town and that e is the fraction of recyclable paper generated by the town. Derive an LP model to minimize total cost y, the total cost of processing and disposal for the network, in terms of the control variables x_2 and x_3.

2.1.6 Two towns, a recycling center, a landfill, and a marketplace for recyclable materials comprise an MSW district. Assume that towns 1 and 2 generate annually 1000 and 2000 bulk tons of MSW, respectively. The total mass for each town consists of 30% and 40% of recyclable materials, respectively.

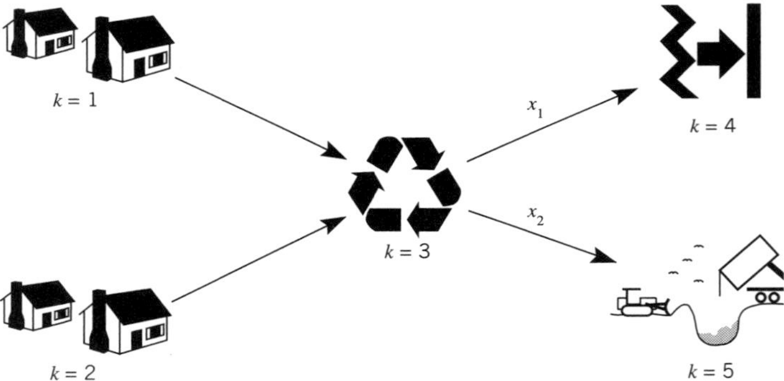

The tipping fee for the recycling center is \$30 per ton. This figure includes all shipping costs from the towns. The shipping fee from recycling center to marketplace for recyclable material is \$50 per ton. The buying price of recyclable materials is \$75 per ton. The landfill tipping fee plus transport is \$45 per ton.

(a) Define the control variables to be

x_1 = mass of recyclables

and

x_2 = mass of remaining MSW after the recyclables are removed

Derive a minimum cost optimization model.

(b) Determine the optimum allocation of resources $\mathbf{x}^*$ and y^* by graphical means. Identify the feasible region.

(c) Use the method of substitution to find $\mathbf{x}^*$ and y^* in part (b).

2.1.7 A town that currently disposes all its MSW in a landfill is considering a volunteer recycling program for metals. Recycling bins will be placed at convenient locations in the town to promote the program. Let the control variables be defined as

$$x_1 = \text{mass of material disposed of in the landfill,}$$

and

$$x_2 = \text{mass of recycled metals}$$

The landfill tipping fee is \$50 per ton, and the selling price of recycled metals is \$100 per ton. The residents of the town generate 10,000 tons of MSW per year. Of this, 1000 tons are recyclable metals. Assume that 100% of the population will volunteer to recycle if it is financially beneficial to them. Assume also that all shipping fees are reflected in the tipping fee and selling price of the metal.

(a) Draw a network diagram for this system.

(b) Derive an LP model to minimize total cost y in terms of x_1 and x_2.

(c) Use a graphical method to solve for the optimum solution, $\mathbf{x}^* = (x_1^*, x_2^*)$ and y^*. Show the feasible region.

(d) Use the method of substitution to solve for the optimum solution $\mathbf{x}^* = (x_1^*, x_2^*)$ and y^*.

2.1.8 A town that currently burns its solid waste in a mass-burn incinerator is considering removing aluminum before it enters the incinerator. This plan is considered economically and ecologically attractive because it reduces the mass of material that is placed in a landfill and saves natural resources. In addition, a profit is made from the sale of recycled aluminum. The price of recycled aluminum is assumed to affect the choice between:

A. Recycling aluminum and burning the remaining material, and

B. Mass burning without recycling.

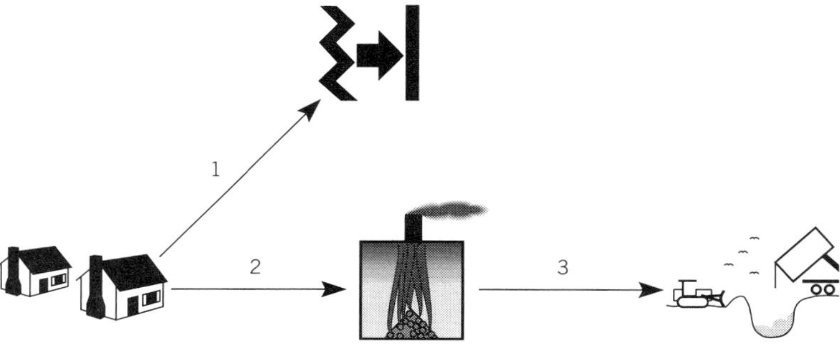

The town has a population of 10,000 and generates 0.7 ton of MSW per person per year. Fifteen percent of the MSW is aluminum. The incinerator, landfill, and aluminum marketplace have a capacity of 8000, 2000, and 5000 tons/year, respectively. The distance from the town to the incinerator and marketplace is 10 and 15 miles, respectively. The shipping cost from the town to the incinerator and the marketplace is \$1 per ton-mile. A 100% volunteerism rate is assumed for the incinerator and marketplace. The incinerator tipping fee is \$45/ton. The distance, shipping cost, and tipping fee for the landfill are assumed to be zero. Assume that 30% of the MSW is ash.

(a) Define control variables and derive an optimization model to minimize total cost y. The selling price of aluminum is \$20 per ton.

(b) Find the optimum solution $\mathbf{x}^*$ and y^* by graphical means.

Indicate the feasible region.

(c) Find the optimum solution $\mathbf{x}^*$ and y^* by the method of substitution.

2.1.9 A city generates no more than 500 tons of aluminum per year. It has two alternatives:

A. Return aluminum products to grocery stores located at selected sites in the city, or

B. Dispose of aluminum products along with other waste products in a curbside recycling program. Aluminum is separated and sold at the marketplace.

The following network diagram depicts the two options. It shows alternative A with a single icon where all grocery stores are assumed to be located at one place, marked as "Grocery."

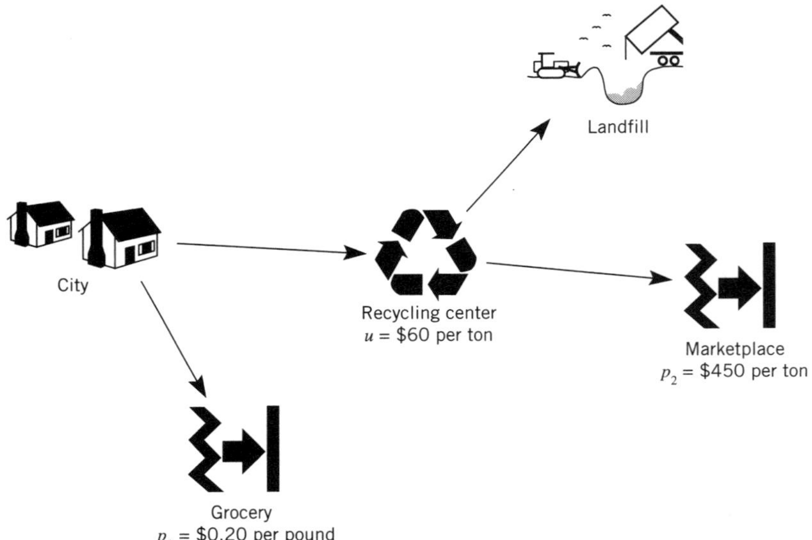

For alternative A, the individual receives \$0.20 per pound for recycled aluminum. Since the individual is assumed to shop at the same time he or she recycles, no shipping or other costs are assumed. A maximum of 25% of the population will volunteer to recycle at the grocery. Assume 2000 pounds = 1 ton.

For alternative B, the aluminum products are separated at a recycling center and sent to the market place. The cost of collecting and processing aluminum is assumed to be \$60 per ton. The recycling center receives \$450 per ton.

Assume that all citizens know that the entire city benefits from recycling with either option. In option A, the benefits from recycling are received as direct payments to the individual. In option B, the benefits are received indirectly through reducing residential taxes or user fees from the sale of recycled aluminum.

(a) Derive a maximum profit model in terms of the control variables:

x_1 = amount of aluminum recycled at the grocery stores, (tons per year)
x_2 = amount of aluminum recycled through the recycling center (tons per year)
z = total profits by all citizens of the city from options A and B

(b) Find the optimum solution $\mathbf{x}^*$ and z^* by graphical means. Show the feasible region.

(c) Find the optimum solution $\mathbf{x}^*$ and z^* for part (b) by the method of substitution.

2.1.10 A construction site requires at least 100 tons of a sand-gravel mixture consisting of a minimum of 75% sand. Material may be obtained at either of two gravel pits.

% Sand		Shipping Distance (miles)	
Pit 1	Pit 2	Pit 1	Pit 2
80	60	20	10

(a) Draw a network diagram and derive an LP model to minimize total travel distance y for the required sand-gravel mixture and material properties.

Let

$$x_1 = \text{amount of material taken from pit 1,}$$

and

$$x_2 = \text{amount of material taken from pit 2}$$

(b) Use a graphical method to solve for the optimum solution, $\mathbf{x}^* = (x_1^*, x_2^*)$ and y^*. Show the feasible region.

2.1.11 An aggregate mix of sand and gravel must contain no less than 20% nor more than 30% gravel. The in situ soil (soil at the site) contains 40% gravel and 60% sand. Pure sand may be purchased and shipped to the site at a unit cost of \$5 per yd^3(cubic yard). A total supply of 1,000 yd^3 is required. There is no charge to use in situ soil.

(a) Draw a network diagram, define control variables, and formulate a linear programming model to minimize total cost.

(b) Draw and identify the feasible solution on a graph.

(c) Solve for the optimum solution by a graphical method. That is: $\mathbf{x}^*$ and y^*.

2.1.12 A liquid mixture must contain no less than 20% and no more than 30% of a chemical. Material 1, which may be purchased for \$50 per gallon, contains 40% of this chemical. Material 2, which may be purchased at \$20 per gallon, contains 10% of this chemical. A total supply of 1000 gallons is required. The supplies are 600 gallons of material 1 and 500 gallons from material 2.

(a) Draw a network diagram, define control variables $\mathbf{x}$, and formulate a linear programming model to minimize total cost.

(b) Draw and identify the feasible solution on a graph.

(c) Solve for the optimum solution by a graphical method. That is $\mathbf{x}^*$ and y^*.

2.2 SENSITIVITY ANALYSIS

Sensitivity analysis is most important in learning how a system behaves under different conditions. In this section we focus on linear programming models and the effect of varying one or more model parameter values $\mathbf{b}$ and $\mathbf{c}$ on $\mathbf{x}^*$ and y^*. Changes in one or more of these parameters are of practical concern because they often represent values imposed by society. First, we introduce new terminology to aid in the discussion.

Active and Inactive Constraints

The terms *active* and *inactive constraints* are used to identify whether or not a solution $\mathbf{x}_q$ is located on the boundary or the extreme point of a feasible region. For example, the constraint equation $f(\mathbf{x}) \leq b$ is an active constraint at $\mathbf{x}_q$ when the strict equality, $f(\mathbf{x}_q) = b$, holds. Otherwise, it is an inactive constraint at $\mathbf{x}_q$, when the strict inequality, $f(\mathbf{x}_q) < b$, holds.

Consider the minimum cost model from Example 2.3:

$$
\begin{aligned}
&\text{minimize } y = 50x_1 - 50x_2 \\
&x_1 + x_2 \geq 100 \qquad \text{(Town)} \\
&0 \leq x_1 \leq 80 \qquad \text{(Landfill 1)} \\
&0 \leq x_2 \leq 60 \qquad \text{(Landfill 2)}
\end{aligned}
$$

Its optimum solution is shown in Figure 2 as the point, $\mathbf{x}_3 = \mathbf{x}^* = (40, 60)$, where $y^* = -\$1{,}000$. Substituting into the constraint equations gives the following results:

$$
\begin{aligned}
x_1^* + x_2^* = 40 + 60 = 100 &\rightarrow \text{Active} \qquad \text{(Town)} \\
x_1^* = 40 < 80 &\rightarrow \text{Inactive} \qquad \text{(Landfill 1)} \\
x_2^* = 60 = 60 &\rightarrow \text{Active} \qquad \text{(Landfill 2)}
\end{aligned}
$$

Graphically, the optimum solution line y^* intersects the feasible region at the extreme point $\mathbf{x}_3 = \mathbf{x}^*$. The extreme point for $\mathbf{x}^*$ is formed by the constraint equations for landfill 2 and town. As a result, the constraint equations for landfill 2 and town are called active constraints. Since y^* does not intersect the landfill 1 line at $\mathbf{x}_3$, it is an inactive constraint as shown in Figure 2.

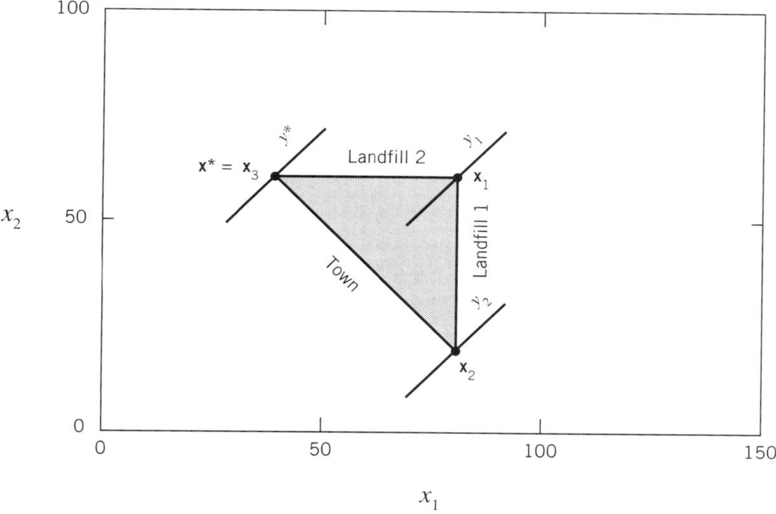

Figure 2 Minimum cost solution of Example 2.3.

Using the same graphical approach, we can classify the constraints at points $\mathbf{x}_1$ and $\mathbf{x}_2$.

Location	Town	Landfill 1	Landfill 2
$\mathbf{x}_1$	inactive	active	active
$\mathbf{x}_2$	active	active	inactive

The point $\mathbf{x}_4 = (100, 0)$ is active for $x_1 + x_2 \geq 100$; however, it is neither a feasible solution nor a candidate for an optimum solution. For our purposes, only active and inactive constraints associated with the feasible region are interesting in a sensitivity analysis.

Consider the effect of changes in one or more values in $\mathbf{b}$ at $\mathbf{x}^*$.

- A change in b_i for an active constraint will cause a parallel shift in the constraint equation and, in turn, a shift in the location of $\mathbf{x}^*$ and the magnitude of y^*.
- In contrast, a small change in b_i of an inactive constraint will shift the inactive constraint line, but it will not affect either $\mathbf{x}^*$ or y^*.

Consider the effect of changes in one or more values in $\mathbf{c}$ at $\mathbf{x}^*$.

- Small changes in $\mathbf{c}$ will cause a change in the slope of y^* and in the magnitude of y^*. The location of $\mathbf{x}^*$ will be unchanged.
- Large changes in $\mathbf{c}$ could cause $\mathbf{x}^*$ to move from one extreme point to another one. The value of y^* will also change.

EXAMPLE 2.4 *A Minimum Cost System Versus a Minimum Mass Disposal System*

Suppose the town in Example 2.3 learned that it could increase the amount of material it currently incinerates by increasing the amount of ash it currently permits to be disposed of in the landfill from 15 to 20 tons per year. In order to accept extra waste for incineration, it must employ additional ash treatment steps. It will increase the cost of ash disposal from $80 to $100 per ton.

Will the town increase its profits if it allows more ash to be disposed of in the landfill?

SOLUTION

The question leads to changes in both a resource constraint parameter and a unit cost parameter in the objective function. Making the appropriate substitutions, we find that the original objective function, $y = 50x_1 - 50x_2$, becomes

$$\text{minimize } y = 50x_1 - 45x_2$$

Note that the increase in treatment cost causes the unit profit for energy recovery to be reduced from $50 to $45 per ton. The landfill 2 constraint becomes $0.25\,x_2 \le 20$, or $x_2 \le 80$.

The model is

$$\begin{aligned} \text{minimize } y = 50x_1 - 45x_2 & \\ x_1 + x_2 \ge 100 & \quad \text{(Town)} \\ x_1 \le 80 & \quad \text{(Landfill 1)} \\ x_2 \le 80 & \quad \text{(Landfill 2)} \\ x_1 \ge 0 & \\ x_2 \ge 0 & \end{aligned}$$

The effect of the changes is shown in the following diagram. The original solution is superimposed to show that the feasible region has changed. The slope of the objective line is changed slightly.

The optimum solution becomes $\mathbf{x}^* = (20, 80)$ with $y^* = -\$2600$ per year. The profit will increase from $1000 to $2600 per year. The amount of material that is incinerated increases from 60 to 80 tons per year. The additional expense for added treatment is outweighed by the added profits from incinerating more solid waste. In addition, the total amount of material dumped in the landfill decreases from 55 to 40 tons per year.

Keep in mind that these changes are relatively *small* parameter changes. Large parameter changes in either active or inactive constraints could dramatically affect the feasible region and solution. In the extreme, the region can be made infeasible (a null set), and so no solution can exist. Large changes in the unit cost parameters can cause the slope of objective function to shift the location of $\mathbf{x}^*$ from one extreme point to another.

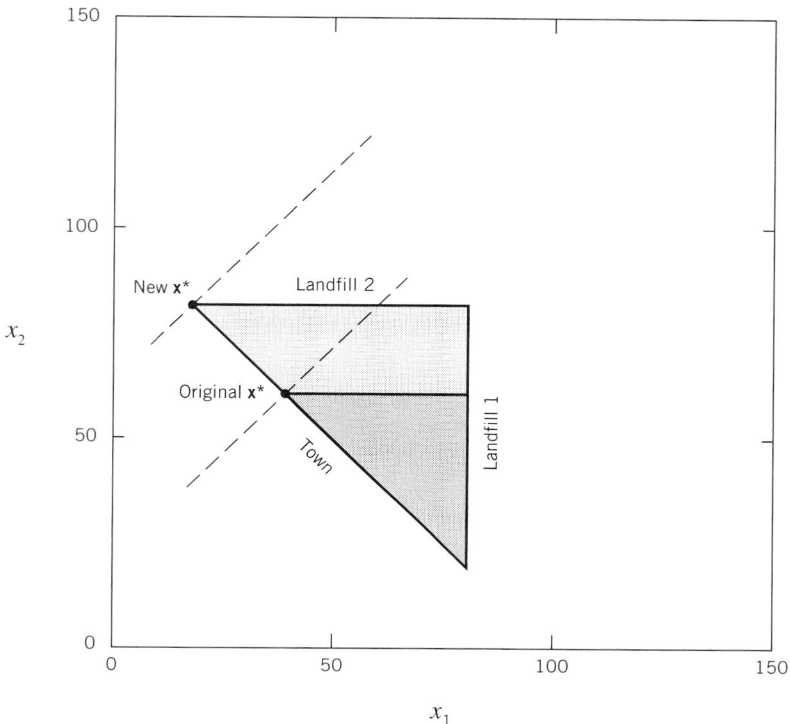

QUESTION FOR DISCUSSION

2.2.1 Define the following terms:
active constraint inactive constraint

QUESTIONS FOR ANALYSIS

2.2.2 The purpose of this example is to study the effects of changing a unit cost parameter in the objective function on $\mathbf{x}^*$ and y^*.

(a) Use a graphical solution to solve for $\mathbf{x}^*$ and y^*.

$$\begin{aligned} \text{minimize } y &= x_2 \\ 3x_1 + 4x_2 &\geq 9 \quad \text{(A)} \\ 5x_1 + 2x_2 &\leq 8 \quad \text{(B)} \\ 3x_1 - x_2 &\leq 0 \quad \text{(C)} \\ x_1 &\geq 0 \\ x_2 &\geq 0 \end{aligned}$$

(b) Identify the active and inactive constraints for $\mathbf{x}^*$ on the graph by equation letter.

(c) Perform a sensitivity analysis by changing $c_1 = 0$ to 4. Determine $\mathbf{x}^*$ and y^*

2.2.3 The purpose of this example is to study the effects of changing a resource constraint on $\mathbf{x}^*$ and y^*.

(a) Use a graphical solution to solve for $\mathbf{x}^*$ and y^*.

$$\begin{aligned} \text{minimize } y = -5x_1 + 7x_2 & \\ x_1 + x_2 \geq 2 & \quad \text{(A)} \\ -2x_1 + x_2 \geq 0 & \quad \text{(B)} \\ -x_1 + 3x_2 \geq 0 & \quad \text{(C)} \\ 4x_1 - x_2 \geq 0 & \quad \text{(D)} \\ x_1 \geq 0 & \\ x_2 \geq 0 & \end{aligned}$$

(b) Identify the active and inactive constraints for $\mathbf{x}^*$ by equation letter.

(c) Perform a sensitivity analysis by changing $b = 0$ of equation (C) to 3. Determine $\mathbf{x}^*$ and y^*.

(d) Perform a sensitivity analysis by changing $b = 3$ of equation (C) to 6. Determine $\mathbf{x}^*$ and y^*.

2.2.4

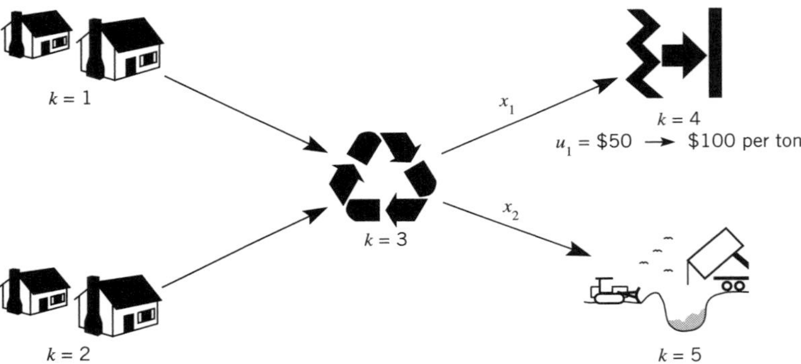

Determine the effects of shipping cost on the recycling program of Problem 2.1.6. In the original problem statement, the shipping fee from recycling center to marketplace is given as u_1 = $50 per ton. If this fee is doubled, how does it affect $\mathbf{x}^*$ and y^*?

(a) Formulate LP models for u_1 = $50 and u_1 = $100 per ton. Solve graphically and compare $\mathbf{x}^*$ and y^* in each case. Use the information given in Problem 2.1.6.

(b) Determine the shipping fee u_1 where it is equally cost effective to have either a combined effort of recycling and shipping MSW to landfill or a singular effort of shipping all MSW materials to the landfill.

2.2.5 Determine the effect of volunteerism on the recycling program of Problem 2.1.7. In the original problem it is assumed that 100% of the people will volunteer to recycle if it is cost effective for them to do so.

Formulate three LP models in which the maximum percentage of people who will volunteer to recycle is 100%, 50% and 0%. Solve graphically and compare $\mathbf{x}^*$ and y^* for each case. Use the information given in Problem 2.1.7.

2.2.6 Scenario A in the following table is considered in Problem 2.1.10. The original problem is to determine which pits should be used to minimize total shipping distance and, at the same time, to supply a construction site with at least 100 tons of a sand-gravel mixture consisting of a minimum of 75% sand. In scenarios B and C, the effects of shipping distance and material properties at each pit are considered, respectively.

	% Sand		Shipping distance, (miles)	
Scenario	Pit 1	Pit 2	Pit 1	Pit 2
A	80	60	20	10
B	80	60	10	10
C	60	60	20	10

(a) Compare scenarios A and B. Formulate appropriate LP models for each scenario and solve graphically.

(b) Formulate an LP model for scenario C. Discuss any difficulties associated with formulating this model.

2.3 THE SIMPLEX METHOD

In Sections 2.1 and 2.2, the fundamental principles of optimization were introduced. Since problems were formulated as 1-D and 2-D LP models, graphical methods of analysis were used. For problems of greater dimension, analytical methods of analysis are required. The simplex method and an electronic computer will be used for this purpose. However, in order to obtain a basic understanding of how this iterative scheme finds the optimum solution, we will use graphical methods.

The simplex method for finding the optimum solution, $\mathbf{x}^*$ and y^*, of a linear programming model consists of the following steps:

1. Establish an initial candidate solution $\mathbf{x}^0$.
2. Test $\mathbf{x}^0$ to determine if it is the optimum solution $\mathbf{x}^*$.
3. If the test of step 2 fails, establish a new candidate solution $\mathbf{x}^1$, assign $\mathbf{x}^0 = \mathbf{x}^1$, and repeat steps 2 and 3 until $\mathbf{x}^*$ is found.

All candidate solutions of $\mathbf{x}^0$ and $\mathbf{x}^1$ are extreme points and feasible solutions.

For finding a minimum, the simplex method uses a method of steepest descent. If a candidate solution of $\mathbf{x}^0$ is not the location of the optimum, then a new candidate solution of $\mathbf{x}^1$ is established by comparing the relative magnitudes of the slope of the objective function from $\mathbf{x}^0$ to all extreme points in the vicinity of $\mathbf{x}^0$. The extreme point that corresponds with the largest negative slope is selected as $\mathbf{x}^1$. It replaces $\mathbf{x}^0$ and the optimality test is repeated. This iterative procedure is repeated until $\mathbf{x}^*$ is found.

Figure 3 is used to illustrate the procedure. For this 2-D problem, there are only three candidate solutions for $\mathbf{x}^*$, $\mathbf{x}_1$, $\mathbf{x}_2$, and $\mathbf{x}_3$. For purposes of illustration, assume in step 1 that $\mathbf{x}^0 = \mathbf{x}_2$ is assigned. This is a satisfactory starting point because $\mathbf{x}_2$ is a feasible solution and an extreme point. In step 2, we test to determine if $\mathbf{x}^0 = \mathbf{x}_2$ is a minimum and find that it is not. In step 3, the candidate points for $\mathbf{x}^1$ in the vicinity of $\mathbf{x}_2$ are $[\mathbf{x}_1, \mathbf{x}_3]$. We calculate the slopes of the lines drawn between $\mathbf{x}_2$ and $\mathbf{x}_1$ and between $\mathbf{x}_2$ and $\mathbf{x}_3$. After comparing the two slopes, we find that the greater negative slope is between $\mathbf{x}_2$ and $\mathbf{x}_3$. We assign $\mathbf{x}^0$ to be equal to $\mathbf{x}_3$, repeat step 2, and find that $\mathbf{x}_3$ is the location of the minimum or $\mathbf{x}^*$.

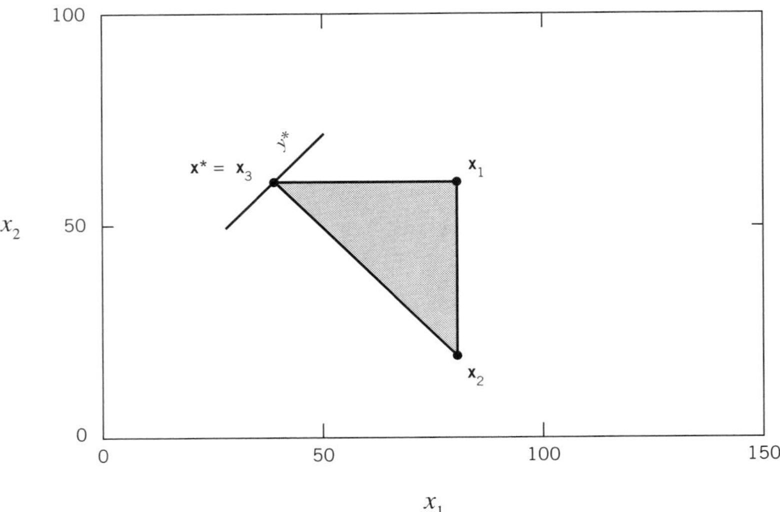

Figure 3 The minimum cost solution of Example 2.3.

Now, let us investigate other features of the simplex method.

First, since $\mathbf{x}^*$ can never lie within the interior of the feasible region, the search for $\mathbf{x}^*$ is always limited to extreme points. This eliminates all possible interior points from consideration.

Second, the simplex method is not an exhaustive search method. Not all extreme points of the feasible region need to be evaluated to find $\mathbf{x}^*$, but only those points that lie on a path of steepest descent. While efficient, the procedure may not minimize the number of extreme points that are evaluated.

Third, the test for optimality is incorporated into the method of steepest descent. Imagine for the moment that the initial candidate for $\mathbf{x}^*$ is $\mathbf{x}_2$ and that all other information about other extreme points in Figure 3 is not available to us. If we can proceed down a path of steepest descent, then we will move from $\mathbf{x}_2$ to $\mathbf{x}_3$. The direction of the move is not altered until another extreme point is reached. In this case, $\mathbf{x}^* = \mathbf{x}_3$ is found. If we attempt to move to a new extreme point from $\mathbf{x}_3$ along a path of maximum descent, it is impossible because we restrict our search to the feasible region and extreme points only. We can only conclude that $\mathbf{x}_3$ is the location of $\mathbf{x}^*$.

For these three reasons, the simplex method is considered an efficient method of analysis. However, if an optimum solution is all points along a boundary line, the simplex method will not identify all these points as being members of $\mathbf{x}^*$. It can only find one point, an extreme point on the optimum boundary line. Since this is a rare event in practical application, this flaw of the simplex method is considered a minor one.

Convex Sets

Another important requirement of the simplex method for finding $\mathbf{x}^*$ is that the feasible region must form a convex set. The feasible region of Figure 3 forms a convex set, but the region given in Figure 4 is an example of a nonconvex set.

Formally, any arbitrary line joining two points that fall within or on the boundary of the feasible region must lie within or on the boundary of the feasible region if the set is to be convex. Since the dashed line shown in Figure 4 fails this requirement, the region is classified as a nonconvex set.

Convex sets are an important consideration in any search for a global optimum with a numerical method like the simplex method. If a constraint set is not convex, generally there is no guarantee that a global optimum has been found.

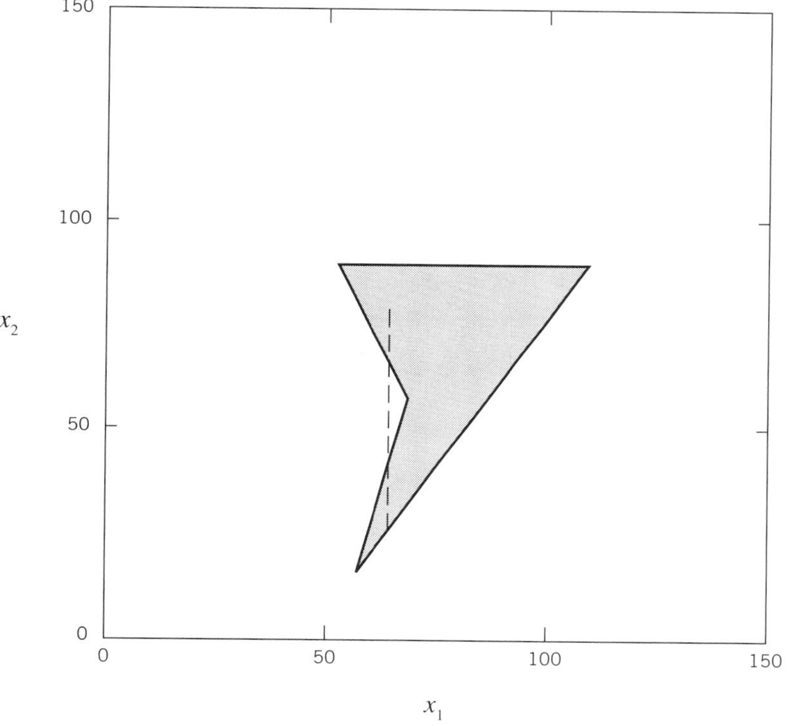

Figure 4 A nonconvex set.

EXAMPLE 2.5 *A Minimum Cost Aggregate Mix Model*

A construction site requires 100 yd³ of aggregate. The aggregate mixture must contain a minimum of 30% sand and no more than 60% gravel or 10% silt. Three sources of aggregate are available. Source 3 meets this specification, but it is the most expensive. By mixing an appropriate amount of material from each source, it is possible to meet the specification and reduce total cost at the same time.

Source	1	2	3
% Sand	5	30	100
% Gravel	60	70	0
% Silt	35	0	0
Unit cost ($ per yd^3)	2	10	8

(a) Formulate an LP model to determine the minimum cost solution y^* and the optimum allocation of material $\mathbf{x}^*$.

(b) Determine the effect of the unit cost c_1 on the y^* and $\mathbf{x}^*$.

SOLUTION

The network diagram for this problem is:

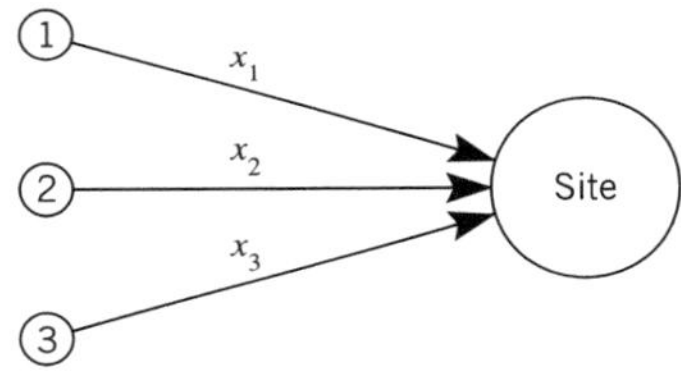

(a) Define the control variables as

x_1, x_2, x_3 = amounts of material in yd^3 taken from sources 1, 2, and 3, respectively (L^3)

The objective function to minimize total cost y is

$$\text{minimize } \underset{(\$)}{y} = \underset{(\$/L^3)(L^3)}{2x_1} + \underset{(\$/L^3)(L^3)}{10x_2} + \underset{(\$/L^3)(L^3)}{8x_3}$$

The aggregate mixture requirements are introduced as a constraint set. A constraint equation for sand, gravel and silt will be derived.

The aggregate must contain a minimum of (0.3) (100) = 30 cubic yards of sand. The total amount or supply of sand in the aggregate in yd^3, written in terms of the control variables, is the sum of the individual supply of sand from each source, or

$$s_s = 0.05x_1 + 0.3x_2 + 1.0x_3$$

Since $s_s \geq 30$, the constraint equation for sand is

$$0.05x_1 + 0.3x_2 + 1.0x_3 \geq 30 \quad \text{(Sand)}$$

Using the same approach, we find that the supplies of gravel s_g and silt s_{sl} are

$$s_g = 0.6x_1 + 0.7x_2 \quad \text{(Gravel)}$$
$$s_{sl} = 0.35x_1 \quad \text{(Silt)}$$

Since $s_g \leq 60$ yd^3 and $s_{sl} \leq 10$ yd^3, the following constraint equations are obtained:

$$0.6x_1 + 0.7x_2 \leq 60 \quad \text{(Gravel)}$$
$$0.35x_1 \leq 10 \quad \text{(Silt)}$$

The total aggregate supply of 100 yd^3 must be equal to the sum of the amounts supplied from each source, or

$$x_1 + x_2 + x_3 = 100 \quad \text{(Supply)}$$

The complete LP model is

$$\begin{aligned}
&\text{minimize } y = 2x_1 + 10x_2 + 8x_3\\
&0.05x_1 + 0.3x_2 + 1.0x_3 \geq 30 \quad \text{(Sand)}\\
&0.6x_1 + 0.7x_2 \leq 60 \quad \text{(Gravel)}\\
&0.35x_1 \leq 10 \quad \text{(Silt)}\\
&x_1 + x_2 + x_3 = 100 \quad \text{(Supply)}\\
&x_1 \geq 0,\ x_2 \geq 0,\ x_3 \geq 0
\end{aligned}$$

The simplex method returns the following optimum solution,

$$y^* = \$628.60$$

where

$$x_1^* = 26.6 \text{ yd}^3,\ x_2^* = 0 \text{ yd}^3,\ x_3^* = 71.4 \text{ yd}^3$$

(b) The results of the sensitivity analysis show that for $c_1 \leq \$7$ per yd^3 cost savings are achieved by purchasing and mixing material from sources 1 and 3. For $c_1 > \$7$ per yd^3, the minimum cost solution is obtained by purchasing all material from source 3.

c_1 (\$ per yd^3)	x_1^* (yd^3)	x_2^* (yd^3)	x_3^* (yd^3)	y^* (\$)
0	28.6	0	71.4	571
1	28.6	0	71.4	600
2	28.6	0	71.4	629
3	28.6	0	71.4	657
4	28.6	0	71.4	686
5	28.6	0	71.4	714
6	28.6	0	71.4	743
7	28.6	0	71.4	771
8	0	0	100	800
9	0	0	100	800
10	0	0	100	800

QUESTION FOR DISCUSSION

2.3.1 (a) Give three reasons why the simplex method is considered an efficient method of analysis.

(b) Give three shortcomings or limitations of the method.

QUESTIONS FOR ANALYSIS

Instructions: All problems in this section should be solved with the simplex method and the aid of a computer.

2.3.2 (a) Use the simplex method to solve for **x*** and y^*:

$$\begin{aligned} \text{minimize } y &= x_2 \\ 3x_1 + 4x_2 &\geq 9 \quad \text{(A)} \\ 5x_1 + 2x_2 &\leq 8 \quad \text{(B)} \\ 3x_1 - x_2 &\leq 0 \quad \text{(C)} \\ x_1 &\geq 0 \\ x_2 &\geq 0 \end{aligned}$$

This LP model is the same one given in Problem 2.2.2.

(b) Perform a sensitivity analysis. Change c_1 from 0 to 4 in the LP objective function. Determine **x*** and y^* for each case.

2.3.3 (a) Use the simplex method to solve for **x*** and y^*.

$$\begin{aligned} \text{minimize } y &= -5x_1 + 7x_2 \\ x_1 + x_2 &\geq 2 \quad \text{(A)} \\ -2x_1 + x_2 &\geq 0 \quad \text{(B)} \\ -x_1 + 3x_2 &\geq 0 \quad \text{(C)} \\ 4x_1 - x_2 &\geq 0 \quad \text{(D)} \\ x_1 &\geq 0 \\ x_2 &\geq 0 \end{aligned}$$

This LP model is the same one given in Problem 2.2.3.

(b) Perform a sensitivity analysis. Change the resource constraint equation (C) from 0 to 6. Determine **x*** and y^* for each case.

2.3.4 A sand and gravel mixture must contain a minimum of 42.5% sand and less than 57.5% gravel. The total aggregate supply of 100 yd^3 is needed. The percentage of sand and gravel and the unit cost from each source are:

Source	1	2	3
% Sand	35	40	45
% Gravel	65	60	55
Unit cost, ($ per yd^3)	6	9	11

(a) Formulate an LP model to minimize cost. Draw a network diagram to assist you in the model formulation. Use a simplex method to solve for **x*** and y^*.

(b) Perform a sensitivity analysis. Change the unit cost for source 2 from c_2 = \$9 to \$6 per yd^3.

(c) Perform another sensitivity analysis. Change the resource constraint for gravel from less than 57.5% to less than 55%. Let c_2 = \$9 per yd^3.

(d) Compare the answers from (a), (b), and (c) in terms of the minimum cost y^*, the supplier, and the percentage of sand and gravel for the optimum solution $\mathbf{x}^*$.

2.3.5 Use the data supplied in the table for Problem 2.3.4. The total aggregate supply of 100 yd^3 is needed.

(a) Formulate a minimum cost LP model to obtain an exact 50–50% mixture of sand and gravel. It should be obvious that a 50–50% mixture cannot be obtained; however, use a simplex method to show that an infeasible solution exists for this mix specification.

(b) Perform a feedback analysis by reformulating and solving the problem from (a) again. The exact mixture specification may be obtained by (1) shifting the mixture, or (2) adding pure sand. The unit cost to remove excess gravel is \$20 per yd^3. The cost of pure sand is \$25 per yd^3. Formulate and solve a minimum cost LP model. Use a simplex method to solve for $\mathbf{x}^*$ and y^*.

2.3.6

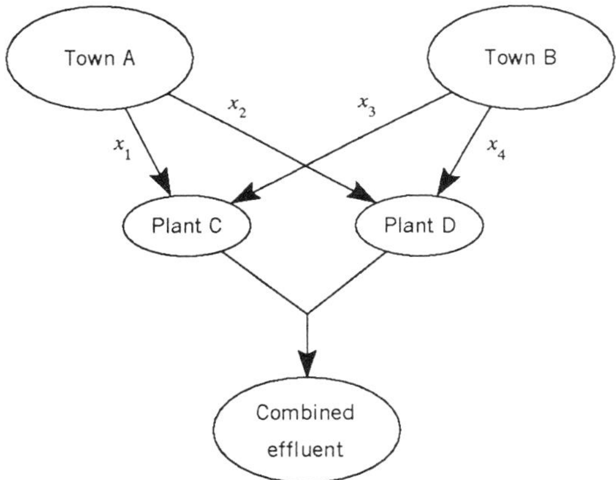

Towns A and B produce 3 and 2Mgal per day of wastewater, respectively. The five-day biological oxygen demand (BOD) level of the wastewater is 200mg per liter. Towns A and B belong to a regional wastewater treatment cooperative that operates two treatment plants. Plants C and D have a capacity of 3 and 4Mgal and BOD removal efficiencies of 90% and 80%, respectively. The cooperative has the authority to assign wastewater flow to either plant. The objective is to minimize wastewater BOD effluent from the combined flows of the two plants.

(a) Formulate an LP model to allocate the flow to minimize the BOD concentration in the combined effluent. Use a simplex method to find the optimum flows.

(b) The BOD standard for the combined flows should not exceed 30mg per liter. Suppose that the flow through plant C will be reduced to 2Mgal per day for plant repair. Can the BOD standard be satisfied?

REFERENCES

Au T., and T. Stelson (1969). *Introduction to Systems Engineering, Deterministic Methods*, Addison-Wesley, Reading, MA.

Haith, D. A. (1982). *Environmental Systems Optimization*, John Wiley & Sons, New York.

Ortolano, L. (1984). *Environmental Planning and Decision Making*, John Wiley & Sons, New York.

Ossenbruggen, P. J. (1984). *Systems Analysis for Civil Engineers*, John Wiley & Sons, New York.

Stark, R. M., and R. L. Nichols (1972). *Mathematical Foundations for Design, Civil Engineering Systems*, McGraw-Hill, New York.

Wagner, H. (1969). *Principles of Operations Research*, Prentice-Hall, Englewood Cliffs, NJ.

Chapter 3

Linear and Network Models

Every method is imperfect.

Charles-Jean-Henri Nicolle (1932)

In Chapters 1 and 2, network diagrams are employed to conceptualize problems and to assist in their model formulations. In this chapter, once again we utilize network diagrams and the simplex method for formulating and solving LP models. Some problems, such as critical path scheduling and transportation problems, have unique properties that are well suited to network analysis and LP model formulation. In Sections 3.1 and 3.2, we study different types of network models, and in Section 3.3, we formulate and solve more general types of LP models.

3.1 NETWORK MODELS

As we have discovered, a network consists of a set of nodes and a set of links connecting various pairs of nodes. Each link has a specified direction. The links show the direction of flow and can be used to indicate the amount of material being shipped between two nodes. The optimum solution $\mathbf{x}^*$ is one or more paths consisting of a set of connected links between source, intermediate, and sink nodes.

Networks are used for a wide range of applications, and their control variables have various meanings. However, certain problems arise when control variables are assigned to be integer values in lieu of continuous values that have been used in the book up to this point. For example, when looking for the shortest route between source and finish nodes in a network, a zero-one control variable, $x_j = 0$ or 1, is used to designate or reject a link as being part of the shortest or optimum path. Models are classified by control variable type.

- LP models: all link control variables are $x_j \geq 0$ where x_j is defined to be a continuous variable over a nonnegative interval.
- Integer programming models: all link control variables are $x_j = 0, 1, 2, \ldots$.
- Zero-one programming models: all link control variables are $x_j = 0$ or 1.
- Mixed programming models: some links are defined to be integer control variables, and others, continuous control variables.

The search procedure employed by the simplex method assumes that all control variables are of the continuous type. Recall that the overall procedure follows a regimented pattern of examining candidate extreme points. No restriction is placed on the simplex method to limit these extreme points of **x** to be integer values.

The simplex method can be used to successfully solve some zero-one, integer, and mixed model problems. Owing to a unique set of properties of these formulations, the simplex method will lead to an optimum solution $\mathbf{x}^*$ satisfying the condition of integer values. If the solution to the problem leads to an optimum solution $\mathbf{x}^*$ of noninteger values, then rounding values of $\mathbf{x}^*$ from continuous to integer numbers may be possible. Be warned: it may lead to an improper location of $\mathbf{x}^*$ and an unsatisfactory result. The shortest route, critical path scheduling, and transportation and transshipment models that are formulated in this chapter as zero-one and integer programming problems can be successfully solved with the simplex method. However, for applications with a large number of variables in **x**, special computer algorithms for integer, zero-one, and mixed programming problems may prove to be more efficient than the simplex method.

Consider the following 2-D integer programming problem that is solved graphically, assuming the control variables are integers, and then solved a second time, assuming the variables are continuous variables.

EXAMPLE 3.1 *A Truck Purchase*

Each day a trucker must ship a minimum of 70 yd^3 per day a distance of 200 miles. Two truck designs are to be considered. The trucker will spend no more than $300,000 to purchase trucks.

Alternatives	Operating Cost (in $ per mile)	Capital Cost	Truck Capacity (in yd^3)
A	$1.50	$115,000	20
B	$1.75	$85,000	30

(a) Determine the optimum number of truck types to minimize total operating cost.

(b) Reformulate and solve the integer programming model of (a) as an LP model with continuous variables. If necessary, roundoff the solution to obtain an integer value solution and then discuss the ramifications of rounding.

SOLUTION

(a) We will formulate an integer programming model because the number of trucks purchased are whole numbers. Define the control variables to be

$$x_1 = \text{number of type A trucks purchased}$$
$$x_2 = \text{number of type B trucks purchased}$$

The objective function to minimize operating cost y, measured in \$ per mile, is

$$\text{minimize } y = \$1.50\, x_1 + \$1.75\, x_2$$

The requirement that 70 cubic yards must be shipped each day is introduced as the constraint equation:

$$20\, x_1 + 30\, x_2 \geq 70$$

The budget limitation is introduced as the constraint equation:

$$\$115{,}000\, x_1 + \$85{,}000\, x_2 \leq \$300{,}000$$

The integer programming model is

$$\begin{aligned} &\text{minimize } y = \$1.50\, x_1 + \$1.75\, x_2 \\ &20\, x_1 + 30\, x_2 \geq 70 && \text{(a)} \\ &\$115{,}000\, x_1 + \$85{,}000\, x_2 \leq \$300{,}000 && \text{(b)} \\ &x_1 \geq 0, 1, 2, 3, \ldots \\ &x_2 \geq 0, 1, 2, 3, \ldots \end{aligned}$$

The following figure shows the feasible region consisting of two points, (0, 3) and (1, 2). The light gray lines are constructed from equations (a) and (b) and placed on the diagram to help identify the feasible region. The light gray lines are not part of the feasible region.

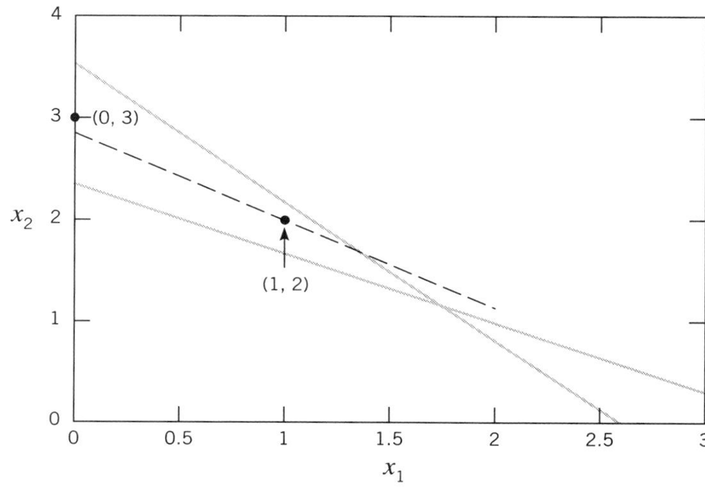

The minimum operating cost line must pass through either one of these points. The optimum solution is $\mathbf{x}^* = (1, 2)$, or $x_1^* = 1$ type A truck, $x_2^* = 2$ type B trucks and $y^* = \$5$ per mile per day. The daily operating cost is

$$(200 \text{ miles})(\$5 \text{ per mile per day}) = \$1000 \text{ per day}$$

(b) The model formulation from (a) is the same, but the nonnegative constraints are changed from

$$x_1 \geq 0, 1, 2, 3, \ldots \rightarrow x_1 \geq 0,$$

and

$$x_2 \geq 0, 1, 2, 3, \ldots \rightarrow x_2 \geq 0$$

The feasible region is shown in the graph as the shaded region.

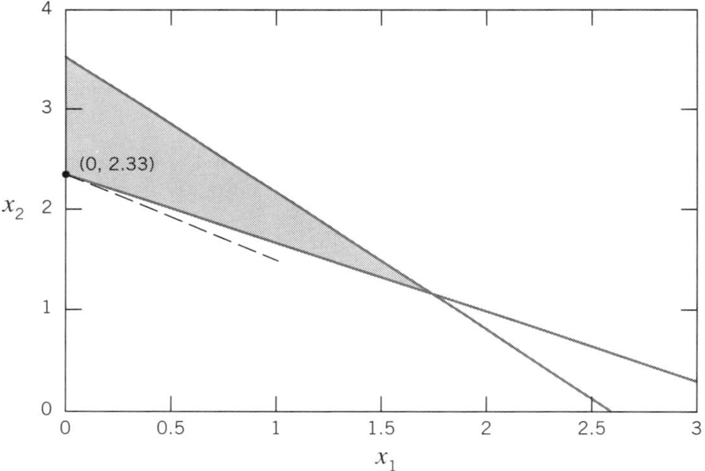

The optimum solution for this formulation is at the extreme point, $\mathbf{x}^* = (0, 2.33)$, or $x_1^* = 0$ type A trucks, $x_2^* = 2.33$ type B trucks, and $y^* = \$3.50$ per mile. This solution makes no sense because a whole number of trucks must be purchased.

Rounding off the solution gives $x_1^* = 0$ type A trucks and $x_2^* = 2$ type B trucks. Rounding only exacerbates the difficulty with this solution. The budgetary constraint of $\$115{,}000\, x_1^* + \$85{,}000\, x_2^* \leq \$300{,}000$ is satisfied, but the supply constraint of $20\, x_1^* + 30\, x_2^* \geq 70$ is not! While it is possible to calculate the operating cost, it is a meaningless figure.

Because the simplex method assumes that all LP model formulations have continuous variables, using it in this example and then rounding its optimum solution to integer values will lead to the same errors as illustrated using a graphical solution method. For this model, an integer programming algorithm is needed.

Some Network Model Formulations

In this section, the discussion is limited to multinodal networks with single start and finish nodes as shown in Figure 1. This network diagram will be used to find an optimum path or solution x^* that defines the shortest path and longest path between the start (source) and finish (sink) nodes and network capacity. The choice of control variables, whether they are continuous or integer values, will be dictated by the problem. The utility of a network diagram with a single start-finish nodal pair will be demonstrated by examples.

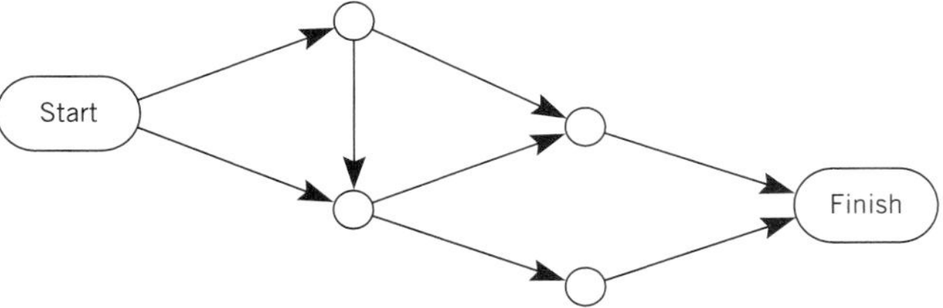

Figure 1 A network diagram with a single start-finish nodal pair.

EXAMPLE 3.2 A Shortest Route Problem

A roadway network is comprised of local streets, arterial roads, and expressways. The characteristics of each link of the network are summarized in the following table.

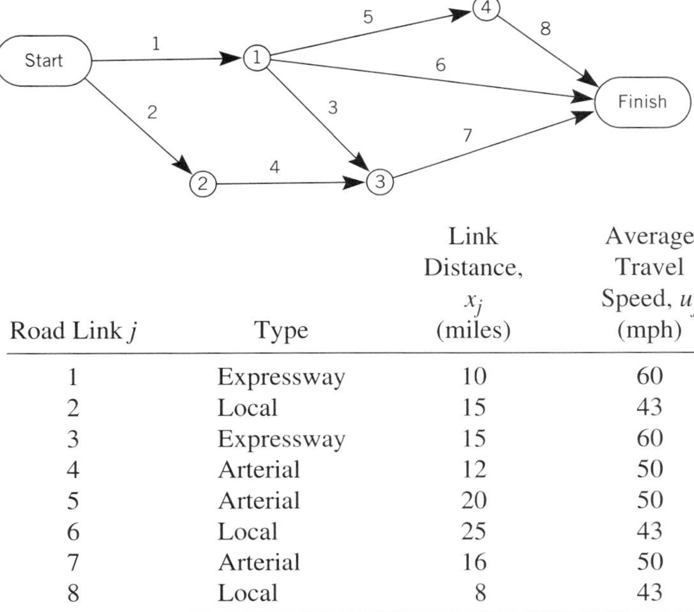

Road Link j	Type	Link Distance, x_j (miles)	Average Travel Speed, u_j (mph)
1	Expressway	10	60
2	Local	15	43
3	Expressway	15	60
4	Arterial	12	50
5	Arterial	20	50
6	Local	25	43
7	Arterial	16	50
8	Local	8	43

Formulate a zero-one programming model to determine the shortest route from the start to finish nodes of the network.

SOLUTION

Define a zero-one control variable for each link of the network:

$$x_j = \begin{cases} 1 \text{ or link } j \text{ is assigned to be on the shortest path.} \\ 0 \text{ or link } j \text{ is not on the shortest path.} \end{cases}$$

The total travel time y is the sum of individual travel times for the path connecting the start and finish nodes. The travel time, in hours, for link j is equal to travel distance over the link divided by the average speed,

$$t_j = \frac{x_j}{u_j}$$

where

$$j = 1, 2, \ldots, 8$$

The objective function is

$$\text{minimize } y = \frac{10}{60}x_1 + \frac{15}{43}x_2 + \frac{15}{60}x_3 + \frac{12}{50}x_4 + \frac{20}{50}x_5 + \frac{25}{43}x_6 + \frac{16}{50}x_7 + \frac{8}{43}x_8$$

The mass conservation law will be applied at the start, finish, and each intermediate node. The start and finish nodes are assigned links and a value of 1.

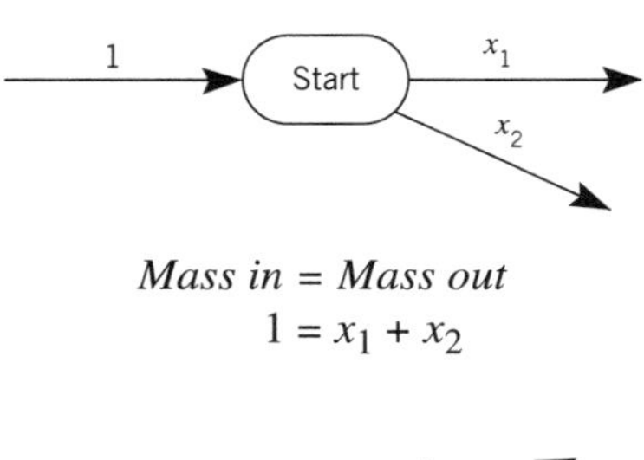

Mass in = Mass out

$$1 = x_1 + x_2$$

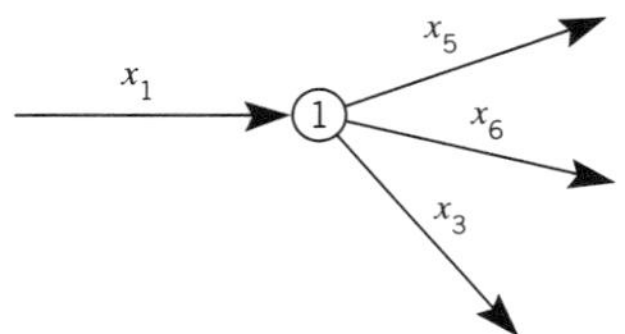

Mass in = Mass out

$$x_1 = x_5 + x_6 + x_3$$

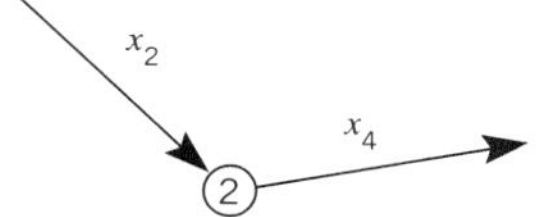

Mass in = Mass out

$$x_2 = x_4$$

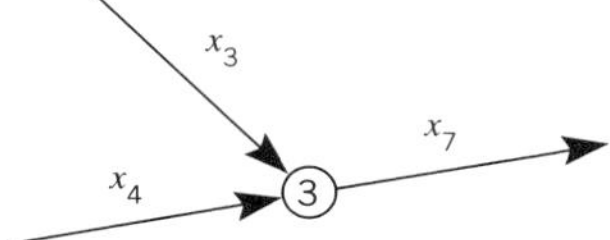

Mass in = Mass out

$$x_3 + x_4 = x_7$$

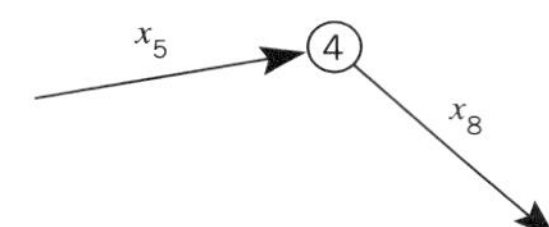

Mass in = Mass out

$$x_5 = x_8$$

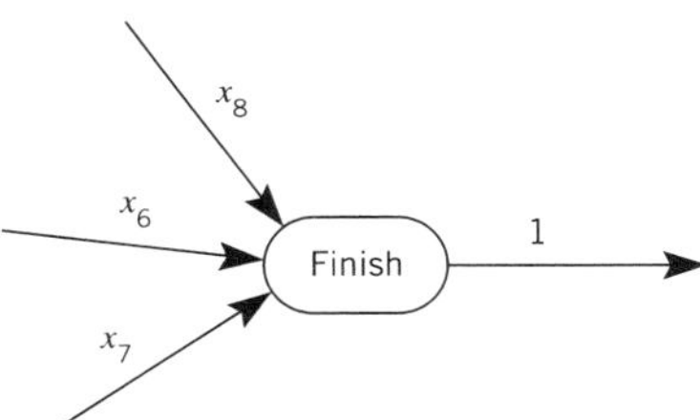

Mass in = Mass out

$$x_6 + x_7 + x_8 = 1$$

The final model is a zero-one programming model,

$$\text{minimize} \quad y = \frac{10}{60}x_1 + \frac{15}{43}x_2 + \frac{15}{60}x_3 + \frac{12}{50}x_4 + \frac{20}{50}x_5 + \frac{25}{43}x_6 + \frac{16}{50}x_7 + \frac{8}{43}x_8$$

$$\begin{array}{ll}
x_1 + x_2 = 1 & \text{(Start)} \\
x_1 - x_5 - x_6 - x_3 = 0 & \text{(Node 1)} \\
x_2 - x_4 = 0 & \text{(Node 2)} \\
x_3 + x_4 - x_7 = 0 & \text{(Node 3)} \\
x_5 - x_6 = 0 & \text{(Node 4)} \\
x_6 + x_7 + x_8 = 1 & \text{(Finish)} \\
x_j = 0, 1 \text{ for j} = 1, 2, \ldots, 8 &
\end{array}$$

The optimum solution is $y^* = 0.74$ hour with $x_1^* = x_3^* = x_7^* = 1$ and all other control variables equal to zero. The shortest path is

$$1 \rightarrow 3 \rightarrow 7$$

As shown here, all shortest route problems will consist of a single path connecting the start and finish nodes. For some shortest route network problems, there may not be a unique solution. The simplex and zero-one programming methods may only identify one of them.

EXAMPLE 3.3 *Network Capacity*

Consider the highway network of Example 3.2. The flow capacity of each link is summarized in the following table.

Road j	Type	Number of Lanes, n_j	Flow Capacity, q_j (vehicles per hour per lane)
1	Expressway	3	2000
2	Local	1	1900
3	Expressway	3	2000
4	Arterial	2	2000
5	Arterial	2	2000
6	Local	1	1900
7	Arterial	2	2000
8	Local	1	1900

Determine the capacity of the network by formulating an LP model with continuous control variables.

SOLUTION

Define a continuous control variable for each link of the network, where

$$x_j = \text{flow in vehicles per hour on link } j$$

The maximum flow z is the same at the start or at the finish nodes,

$$z = x_1 + x_2 = x_6 + x_7 + x_8$$

The mass conservation law is applied at all intermediate nodes. The formulations of these relationships are shown in Example 3.2.

The capacity for each link is given by the side constraint,

$$x_j \le n_j q_j$$

The final model is an LP model.

$$\begin{aligned}
&\text{maximize } z = x_1 + x_2 \\
&x_1 - x_5 - x_6 - x_3 = 0 \qquad \text{(node 1)} \\
&x_2 - x_4 = 0 \qquad \text{(node 2)} \\
&x_3 + x_4 - x_7 = 0 \qquad \text{(node 3)} \\
&x_5 - x_8 = 0 \qquad \text{(node 4)} \\
&x_1 \le 6000 \\
&x_2 \le 1900 \\
&x_3 \le 6000 \\
&x_4 \le 4000 \\
&x_5 \le 4000 \\
&x_6 \le 1900 \\
&x_7 \le 4000 \\
&x_8 \le 1900 \\
&x_j \ge 0 \text{ for } j = 1, 2, \ldots, 8
\end{aligned}$$

The optimum solution or capacity of the network is $z^* = 7800$ vehicles per hour. The optimum flows $\mathbf{x}^*$ at capacity are shown on the network diagram.

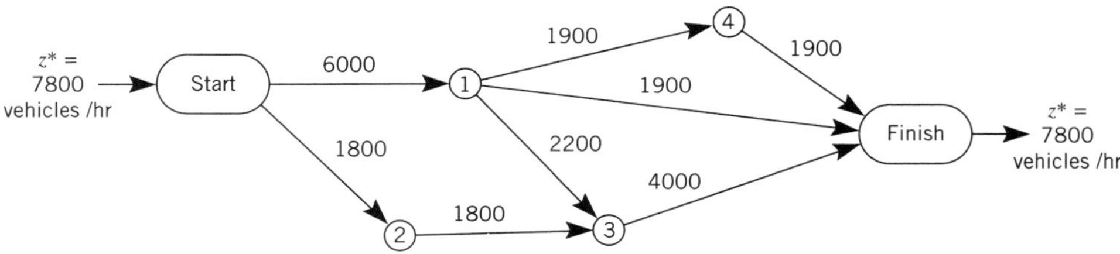

The following shows that most roadways operate below capacity:

$$\begin{aligned}
x_1^* &= 6000 = 6000 \\
x_2^* &= 1800 < 1900 \\
x_3^* &= 2200 < 6000 \\
x_4^* &= 1800 < 4000 \\
x_5^* &= 1900 < 4000 \\
x_6^* &= 1900 = 1900 \\
x_7^* &= 4000 = 4000 \\
x_8^* &= 1900 = 1900 \text{ vehicles per hour}
\end{aligned}$$

This solution is not unique. Note that the following allocation of vehicular flows for links 1, 2, and 4 will also have the same network capacity of $z^* = 7800$ vehicles per hour:

$$x_1^* = 5900 < 6000$$
$$x_2^* = 1900 = 1900$$
$$x_4^* = 1900 < 4000 \text{ vehicles}$$

The simplex method will lead to the determination of one of these solutions.

EXAMPLE 3.4 *A Critical Path Schedule*

A construction project of a single-family house consists of the following tasks, with estimated times to complete each of these tasks.

Task	Description	Estimated Completion Time in Weeks	Predecessor Task
1	Excavate the site	1	start
2	Lay foundation	2	1
3	Erect walls	2	2, 5
4	Install roof	1	3, 6
5	Prefabricate walls	1	start
6	Prefabricate roof	1.5	2, 5
7	Landscape the site	1	3, 6
8	Install the pumping and heating systems	2	3, 6

The order in which the tasks must be completed are indicated in the predecessor task column but are more clearly described by a network diagram. The network shows that some tasks may proceed simultaneously with others.

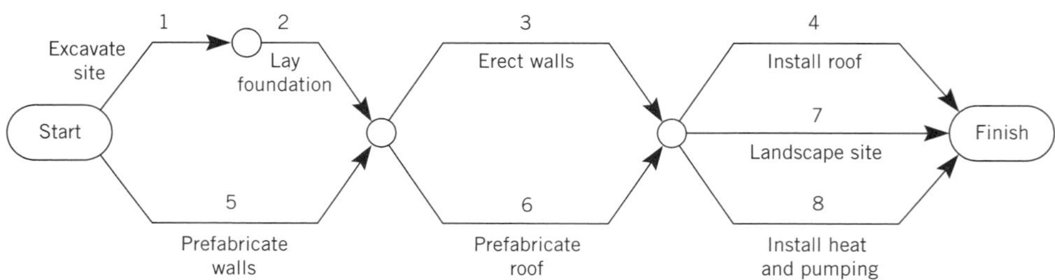

Determine the critical time to complete the house using a zero-one programming model.

SOLUTION

A critical time path is the sequence of tasks that must be begun and be completed on time to avoid delay. The critical time path is defined as the longest possible time path of the network. Define a zero-one control variable for each link of the network,

$$x_j = \begin{cases} 1 \text{ or link } j \text{ is assigned to be on the critical path.} \\ 0 \text{ or link } j \text{ is not on the critical path.} \end{cases}$$

The objective function to maximize the total time is

$$\text{maximize } y = 1x_1 + 2x_2 + 2x_3 + 1x_4 + 1x_5 + 1.5x_6 + 1x_7 + 2x_8$$

The mass conservation law will be applied at each node of the network. The start and finish nodes are assigned links with a value of one.

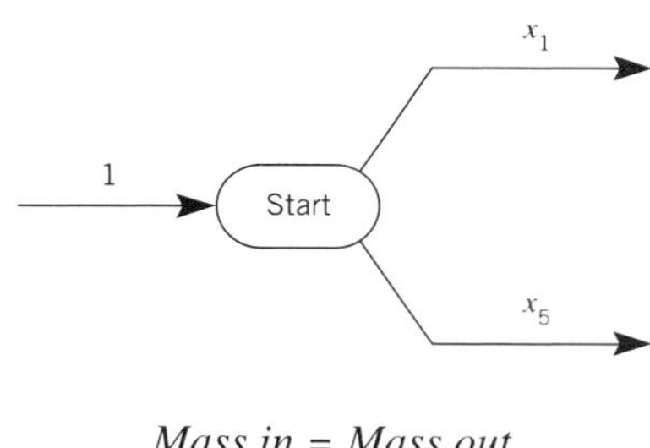

Mass in = Mass out

$$1 = x_1 + x_5$$

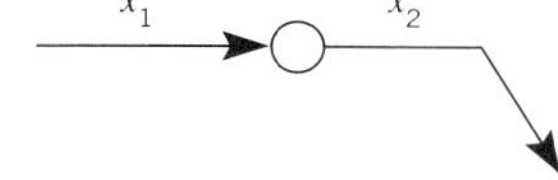

Mass in = Mass out

$$x_1 = x_2$$

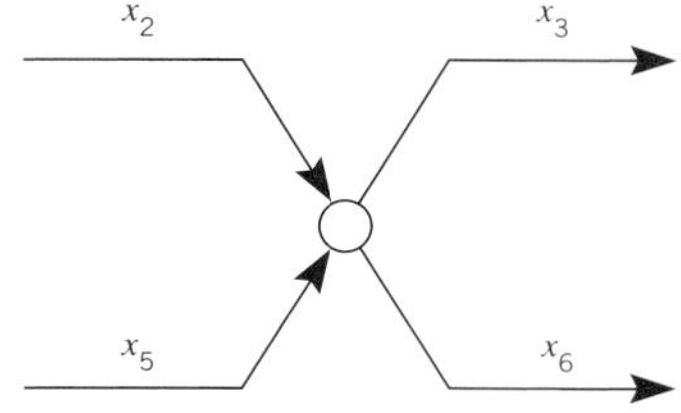

Mass in = Mass out

$$x_2 + x_5 = x_3 + x_6$$

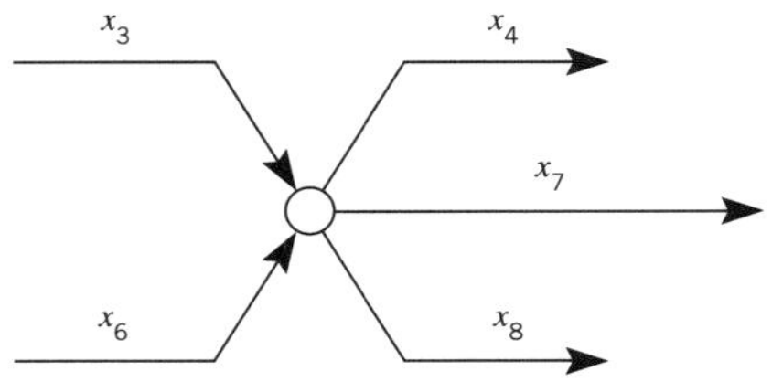

Mass in = Mass out

$$x_3 + x_6 = x_4 + x_7 + x_8$$

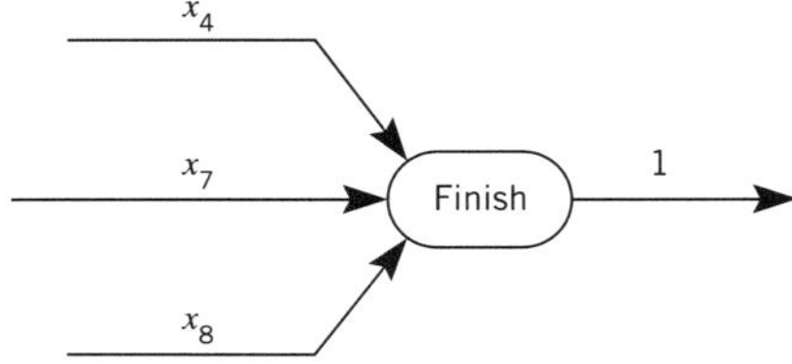

Mass in = Mass out

$$x_4 + x_7 + x_8 = 1$$

The final model is a zero-one programming model:

$$\begin{aligned}
&\text{maximize } y = 1x_1 + 2x_2 + 2x_3 + 1x_4 + 1x_5 + 1.5x_6 + 1x_7 + 2x_8 \\
&x_1 + x_5 = 1 \\
&x_1 - x_2 = 0 \\
&x_2 + x_5 - x_3 - x_6 = 0 \\
&x_3 + x_6 - x_4 - x_7 - x_8 = 0 \\
&x_4 + x_7 + x_8 = 1 \\
&x_j = 0, 1 \text{ for } j = 1, 2, \ldots, 8
\end{aligned}$$

The optimum solution is $y^* = 7$ weeks with $x_1^* = x_2^* = x_3^* = x_8^* = 1$, with all other control variables equal to zero. The critical path is

$$1 \rightarrow 2 \rightarrow 3 \rightarrow 8$$

and the critical tasks are:

excavate the site → lay foundation → erect walls →
install the pumping and heating systems

The tasks that are not on the critical path can begin late or take longer to complete than given by the estimated task completion times and not delay the project. For example, the site excavation and lay foundation tasks are critical. They must start on time and must be completed in a total of three weeks to keep the project on schedule. The parallel, noncritical one-week task of prefabricating the walls must also be completed within this three-week period, but it can start as late as the end of the second week and still remain on schedule. This task has a two-week float time. This time gives the contractor flexibility of scheduling the project and managing work tasks without delaying the project.

QUESTION FOR DISCUSSION

3.1.1 Define the following terms:

network models classified by control variable type

critical time path

QUESTIONS FOR ANALYSIS

Instructions: The problems market with an asterisk (*) are most easily solved with the aid of the simplex method.

3.1.2 (a)Use a graphical method to find for x* and y* for the integer programming model.

$$\begin{aligned} &\text{minimize } y = x_1 \\ &x_1 + x_2 \geq 1 \qquad \text{(A)} \\ &x_2 - 2x_1 \leq 0 \qquad \text{(B)} \\ &x_1 = 0, 1, 2, \ldots \\ &x_2 = 0, 1, 2, \ldots \end{aligned}$$

Show the feasible region.

(b) Assume x_1 and x_2 are continuous variables for the model given in (a), find for **x*** and *y** by graphical means. Roundoff **x***, calculate *y** for the rounded values, and compare it to the results obtained in (a).

3.1.3 (a)Use a graphical method to find for **x*** and *y** for the integer programming model.

$$\begin{aligned} &\text{minimize } y = 2x_1 + 3x_2 \\ &x_1 - x_2 \geq 0 \qquad \text{(A)} \\ &x_2 - x_1 \leq 2 \qquad \text{(B)} \\ &x_1 = 0, 1, 2, \ldots \\ &x_2 = 0, 1, 2, \ldots \end{aligned}$$

Show the feasible region.

(b) Assume x_1 and x_2 are continuous variables for the model given in (a), and find for **x*** and *y** by graphical means. Roundoff **x***, calculate *y** for the rounded values, and compare it to the results obtained in (a).

3.1.4* Determine the shortest route between the start and finish nodes of a highway network. The travel times on the links are

Link k	t_k, Travel Time in Minutes
1	20
6	15
2,3,4,5,7,8	10
9,10	5

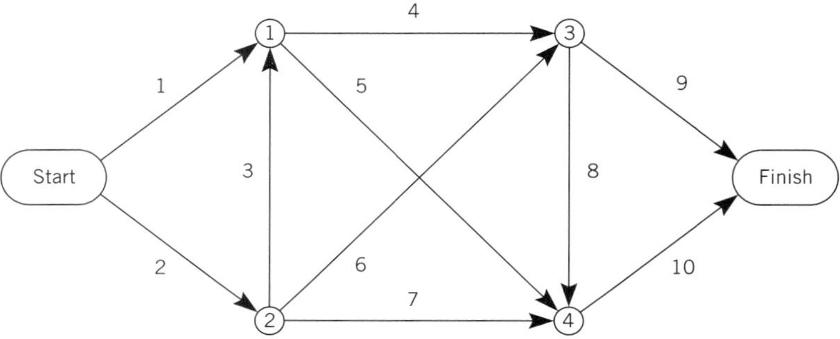

3.1.5* Determine the shortest route between New York and Portland. The travel times in hours between cities are shown on the links.

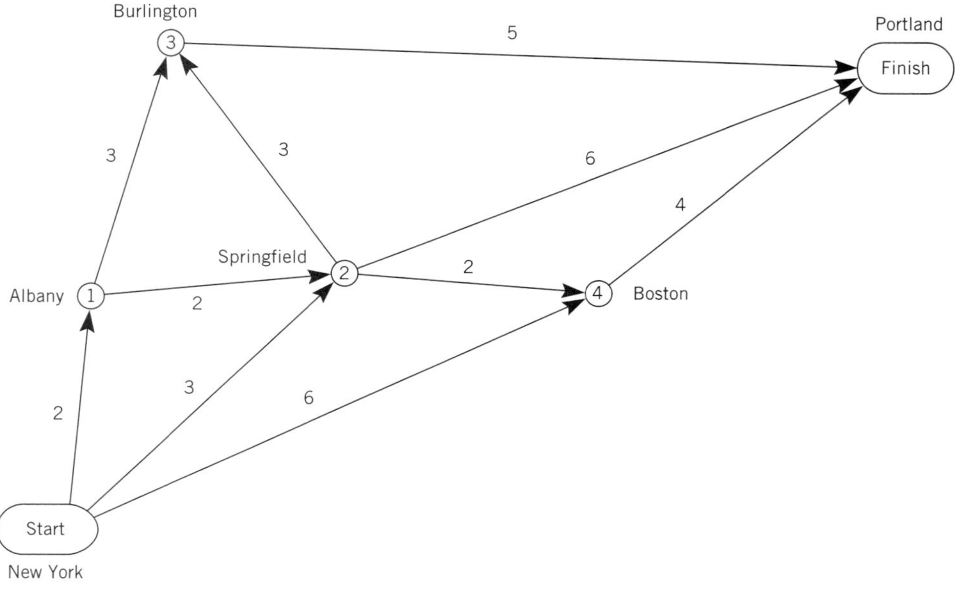

Use the following information to define control variables:

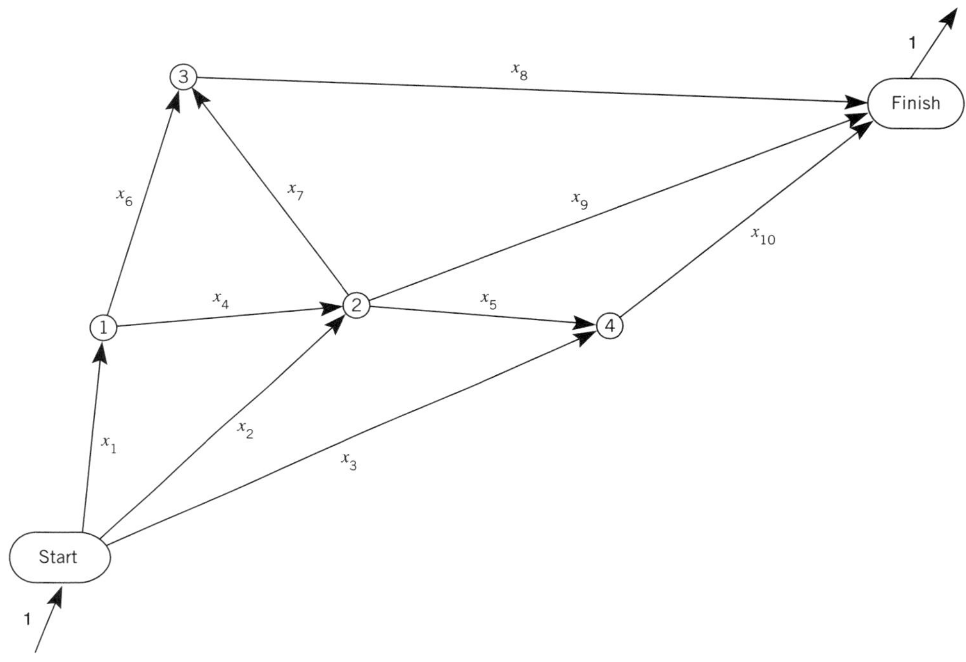

3.1.6* Determine the capacity of the highway network given in Problem 3.1.4. The link capacities are:

Link k	q_k, Capacities (in vehicles per hour)
2,3,5	2000
4,6,7,8	4000
9	6000
1,10	8000

3.1.7* Determine the maximum flow through a piping network. The pipe capacities in mgd (million gallons per day) are shown on the links.

Use the following information to define control variables:

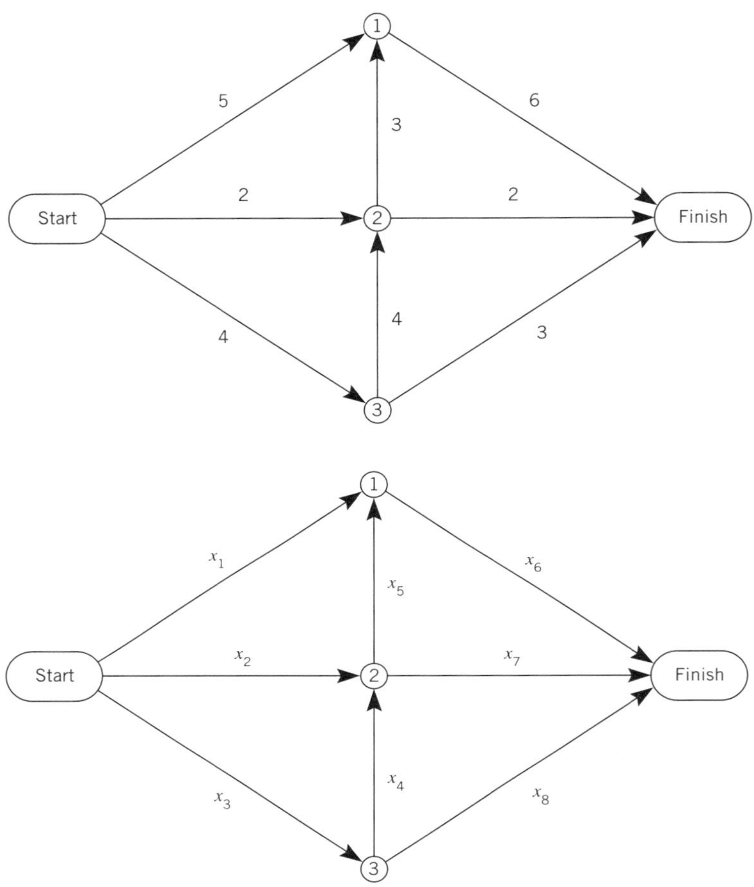

3.1.8* Bidirectional flow between nodes 1 and 2 are modeled with two links 3 and 4. Determine the capacity of the network.

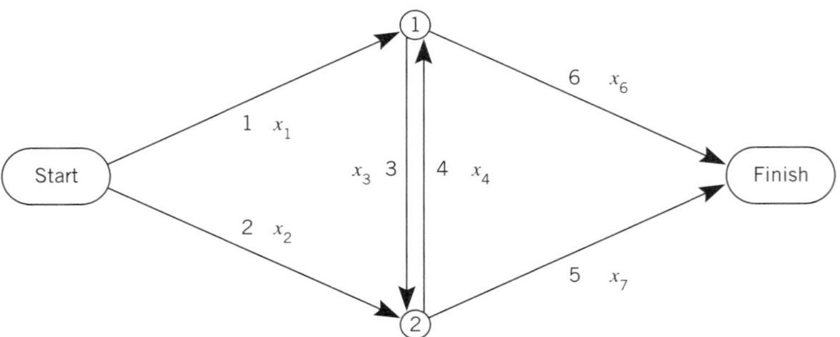

Link j	Link Capacity
1	2000
2	10000
3	6000
4	8000
5	6000
6	8000

3.1.9* There are two crews, one for site excavation and one for pouring concrete.

Task	Description	Estimated Completion Time in Weeks	Predecessor Task
1	Excavate site A	2	start
2	Excavate site B	3	1
3	Excavate site C	1	2, 4
4	Pour concrete at site A	1	1
5	Pour concrete at site B	2	2, 4
6	Pour concrete at site C	3	3, 5

(a) Draw a network flow diagram for scheduling these tasks.

(b) Determine the critical path.

3.1.10* A construction job consists of the following tasks:

Task	Description	Estimated Completion Time in Days	Predecessor Task
1	Clear site	3	start
2	Order and deliver material for foundation	14	start
3	Excavate foundation	5	1, 2
4	Pour concrete foundation	7	3
5	Order and deliver material for structure	14	start
6	Prefabricate structural members	5	4, 5
7	Erect structure	5	6
8	Install water and sewer lines	8	4, 5

(a) Draw a network flow diagram for scheduling these tasks.

(b) Determine the critical path.

3.2 TRANSPORTATION AND TRANSSHIPMENT MODELS

Transportation and transshipment models are special network models that have been thoroughly studied; therefore, they are often called classical models. In this section, we address one basic question: What is the most inexpensive way to assign the shipment materials located in m supply locations to n demand locations? The supply of material at each supply node s_i, the demand for material at each demand node d_j, and the unit transportation cost to ship between each supply and demand link c_{ij} are known.

When the problem is formulated as a transportation problem, we designate supply and demand nodes as being either source or sink nodes, respectively, and we find the direct links between the source and sink nodal pairs that minimize the total shipment cost. When the problem is formulated as a transshipment problem, the objective to minimize total shipment cost is the same but the formulation is less restrictive. The nodes are designated as source, sink, or intermediate nodes. Here, we permit materials to flow through the intermediate or transshipment nodes.

The transportation and transshipment models are important in their own right, but knowledge of their structure and properties gives us insight into formulating and solving more generalized LP models, discussed in Section 3.3.

The Control Variables

Thus far in the book, it has been most convenient to use a single-subscript notation to identify the amount of material to be shipped over a network link. For example, we have defined x_k = the amount of material shipped on link k or between a source node i and a sink node j. In some applications, particularly in formulating transportation and transshipment models, single subscript notation can lead to ambiguity and confusion. In this section, we prefer to use double-subscript notation, or

$$x_{ij} = \text{amount of material shipped on directed link } i \rightarrow j$$

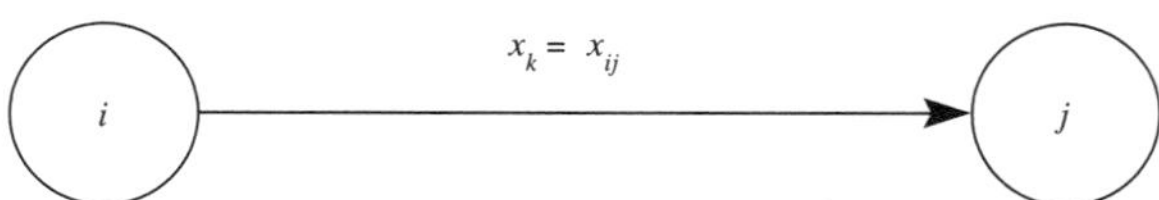

This notation has the added advantage of being able to deal with bidirectional flows between nodes.

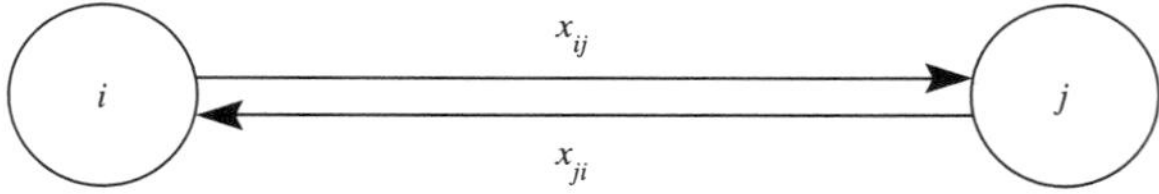

Here, we define

$$x_{ij} = \text{amount of material shipped on directed link } i \rightarrow j$$

and

$$x_{ji} = \text{amount of material shipped on directed link } j \rightarrow i$$

Since $i \neq j$, there is no ambiguity about the direction of flow. Furthermore, it is possible to assign a flow in one direction and not the other, $x_{ij} > 0$ and $x_{ji} = 0$. The unit costs associated with these links are

$$c_{ij} = \text{unit cost on directed link } i \rightarrow j$$

and

$$c_{ji} = \text{unit cost on directed link } j \rightarrow i$$

Note also that c_{ij} and c_{ji} can also be assigned different values.

The Transportation Model

The conceptual format of a transportation model consisting of m sources and n sink nodes is shown in the network diagram of Figure 2. Here we have conveniently arranged all source or supply nodes in the left-hand column and all sink or demand nodes in a right-hand column. Directed arcs or links are drawn between each $i \rightarrow j$ pair. The unit cost and amount of material shipped over this link is designated as c_{ij} and x_{ij}, respectively.

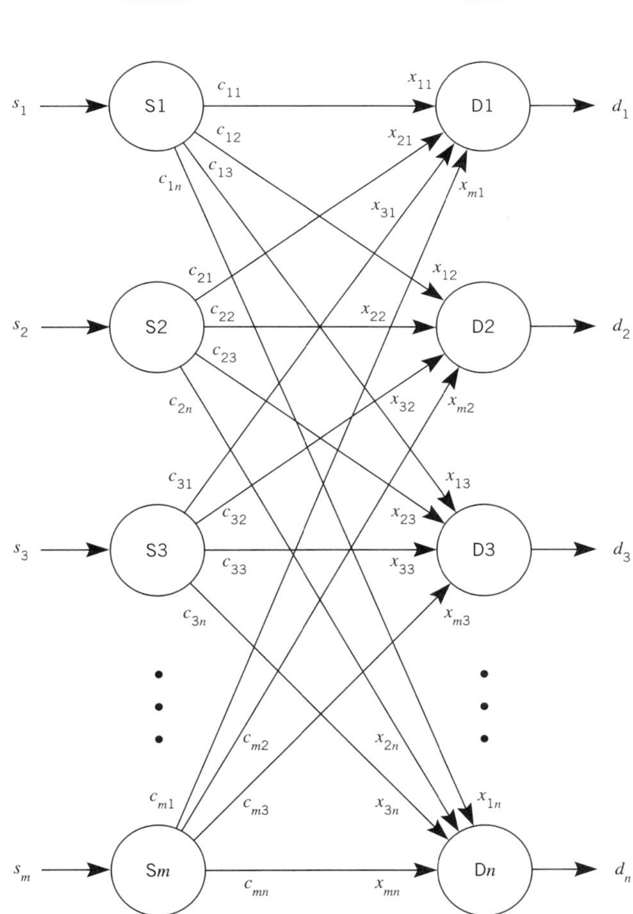

Figure 2 The conceptual format of a transportation model.

The supplies, s_i, and demands, d_j, can be integers or noninteger values. For example, if s_i and d_j represent the number of truckloads of material that are available at each source node and the number of truckloads of material required at each sink node, then the control variables x_{ij} are designated to be integers,

$$x_{ij} = 0, 1, 2, \ldots$$

If one or more of the constraints s_i and d_j are noninteger values, then the problem is formulated as an LP model with continuous variables of x_{ij}. In this section, transportation problems are formulated as integer programming models.

The total supply is assumed to be equal to the total demand; thus,

$$\sum_{i=1}^{m} s_i = \sum_{j=1}^{n} d_j$$

With this information, we can formulate an integer programming model. First, consider the objective function to minimize total transportation cost. The cost to transport over any $i \rightarrow j$ pair is $c_{ij}x_{ij}$. The total cost for the entire network is the sum of these individual costs, or

$$y = \sum_{i=1}^{n} \sum_{j=1}^{m} c_{ij} x_{ij}$$

We have implied that the unit cost is a monetary measure in \$ per truckload. We have the freedom to assign the unit cost to be a travel time where c_{ij} is a unit travel time to ship over an $i \rightarrow j$ link per truckload. Here, since the objective function is a nonmonetary measure of effectiveness, the total cost function is called a merit function.

Second, the flow of materials over the entire network must be formulated as a constraint set. The mass conservation law will be applied at each source and sink node.

At each source node i, the following relationship must hold:

Mass in = Mass out (#)

$$\sum_{j}^{n} x_{ij} = s_i \quad \text{for} \quad i = 1, 2, \ldots, m$$

Note that a unit balance exists because each x_{ij} and s_i is measured in the same units, the number (#) of truckloads.

At each sink node i, the following relationship must hold:

Mass in = Mass out (#)

$$\sum_{i}^{m} x_{ij} = d_j \quad \text{for} \quad j = 1, 2, \ldots, n$$

The restriction on the control variable is formally stated by the nonnegative constraint set,

$$x_{ij} \geq 0$$

where

$$x_{ij} = 0, 1, 2, \ldots$$

We should not be misled by the apparently rigid conceptual and mathematical structure of the transportation model. Assignment, scheduling, and other managerial problems can be solved with the classical transportation model. Special computer algorithms are available to solve the integer, zero-one, and mixed programming models.

Assignment Model

Without any loss of generality, suppose we restrict the control variables to be zero-one variables, or

$$x_{ij} = 0 \quad \text{or} \quad 1$$

When this restriction is made on the transportation model, we call the model an assignment model. The following example illustrates how an optimal job assignment problem can be formulated as an assignment model. Since the only difference between an assignment and transportation model is the restriction placed on the control variables, there is no conceptual difference in formulating these two model types. Nonetheless, the problem is solved with aid of a computer and the simplex method. As a result, the restriction that the control variables are integers is relaxed from $x_{ij} = 0, 1, 2, \ldots$ to continuous variables. Owing to the unique model structure of assignment, transportation and transshipment models, the simplex method will lead to integer values of the optimum solution vector $\mathbf{x}^*$.

EXAMPLE 3.5 ***An Assignment Model***

The time in hours to complete jobs 1, 2, and 3 by machine types 1, 2, and 3 is indicated in the following table.

Machine	Job 1	Job 2	Job 3
1	2	1	1.5
2	3	2	2.5
3	2.5	3	1.5

Formulate and solve for the optimum machine assignments to minimize total work time.

SOLUTION

The network diagram is

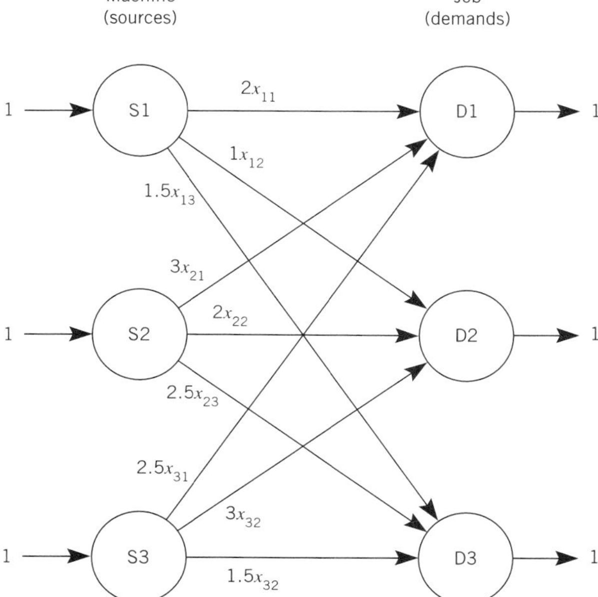

The control variables $x_{ij} = 1$ when machine i is assigned to job j, and $x_{ij} = 0$ when machine i is not assigned to job j.

The objective function is to minimize total work time. For convenience, the cost of assigning machine i to job j is shown on the $i \rightarrow j$ links of the network diagram. The total work time is the sum of the individual work times, or

$$\text{minimize } y = 2x_{11} + 1x_{12} + 1.5x_{13} + 3x_{21} + 2x_{22} + 2.5x_{23} + 2.5x_{31} + 3x_{32} + 1.5x_{33}$$

Mass balance relationships are used at each source (machine) and sink (job) node.

Mass in = Mass out (unitless)

Source Nodes

$$x_{11} + x_{12} + x_{13} = 1 \qquad \text{(S1)}$$
$$x_{21} + x_{22} + x_{23} = 1 \qquad \text{(S2)}$$
$$x_{31} + x_{32} + x_{33} = 1 \qquad \text{(S3)}$$

Sink Nodes

$$x_{11} + x_{21} + x_{31} = 1 \qquad \text{(D1)}$$
$$x_{12} + x_{22} + x_{32} = 1 \qquad \text{(D2)}$$
$$x_{13} + x_{23} + x_{33} = 1 \qquad \text{(D3)}$$

The model is solved by the simplex method. The optimum solution is

$$x_{12}^* = 1,\ x_{21}^* = 1,\ x_{33}^* = 1 \text{ and all other } x_{ij}^* = 0$$
$$y^* = 5.5 \text{ hours}$$

The optimum assignment is

Machine Job

1 S1 D1 1

1 hour

3 hours

1 S2 D2 1

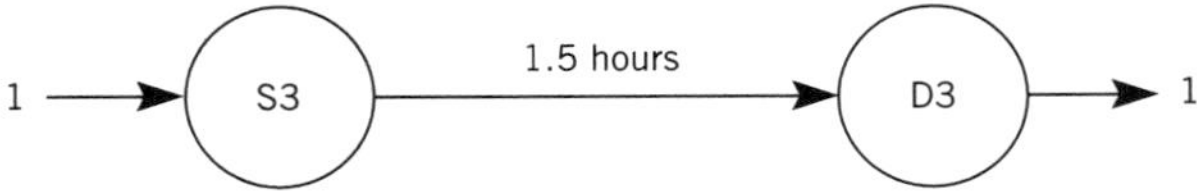

The Transshipment Model

The transshipment model is more general than the transportation model. The transshipment model permits the use of intermediate or transshipment nodes for transport of material through them. Our primary aim at this point is to demonstrate the similarities between formulating transportation and transshipment models.

The transshipment model, like the transportation model, can be formulated as an integer programming model or an LP model. In this discussion, problems will be formulated as integer programming models in which supply and demand are integer values. The major difference between the two models is that in a transshipment model formulation material may flow through nodes. These nodes are called transshipment nodes. In spite of this difference, the network diagram (see Figure 3) and formulation procedure are basically the same for both model types.

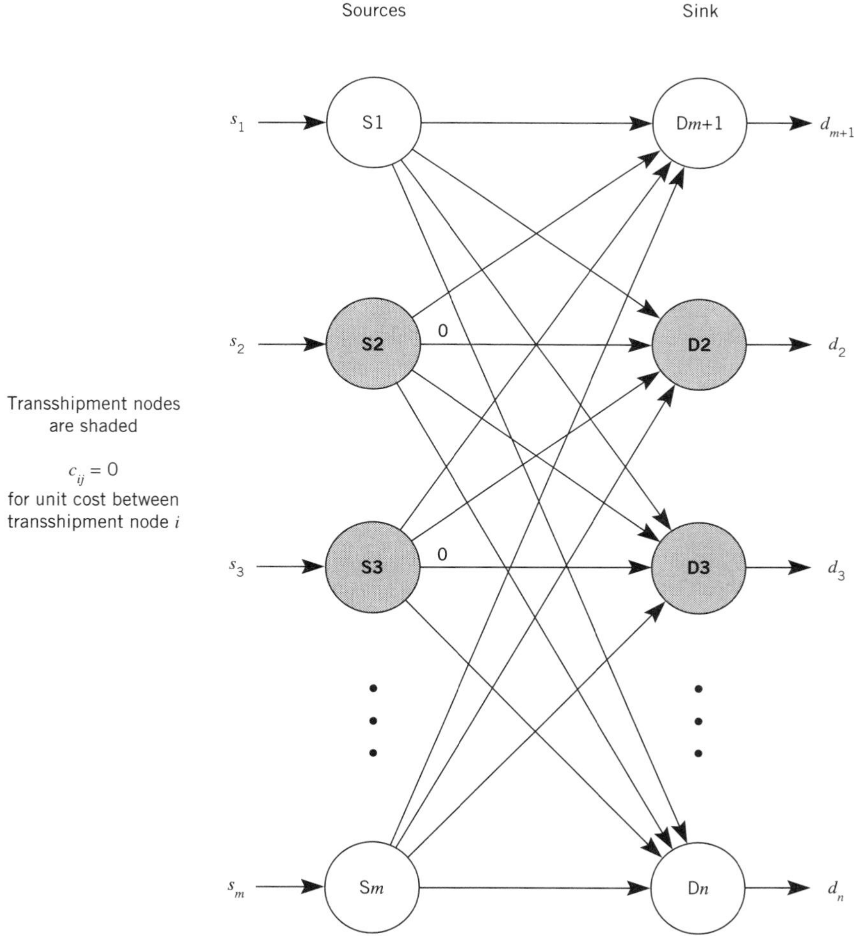

Figure 3 A transshipment model.

Unlike the transportation model, where nodes are designated as either source or sink nodes and numbered accordingly ($i = 1, 2, \ldots, m$ sources, $j = 1, 2, \ldots, n$ sinks), the nodes in a transshipment problem are numbered in sequence, where n = total number of nodes. The network flow diagram of source and sink nodes is most helpful for identifying the x_{ij} control variables. In addition, the transshipment nodes are identified and introduced into the network flow model as both source and sink nodes.

The objective function for the transshipment model is

$$\text{minimize } y = \sum_i \sum_j c_{ij} x_{ij}$$

Since it is assumed that there is no transport cost between transshipment nodes, we assign $c_{ii} = 0$.

The mass balance constraints are

$$\sum_j x_{ij} = s_i \text{ at source node } i$$

and

$$\sum_i x_{ij} = d_j \text{ at sink node } j$$

$$x_{ij} \geq 0$$

where

$$x_{ij} = 0, 1, 2, \ldots$$

Furthermore, it is assumed that the total supply equals the total demand:

$$\sum_i s_i = \sum_j d_j$$

In order to ensure that the search for an optimum solution remains in the feasible region, all transshipment control variables are initially assigned an added value equal to total supply,

$$\psi = x_{ii} = \sum_{i=1}^{m} s_i$$

The ψ value is added to the right-hand side of each transshipment source and sink mass balance equation.

$$\sum_j x_{ij} = s_i + \psi \text{ at transshipment source node } i$$

$$\sum_i x_{ij} = d_j + \psi \text{ at transshipment sink node } j$$

The following example illustrates how to formulate and solve a problem as a transshipment model. The problem is simple enough to be solved by inspection. Our efforts will focus on obtaining a basic understanding of model formulation.

EXAMPLE 3.6 *A Transshipment Model for MSW Collection*

A map of a solid waste district shows an incinerator located in zone 1 and the neighborhoods that supply it as zones 2, 3, 4, and 5. There is a total of 12 collection vehicles. The supply of vehicles by neighborhood is

j, Neighborhood	2	3	4	5
s_j, Supply	4	3	1	4

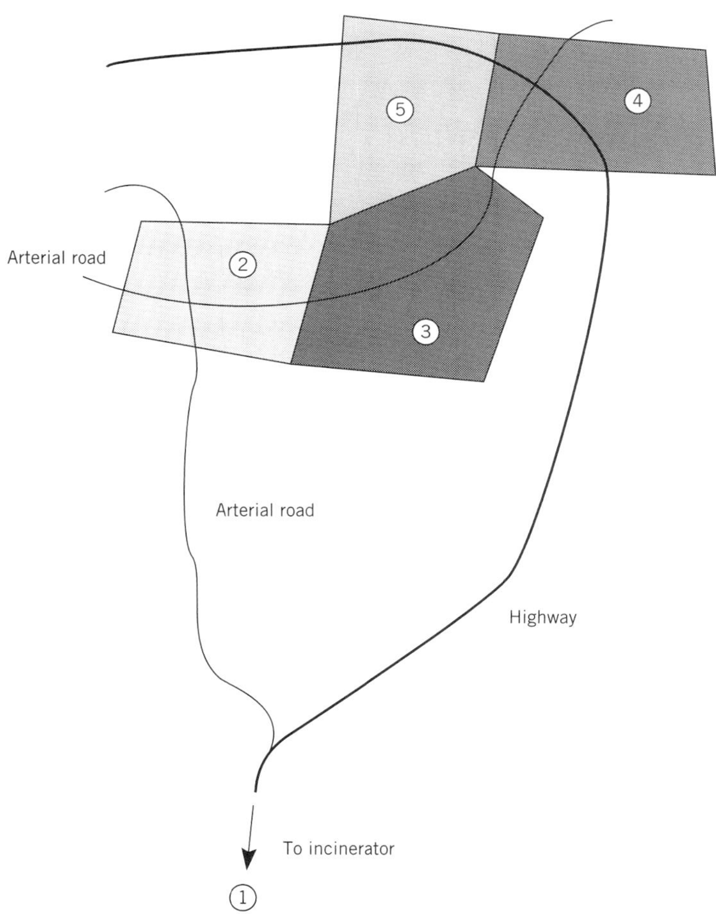

The travel times in minutes between zone centroids using local roads and highways are

	To Zone j				
From Zone i	1	2	3	4	5
1	0	45	∞	20	∞
2	45	0	10	∞	∞
3	∞	10	0	5	∞
4	20	∞	5	0	5
5	∞	∞	∞	5	0

Formulate and solve a transshipment model to minimize the total time to supply the incinerator with MSW.

SOLUTION

Our task is to make a minimum travel time assignment of vehicles. The following network diagram shows the possible flow of trucks through the region.

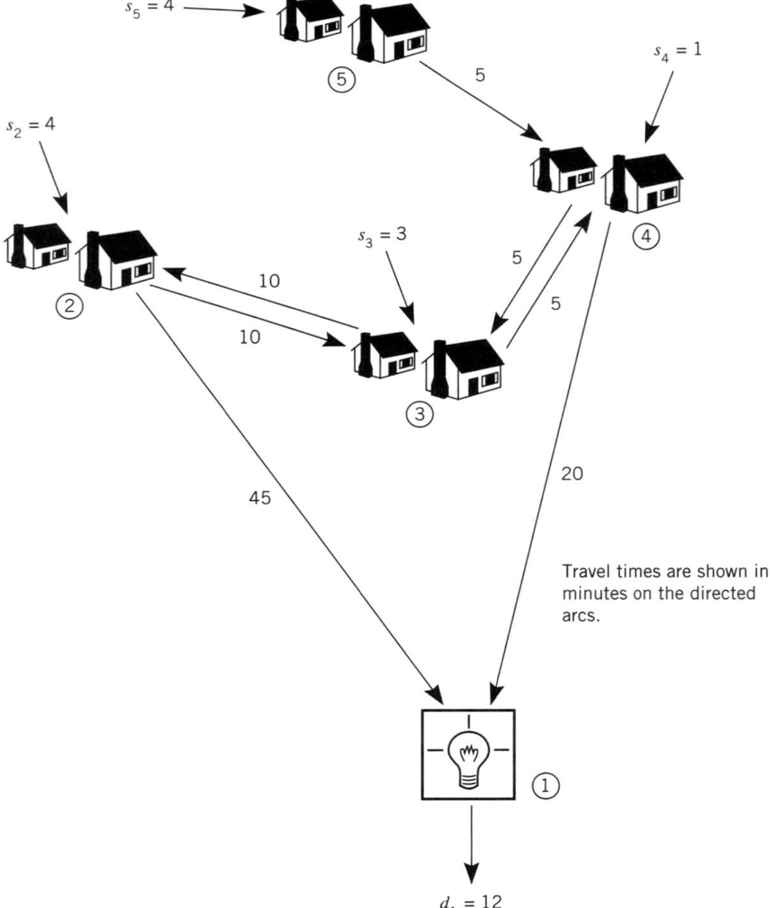

The network shows the roads as directed arcs. The unit costs of travel along these links are shown to be $c_{21} = 45$, $c_{41} = 20$, $c_{23} = c_{32} = 10$, and $c_{34} = c_{43} = c_{54} = 5$ minutes. The number of trucks assigned to each neighborhood are $s_2 = 4$, $s_3 = 3$, $s_4 = 1$, and $s_5 = 4$ trucks, respectively. Since all trucks will arrive at zone 1, it is treated as a demand node with $d_1 = 12$ trucks. Thus, the total-supply-equals-total-demand requirement for a transshipment model is satisfied.

The following network flow diagram shows zone 5 as a supply node and zone 1 as a demand node. Since all other zones permit vehicles to travel through them, they are designated as transshipment nodes. The unit travel times for transshipment nodes are $c_{22} = c_{33} = c_{44} = 0$.

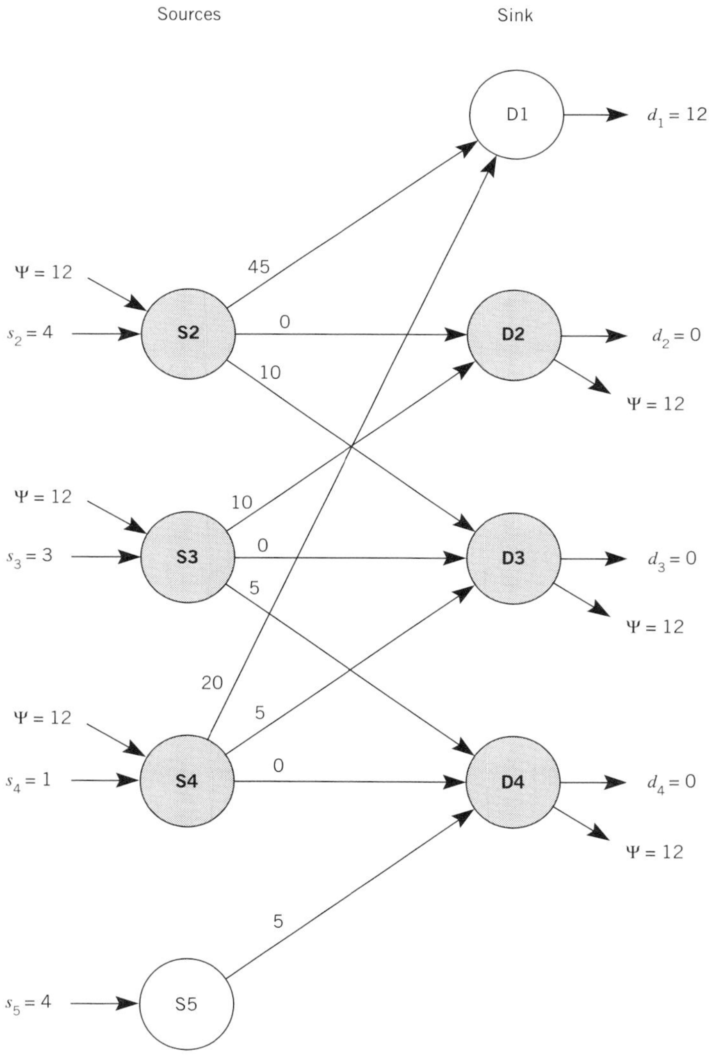

In order to search for the optimum solution and to remain in the feasible region, the transshipment source and sink equations are modified with the following sum,

$$\psi = x_{22} = x_{33} = x_{44} = \sum_i s_i = 4+3+1+4 = 12$$

In order to satisfy the mass balance conditions, the value of $\psi = 12$ is added to each transshipment source and sink nodes 2, 3, and 4, as shown in the network diagram.

The objective function to minimize total travel time y is

$$\text{minimize } y = 45x_{21} + 10x_{23} + 10x_{32} + 5x_{34} + 5x_{43} + 5x_{54} + 20x_{41}$$

The mass balance relationship is satisfied at source and sink node.

$$\textit{Mass in} = \textit{Mass out}(\#)$$

Source nodes

$$x_{21} + x_{22} + x_{23} = 16 \qquad \text{(S2)}$$
$$x_{32} + x_{33} + x_{34} = 15 \qquad \text{(S3)}$$
$$x_{41} + x_{43} + x_{44} = 13 \qquad \text{(S4)}$$
$$x_{54} = 4 \qquad \text{(S5)}$$

Sink Nodes

$$x_{21} + x_{41} = 12 \qquad \text{(D1)}$$
$$x_{22} + x_{32} = 12 \qquad \text{(D2)}$$
$$x_{23} + x_{33} + x_{43} = 12 \qquad \text{(D3)}$$
$$x_{34} + x_{44} + x_{54} = 12 \qquad \text{(D4)}$$

The optimum solution is

$$x_{22}^* = 12$$
$$x_{23}^* = 4$$
$$x_{33}^* = 8$$
$$x_{34}^* = 7$$
$$x_{41}^* = 12$$
$$x_{54}^* = 4 \text{ with all other } x_{ij}^* = 0$$

and

$$y^* = 335 \text{ minutes}$$

The optimum paths are shown on the following diagram:

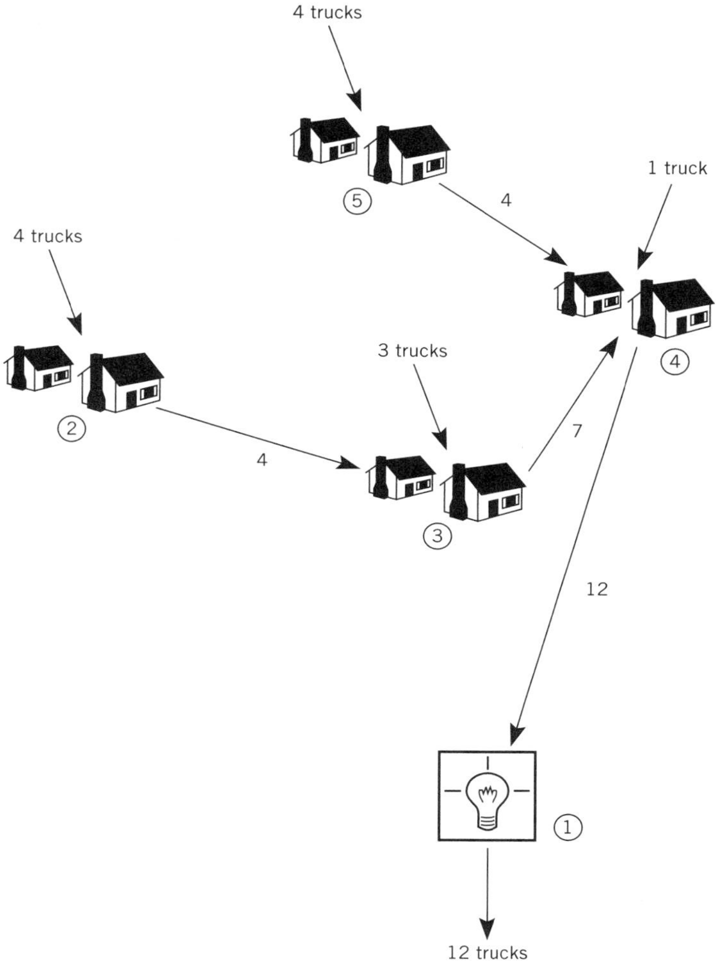

DISCUSSION

This formulation takes advantage of the special features of the transshipment model. Those who failed to recognize that a problem satisfied the conditions for a transshipment model can successfully proceed to formulate and solve the problem as an integer programming model. This approach is shown in the following illustrative example.

EXAMPLE 3.7 *An Integer Programming Model for MSW Collection*

Formulate the problem described in Example 3.6 as an integer programming problem in lieu of a transshipment model.

SOLUTION

Using the same network diagram as before and duplicated here, we proceed with the model formulation as follows. In this formulation, the control variables x_{22}, x_{33}, and x_{44} are not used.

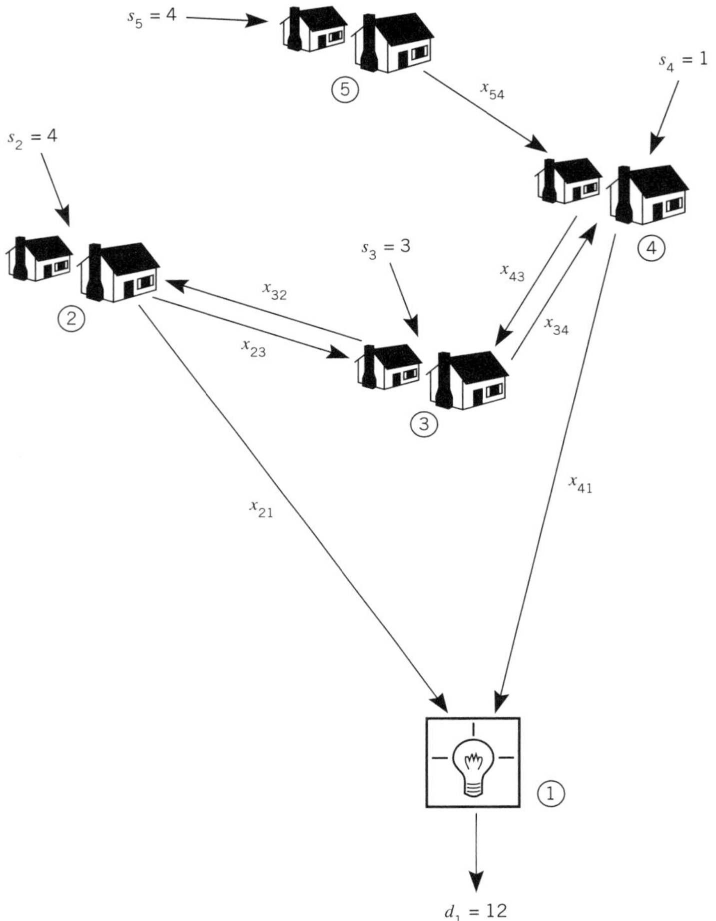

The objective function to minimize the total travel time is

$$\text{minimize } y = 45x_{21} + 10x_{23} + 10x_{32} + 5x_{34} + 5x_{43} + 5x_{54} + 20x_{41}$$

The mass balance relationship is satisfied by isolating each node, $j = 1$ through 5.

Mass in = Mass out (# of trucks)

$$
\begin{array}{ll}
x_{21} + x_{41} = 12 & \text{(Node 1)} \\
4 + x_{32} = x_{23} + x_{21} & \text{(Node 2)} \\
3 + x_{23} + x_{43} = x_{32} + x_{34} & \text{(Node 3)} \\
1 + x_{54} + x_{34} = x_{43} + x_{41} & \text{(Node 4)} \\
4 = x_{54} & \text{(Node 5)}
\end{array}
$$

The model is summarized as follows:

$$
\begin{array}{ll}
\text{minimize } y = 45x_{21} + 10x_{23} + 10x_{32} + 5x_{34} + 5x_{43} + 5x_{54} + 20x_{41} & \\
x_{21} + x_{41} = 12 & \text{(Node 1)} \\
x_{21} + x_{23} - x_{32} = 4 & \text{(Node 2)} \\
-x_{23} + x_{32} + x_{34} - x_{43} = 3 & \text{(Node 3)} \\
-x_{34} + x_{43} - x_{54} + x_{41} = 1 & \text{(Node 4)} \\
x_{54} = 4 & \text{(Node 5)} \\
x_{ij} = 0, 1, 2, \ldots \text{ for } i, j = 1, 2, 3, 4, 5 &
\end{array}
$$

The optimum solution is the same result obtained in Example 3.6.

$$
\begin{aligned}
x_{23}^* &= 4 \\
x_{34}^* &= 7 \\
x_{54}^* &= 4 \\
x_{41}^* &= 12
\end{aligned}
$$

and

$$y^* = 335 \text{ minutes}$$

As a point of interest, all integer and zero-one programming problems are solved with the simplex method, and integer solutions are obtained without rounding. However, special algorithms are more efficient for certain problems.

This and the preceding examples illustrate that different mathematical model structures will give the same result. In other words, there are different ways to address the same problem and be successful.

DISCUSSION: THE ψ VALUE

By comparing the transshipment and integer programming models of Examples 3.6 and 3.7, we can get better insight as to the purpose adding ψ to each transshipment source and sink node. First, consider the control variable x_{22} and node 2. From the transshipment formulation, the mass balance equation for source node 2 is

$$x_{21} + x_{22} + x_{23} = 16 = \psi + 4$$

or

$$(x_{21} + x_{23}) = \psi + 4 - x_{22} \quad \text{(Node 2)}$$

From the integer programming formulation, the mass balance equation for node 2 is

$$x_{21} + x_{23} - x_{32} = 4$$

or

$$(x_{21} + x_{23}) = 4 + x_{32} \qquad \text{(Node 2)}$$

Since the mass balances must be the same node for each model, it follows that

$$(x_{21} + x_{23}) = 4 + x_{32} = \psi + 4 - x_{22}$$

The control variable x_{22} is

$$x_{22} = \psi - x_{32} \qquad \text{(Node 2)}$$

The amount of material transshipped through node 2 is the difference between ψ and the flow of material from the adjacent node 3.

By using the same approach for nodes 3 and 4, the following identities are obtained:

$$x_{33} = \psi - (x_{23} + x_{43}) \qquad \text{(Node 3)}$$

and

$$x_{44} = \psi - (x_{34} + x_{54}) \qquad \text{(Node 4)}$$

Again, the transshipment variables x_{33} and x_{44} are the added value sum ψ minus the flows from the adjacent nodes.

By definition, $\psi > 0$; thus $x_{22} > 0$, $x_{33} > 0$, and $x_{44} > 0$ will always be positive if ψ is assigned to be a large enough value. In the integer programming model, all control variables must be nonnegative. Using

$$\psi = \sum_{all\ i} s_i$$

we are assured that this requirement is met. If we fail to add this quantity, then it is possible that the computer algorithm will lead to an infeasible solution, violating the nonnegative control variable requirement for the simple method.

For example, if we assume $x_{32} > 0$ and $\psi = 0$, then

$$x_{22} = \psi - x_{32} = -x_{32} < 0$$

Clearly, the solution is in the infeasible region and a violation of a basic LP requirement. Since the model is incorrectly formulated, the simplex method will not be able to provide a solution. It is necessary to add an appropriate ψ value to obtain a correct solution.

QUESTIONS FOR ANALYSIS

3.2.1* Service vehicles are stored at nodes 1 and 4 and are needed at nodes 2, 3, and 5. Formulate and solve as a transportation model. The supply and demand of vehicles are shown in the diagram.

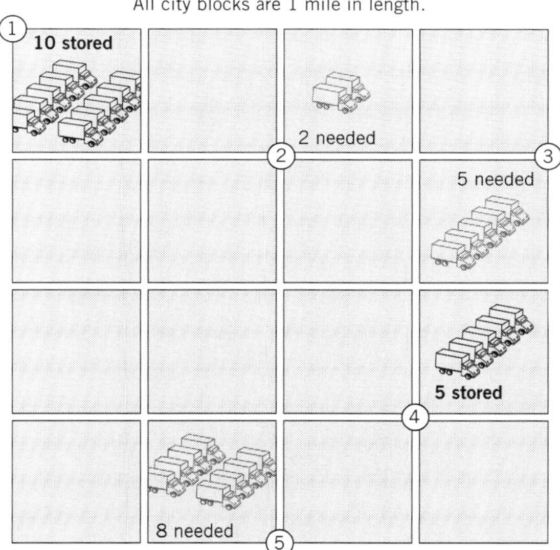

3.2.2* Formulate and solve Problem 3.2.1 as a transshipment model. Treat nodes 2, 3, and 4 as transshipment nodes.

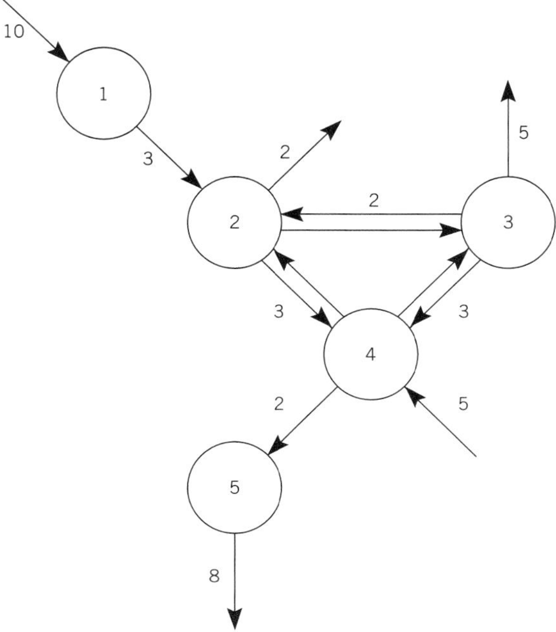

3.2.3* Assume that there is a supply of 15 service vehicles at node 1 and 10 service vehicles at node 4 for the network given in Problem 3.2.1. The total supply exceeds the total demand. Add a dummy demand node 6 and solve as a classical transportation problem.

3.2.4* Formulate and solve Problem 3.2.3 as an LP model. Use inequality constraints to account for excess supply of vehicles at nodes 1 and 4.

3.2.5* Three towns have signed a joint contract agreeing to supply energy recovery plants 1 and 2 with 75M and 60M tons of MSW per year, respectively.

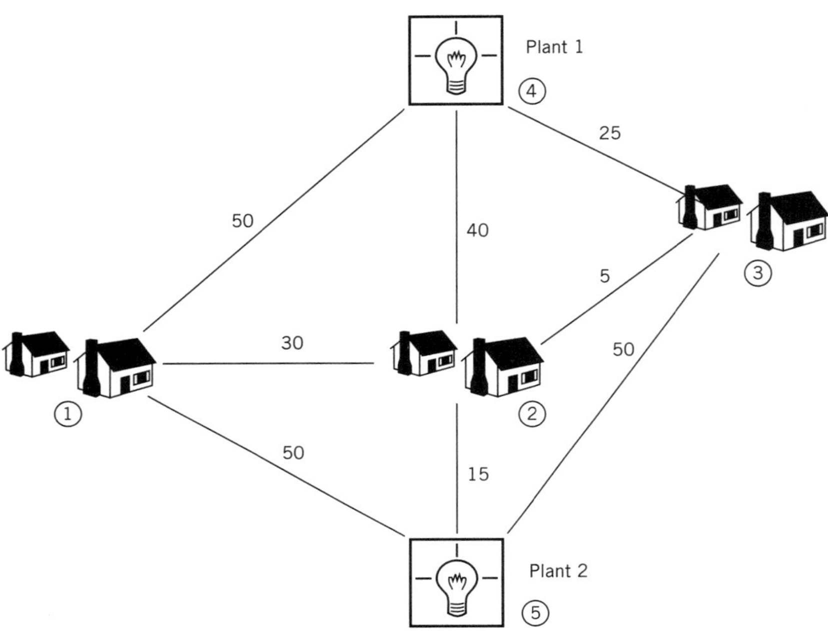

Towns 1, 2, and 3 have agreed to supply annually 50M, 35M, and 50M tons of MSW, respectively. Formulate and solve as a transshipment model to minimize total travel time. The travel times between nodes are shown on the diagram in minutes. Assume bidirectional flows between towns 1 and 2 and between towns 2 and 3.

3.2.6* Utilize the information and data of Example 3.6 from Section 3.2 for this problem. Two incinerators are available which are 20 minutes from zones 2 and 5. Each plant has a capacity of six trucks.

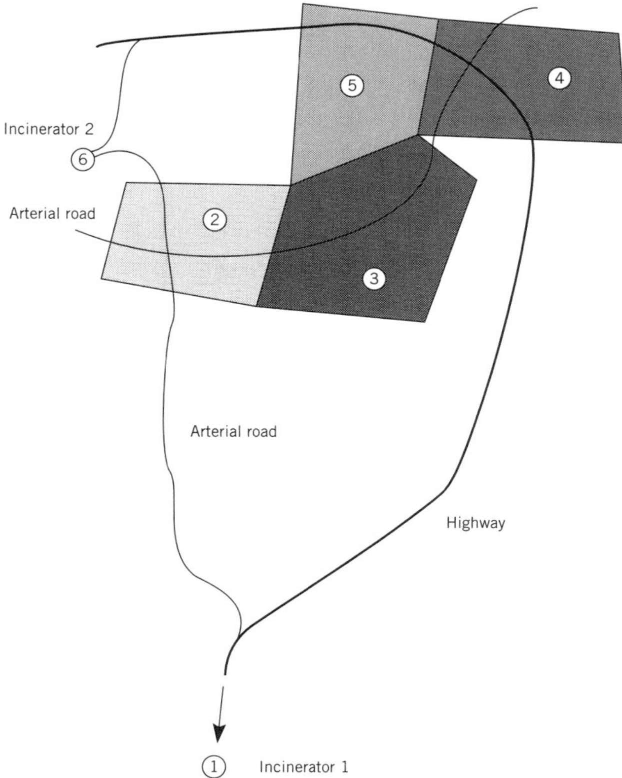

3.2.7* Travel times in minutes are shown on the map (spatial diagram). Towns 1 and 2 generate 10 and 20 truckloads of solid waste per day, respectively. A transfer center is a place where solid waste is transferred for more efficient transport.

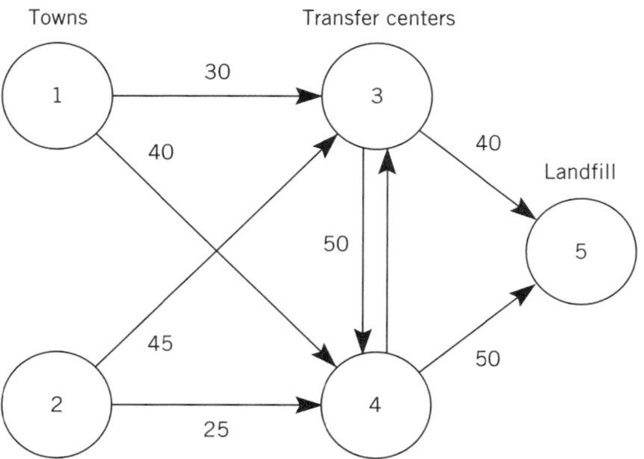

Formulate and solve as a transshipment model.

3.2.8* Find the shortest route between the origin-destination pair 1 → 28. Consider three route alternatives:

A. Expressway between 9 → 21 of length 15 miles.
Use path 1 → 2 → 9 → 21 → 28.

B. Arterial road between 16 → 21.
Use path 1 → 2 → 16 → 21 → 28.

C. Local street route.
Use path 1 → 22 → 28.

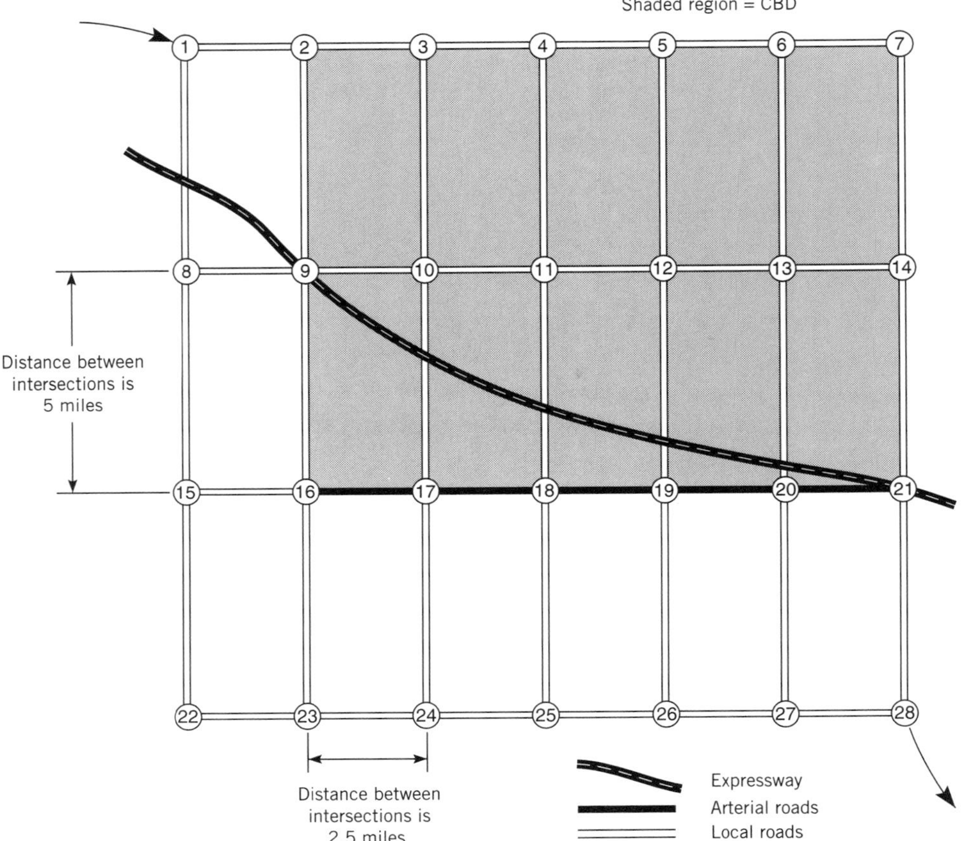

ASSUMPTIONS

- Assume traffic densities of 30, 20, and 0 vehicles per mile for expressway, arterial, and local streets, respectively.
- The entrance and exit nodes for the expressway are located at nodes 9 and 21, respectively.

- There is a two-minute delay at all intersections, including the expressway entrances and exits, except for arterial road B where the traffic signals are timed so there are no intersection delays at intersections 17, 18, 19, 20 and 21.

Formulate and solve as a transshipment model.

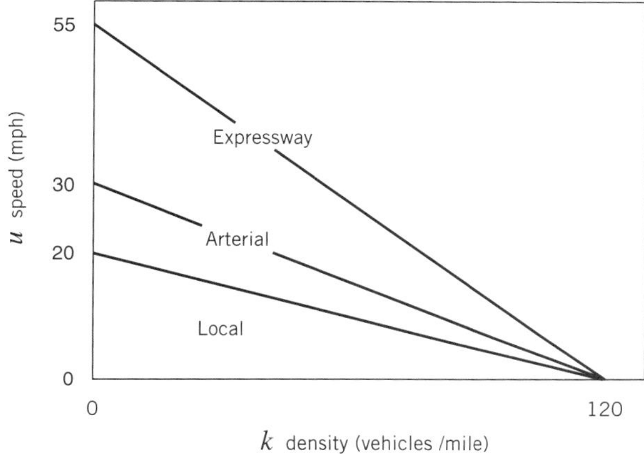

3.3 OTHER LINEAR MODELS

In the preceding sections, we discovered that it is possible to create simple model structures for complex systems. Since model formulation is a creative process, we are not restricted to fixed model forms but are free to formulate and solve problems using different approaches. In this section, we continue to try to find simple model structures and to enhance our understanding of the model formulation process.

In this section, we focus on an MSW management model. This model has analytical features that make it special, particularly in the definition and use of control variables. This model also addresses a number of community and social issues that are relevant to the application of systems modeling.

An MSW Management Model

The following discussion focuses largely on one issue, the use of control variables. MSW is a mixture of materials with particular resources that we want to evaluate, and to do so we must define a set of control variables. In order to demonstrate the importance of the control variable definition, we evaluate two MSW management models. They are similar in two important respects. First, both have the same project objective to minimize the total cost of shipping, processing, and disposing of MSW, and second, both use recycling as an alternative solution. They differ in a major way, however. In one illustration, source separation is an option, but in the other one, it is not allowed. In other words, in the source separation problem the residents are given the option to separate recyclable materials before the material is picked up for them. In the second illustration, residents are not given this option; recyclable materials are separated after MSW pickup.

Once again, MSW management models are used to present general principles of systems analysis. The same principles presented in this book can be used to evaluate other types of systems.

EXAMPLE 3.8 *An MSW Management Model with Source Separation*

Three alternative technologies are under consideration:

A. Landfill disposal

B. Recycling

C. Waste-to-energy incineration

The following information has been gathered:

MATERIAL COMPOSITION				
j	MSW Component	Heat Value (dry mass), h_j (kWh/ton)	Ash Content, a_j, (% of as collected mass)	Moisture Content, m_j (% of as collected mass)
1	Paper	4439	5.4	10.2
2	Remaining	2655	41.9	31.1

MSW GENERATION		
k	Source	Supply, s_k (tons per year)
1	Town	5000

j	MSW Component	Fraction, ω_j
1	Paper	0.46
2	Remaining	0.54

TIPPING FEES AND CONSTRAINTS			
k	Facility	Tipping Fee, u_k (\$/ton)	Constraints, b_k (tons per year)
2	Landfill	40	10,000
3	Recycling center	0	10,000
4	Waste-to-energy	100	10,000

	Selling Prices	
k	Description	Price ($), p_k
5	Paper	150/ton
4	Recovered energy	0.10/kWh

All shipping costs are assumed to be zero, and thermal energy conversion efficiency of the waste-to-energy plant is $e_4 = 60\%$.

The following restrictions are placed on the system:

1. *Policy.* Town government, in its effort to extend the useful life of the landfill, imposes a $b_2 = 10{,}000$ ton per year capacity limit on the total amount of MSW that can be deposited.
2. *Regulatory.* In order to protect public health, the amount of flue gas emitted from the waste-to-energy facility is controlled by restricting the amount of material that is permitted to be burned to be no more than $b_4 = 10{,}000$ tons of MSW per year.
3. *Capacity.* The recycling center has the physical capacity to handle $b_3 = 10{,}000$ tons of MSW per year.
4. *Behavioral.* The number of households that are willing to recycle in a volunteer program is estimated to be $v = 30\%$.

SOLUTION

The following network diagram describes the alternatives under consideration.

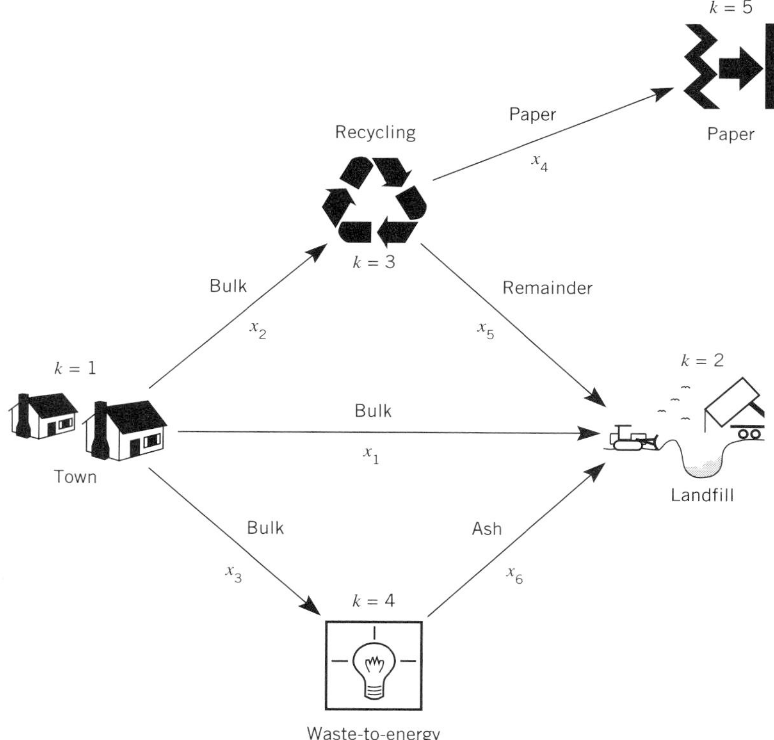

The model formulation process is an evolutionary process. After the control variables are defined, we discover that some of them will be redefined as we proceed. This is a normal process in feedback analysis.

The state of the MSW is altered by separation and incineration. Thus, the fate of the material must be accounted for at each step of the process as shown on the network diagram. The material shipped from the town over links 1, 2, and 3 is bulk mass material. The bulk material entering the recycling and waste-to-energy facilities is processed. The materials entering the marketplace and landfill over links 4, 5, and 6 are paper, the material remaining after paper is removed, and bottom ash, respectively.

The control variables are defined to describe their properties on each link:

$$x_1, x_2, x_3 = \text{bulk mass for links 1, 2, and 3 (M)}$$
$$x_4 = \text{paper mass for link 4 (M)}$$
$$x_5 = \text{remaining bulk mass after removing paper for link 5 (M)}$$
$$x_6 = \text{mass of bottom ash (M) for link 6}$$

These definitions are helpful, but they are inadequate in establishing a mathematical model that will allow MSW to be separated or to change state. For example, paper may be separated and shipped to the recycling center, while the remaining mass is shipped to the landfill. The current definitions of x_1 and x_2 will not permit this shipment. It will be necessary to use two control variables, one for paper and the other for the remaining material. Double-subscript notation of these control variables will allow the link and the material component to be identified:

$$x_{ij} = \text{mass of component } j \text{ of the MSW mixture shipped over link } i$$

where

$$i = \text{link numbers 1, 2, 3, 4, 5, 6}$$

and

$$j = \text{1 for paper and 2 for the remaining MSW mixture}$$

Consider link 1. The amount of paper and remaining material shipped over the link are denoted as x_{11} and x_{12}, respectively. The bulk mass shipped over link 1 is denoted as originally defined x_1. The bulk mass shipped over the link is the sum of components shipped over the link, or

$$x_1 = x_{11} + x_{12}$$

With this definition it is possible to consider removing all the paper at the source and shipping it to a recycling center and shipping the remaining MSW to the landfill. In this case,

$$x_{11} = 0 \quad \text{and} \quad x_1 = x_{12}$$

The total or bulk mass x_1 in this case is all remaining material.

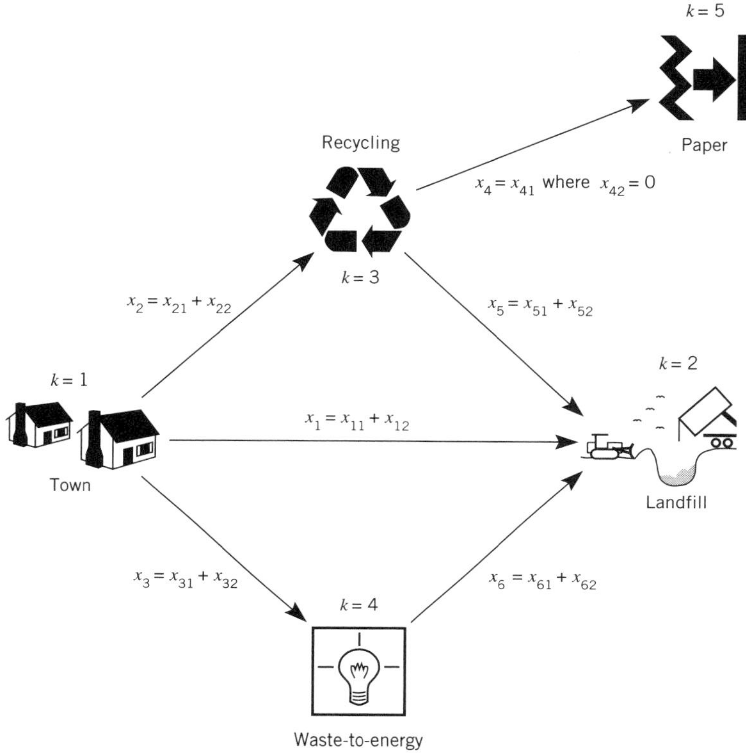

Similar definitions are introduced for links 2 through 5 as shown in the revised network diagram. Note that link 4 is restricted to paper only; thus, $x_{42} = 0$.

Link 6 or control variable x_6 involves the shipment of ash. Information about the ash and water contents of paper and the remaining material entering the waste-to-energy facility will be used to define x_6 in terms of x_{31} and x_{32}.

Double subscripts will be used in the formulation of the final model. The bulk mass or single-subscript notation is used as an aid to formulate the model.

Consider the constraint set. Since material may be separated at town and recycling nodes, these nodes are isolated from the entire system and the mass conservation law is used to derive a constraint set. Empirical relationships will be used at the waste-to-energy node.

The supply of bulk mass s_1 for the town is known. A mass balance at the town node is

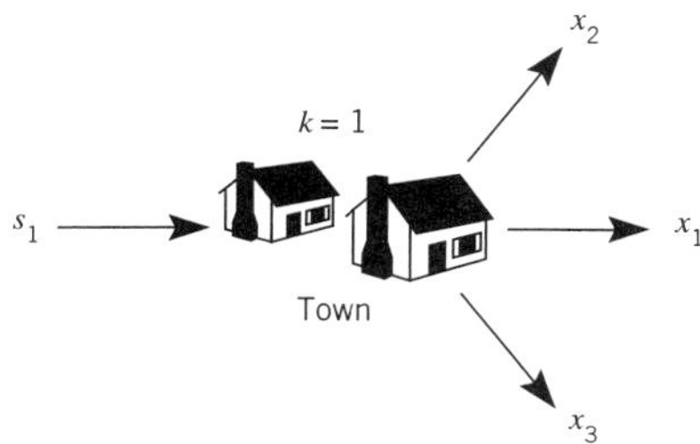

$$\textit{Mass in} = \textit{Mass out}$$
$$s_1 = x_1 + x_2 + x_3 \qquad \text{(Town)}$$
$$(\mathrm{M}) = (\mathrm{M}) + (\mathrm{M}) + (\mathrm{M})$$

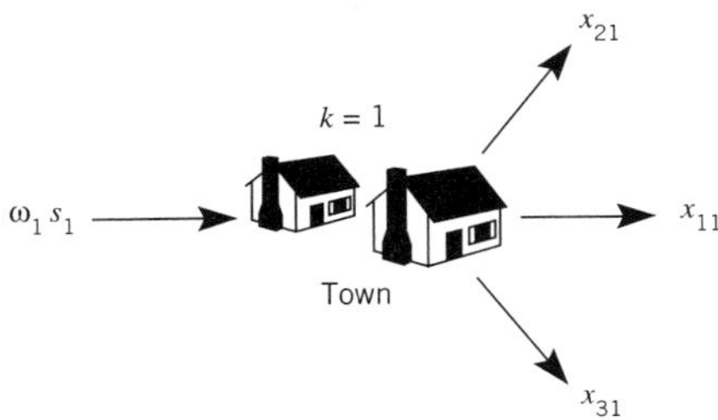

The fraction of paper ω_1 is also known; thus, the mass balance for paper is

$$\textit{Mass in} = \textit{Mass out}$$
$$\omega_1 s_1 = x_{11} + x_{21} + x_{31} \qquad \text{(Paper supply)}$$
$$(\mathrm{M}) = (\mathrm{M}) + (\mathrm{M}) + (\mathrm{M})$$

A mass balance relationship for the remaining material is defined in the same manner:

$$\omega_2 s_1 = (1 - \omega_1) s_1 = x_{12} + x_{22} + x_{32} \qquad \text{(Remaining MSW supply)}$$

These three equations are dependent; thus, only two equations need to be introduced into the constraint set.

The same approach is used at the recycling node.

Recycling

$$\textit{Mass in} = \textit{Mass out}$$
$$x_2 = x_4 + x_5 \qquad \text{(Total mass)}$$
$$x_{21} = x_{41} + x_{51} \qquad \text{(Recycling paper)}$$
$$x_{22} = x_{52} \qquad \text{(Remaining MSW)}$$

Since these equations are dependent, only two of them will be introduced into the optimization model.

The mass conservation law assumes that the mass of the media, MSW in this example, does not change state, degrade, or increase over time. This relationship is suitable for source and recycling nodes where material is separated and transported. In the case of incineration, materials change state. While the conservation law is applicable, it is generally easier to rely on empirical relationships. For example, calorimetric bomb tests are used to determine the amount of ash that is generated from a given mass of material or mixed media.

Calorimetric bomb tests can give the ash content fraction of the bulk mass or a. Thus, the ash can be written in terms of x_3. Assuming all moisture is driven off by the combustion process, the dry weight of ash is

$$x_6 = a(1 - m)x_3$$
$$(\text{M}) = \%(\text{M})$$

This relationship assumes that MSW bulk mass is unseparated. If paper, for example, is separated at the source and shipped to the recycling facility or landfill, this equation will give an erroneous value of ash. It is necessary to express the ash content for paper a_1 and the remaining MSW material, a_2, as

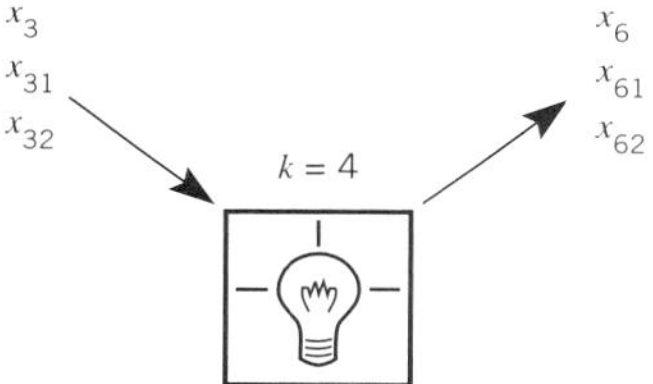

Mass in = *Mass out*

$$x_{61} = a_1(1 - m_1)x_{31} \qquad \text{(Waste-to-energy paper)}$$
$$(\text{M}) = \%\ (\text{M})$$

$$x_{62} = a_2(1 - m_2)x_{32} \qquad \text{(Waste-to-energy remaining MSW)}$$
$$(\text{M}) = \%\ (\text{M})$$

These two constraints dictate the choice of control variables for the entire model. The model will be written in terms of the control variables for paper and remaining MSW. The bulk mass control will not be used. The total ash is defined as

$$x_6 = x_{61} + x_{62} \quad \text{(Total ash)}$$

The policy constraint deals with the material on links 1, 5, and 6 as shown here.

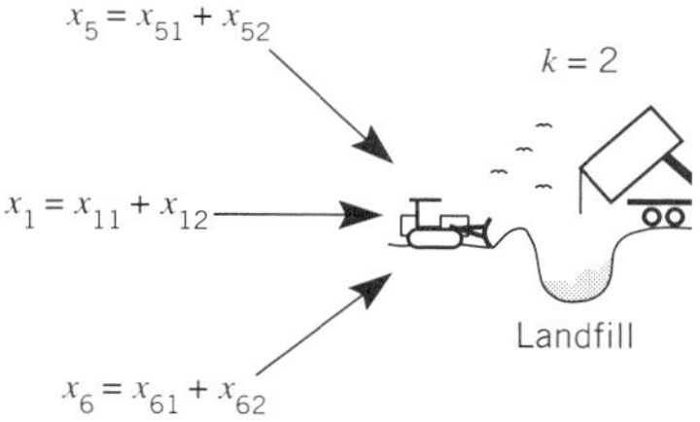

The policy constraint is

$$x_1 + x_5 + x_6 \leq b_2$$

or

$$x_{11} + x_{12} + x_{51} + x_{52} + x_{61} + x_{62} \leq b_2 \qquad \text{(Policy)}$$

The regulatory constraint on air emissions is written in terms of a capacity restriction on the amount of material that may be incinerated in the waste-to-energy facility.

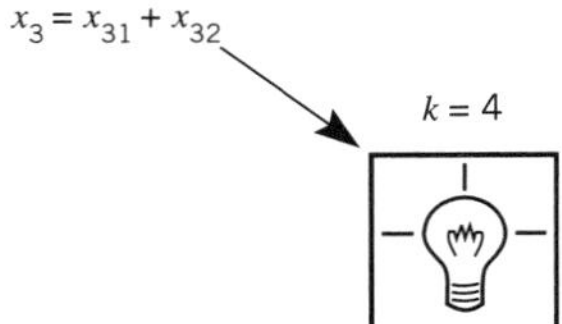

The regulatory constraint is

$$x_3 \leq b_4$$

or

$$x_{31} + x_{32} \leq b_4 \qquad \text{(Regulatory)}$$

The physical capacity of the recycling center limits the amount of material that may be recycled.

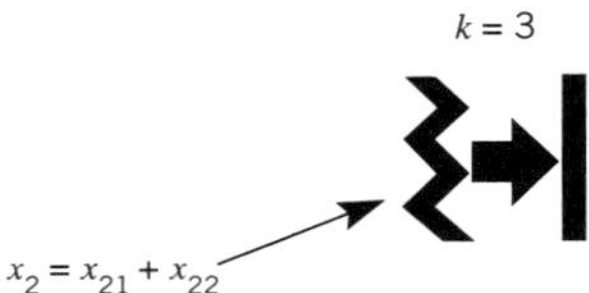

The recycling plant constraint is

$$x_2 \leq b_3$$

or

$$x_{21} + x_{22} \leq b_3 \qquad \text{(Recycling plant capacity)}$$

The behavioral constraint is imposed as a shipping constraint on link 2. The total amount of material that can be shipped from the town is limited to the amount generated by the town, or s_1. Thus, the link 2 capacity constraint for bulk mass is

$$x_2 \leq v s_1 \qquad \text{(Behavorial)}$$

However, individuals have the option either to separate paper in their household or to have the recycling center separate it. The option selected will, of course, depend on minimizing the cost to all households. These two options can be expressed by the following constraint set:

$$x_{21} \leq \nu\omega_1 s_1 \qquad \text{(Link 2 capacity for paper)}$$
$$x_{22} \leq \nu(1 - \omega_1)s_1 \qquad \text{(Link 2 capacity for remaining MSW)}$$

The objective function is to minimize total cost, which is equal to the total shipping costs plus tipping fees minus the profits received from the sale of paper and recovered energy from the waste-to-energy facility. The unit shipping and tipping fees were calculated on the amortized debt of capital investment and operation and maintenance costs. (Amortized debt is discussed in Section 4.2.) Profits from recovery energy were dependent on the properties of the material being incinerated as well as the thermal efficiency of the plant and its energy delivery system. These relationships are used to derive the objective function.

The total shipping cost for link i is the product of the shipping fee, shipping distance, and the bulk mass shipped along the link, or

$$s_i d_i x_i \quad \text{for } i = 1, 2, 3, 4, 5, \text{ and } 6$$

The tipping fees charged will be the product of the tipping fee for the facility and the bulk mass processed or disposed of at the facility:

$$u_k x_i \quad \text{for } i = 1, 2, 3, 5, \text{ and } 6$$

where u_k = tipping fee at facility k with $k = 2, 3, 4$ representing the landfill, recycling center, and waste-to-energy facility, respectively.

The total profit w from the sale of recycled material is the product of the unit selling price and the mass of material sold. For paper, the tipping fee is

$$w_4 = p_5 x_{41}$$

where p_5 = selling price of paper.

The total profit from the sale of recovered heat energy is dependent on the heat value of the material being incinerated and the thermal efficiency of the plant. The total profit is the product of unit profit for recovered energy, heat value, thermal efficiency, and dry mass of the material. A generalized equation is

$$\begin{aligned} w_{ij} &= p_k h_j e_t (1 - m_j) x_{ij} \qquad \text{for } i = 3 \\ (\$) &= (\$/\mathrm{E})(\mathrm{E/M})(\%)(\%)(\mathrm{M}) \end{aligned}$$

where $p_k = p_4$ = selling price of energy (\$/E), $e_k = e_4$ = thermal energy conversion efficiency (%) for the waste-to-energy and delivery system, h_j = energy content of paper, $j = 1$, and remaining material, $j = 2$.

In order to simplify the model, the unit cost for each link i, c_i, or c_{ij}, is determined:

$$\begin{aligned} c_1 &= s_1 d_1 + u_2 \\ c_2 &= s_2 d_2 + u_3 \\ c_{31} &= s_3 d_3 + u_4 - p_3\, h_1\, e_4 (1 - m_1) \\ c_{32} &= s_3 d_3 + u_4 - p_3\, h_2\, e_4 (1 - m_2) \\ c_4 &= s_4 d_4 - p_5 \\ c_5 &= s_5 d_5 + u_2 \\ c_6 &= s_6 d_6 + u_2 \end{aligned}$$

The final model, written in terms of the double-subscript notation for control variables and these unit costs, is

$$\text{minimize } y = c_1(x_{11} + x_{12}) + c_2(x_{21} + x_{22}) + c_{31}x_{31} + c_{32}x_{32}) + c_4 x_{41} + c_5(x_{51} + x_{52}) + c_6(x_{61} + x_{62})$$

subject to the constraints. The minimum cost model is summarized in an array:

x_{11}	x_{12}	x_{21}	x_{22}	x_{31}	x_{32}	x_{41}	x_{51}	x_{52}	x_{61}	x_{62}		b	
1		1		1							=	$\omega_1 s_1$	(Paper supply)
	1		1		1						=	$(1-\omega_1)s_1$	(Remaining MSW supply)
		1				−1	−1				=	0	(Recycling paper)
			1					−1			=	0	(Remaining MSW)
				$-a_1(1-m_1)$							=	0	(Waste-to-energy paper)
					$-a_2(1-m_2)$					1	=	0	(Waste-to-energy remaining MSW)
1	1						1	1	1	1	≤	b_2	(Policy)
		1	1								≤	b_3	(Recycling plant capacity)
				1	1						≤	b_4	(Regulatory)
		1									≤	$v\omega_1 s_1$	(Link capacity for paper)
			1								≤	$v(1-\omega_1)s_1$	(Link capacity for remaining MSW)
c_1	c_1	c_2	c_2	c_{31}	c_{32}	c_4	c_5	c_5	c_6	c_6	=	y	Cost

The array, written in terms of the given data, is

x_{11}	x_{12}	x_{21}	x_{22}	x_{31}	x_{32}	x_{41}	x_{51}	x_{52}	x_{61}	x_{62}		b	
1		1		1							=	2300	(Paper supply)
	1		1		1						=	2700	(Remaining MSW supply)
		1				−1	−1				=	0	(Recycling paper)
			1					−1			=	0	(Remaining MSW)
				−0.0485					1		=	0	(Waste-to-energy paper)
					−0.289					1	=	0	(Waste-to-energy remaining MSW)
1	1						1	1	1	1	≤	10,000	(Policy)
		1	1								≤	10,000	(Recycling plant capacity)
				1	1						≤	10,000	(Regulatory)
		1									≤	690	(Link capacity for paper)
			1								≤	810	(Link capacity for remaining MSW)
c_1	c_1	c_2	c_2	c_{31}	c_{32}	c_4	c_5	c_5	c_6	c_6	=	y	Cost

where the unit costs are:

$$
\begin{aligned}
c_1 &= \$40/\text{ton} \\
c_2 &= \$0/\text{ton} \\
c_{31} &= 100 - (0.1)(4{,}439)(0.6)(1 - 0.102) = -\$139.17/\text{ton} \\
c_{32} &= 100 - (0.1)(2{,}655)(0.6)(1 - 0.311) = -\$9.76/\text{ton} \\
c_4 &= \${-150}/\text{ton} \\
c_5 &= \$40/\text{ton} \\
c_6 &= \$40/\text{ton}
\end{aligned}
$$

The problem is solved by the simplex method. The minimum cost solution is

$$y^* = -\$320{,}000$$

or a profit of $320,000 per year is made. The optimum allocation of material is shown on the network diagram as a combination of recycling and energy recovery.

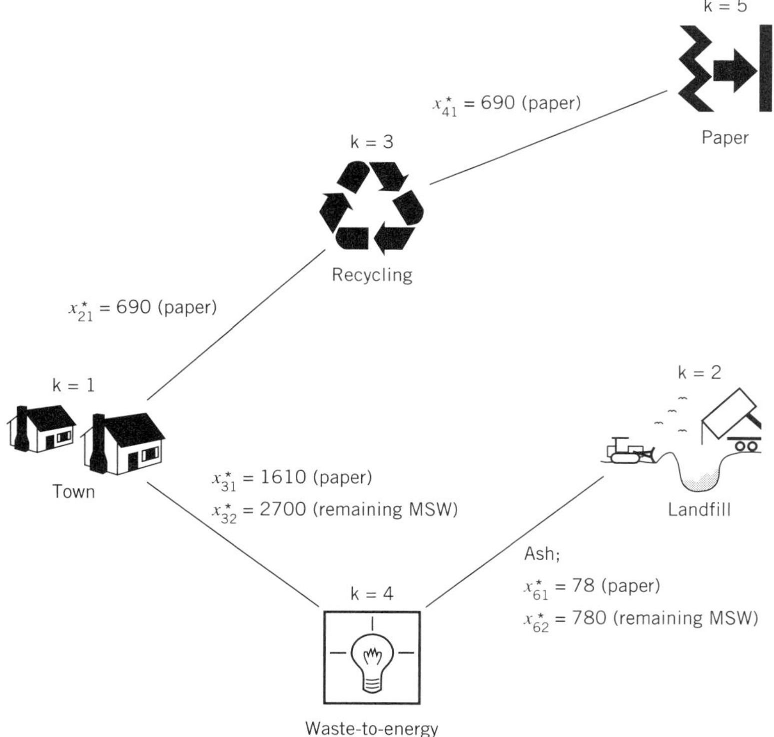

Path 2 → 4 consists of paper only. This solution implies that no separation of material is being performed in the recycling center. In essence, it is not needed because the residents of the town separate it. If this is not a viable option, the problem can be reformulated as a no-source separation model, as illustrated in the next problem.

EXAMPLE 3.9 ***An MSW Management Model with No Source Separation***

Two neighboring towns currently dispose of a total of 20 tons of MSW per year in a landfill. Of the 8000 and 12,000 tons of solid waste a year generated by towns 1 and 2, 25 and 45 percent, respectively, is paper. Since there is a potential to recycle 7400 tons of paper per year, a recycling center that serves both towns is to be considered.

The following network diagram shows the two alternatives:

A. Ship MSW to the landfill without processing.

B. Ship MSW to the recycling center to separate and sell paper.

No curbside collection of recycled paper will be provided. All MSW processing is done at the recycling center.

The following information has been gathered:

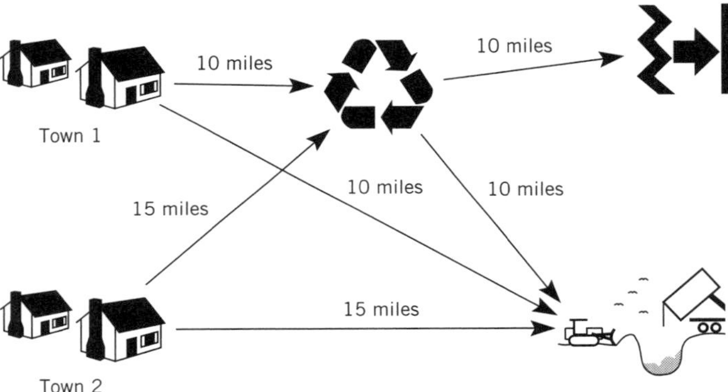

k		Tipping Fee ($ per ton)	Capacity (tons per year)
3	Recycling center	0	15,000
4	Landfill	40	15,000
5	Recycled paper	50 (profit)	5,000

All shipping costs are $1 per ton-mile.

Formulate a mathematical model to minimize total MSW disposal costs. In addition, perform a feedback analysis to determine the effects of source and no-source separation.

SOLUTION

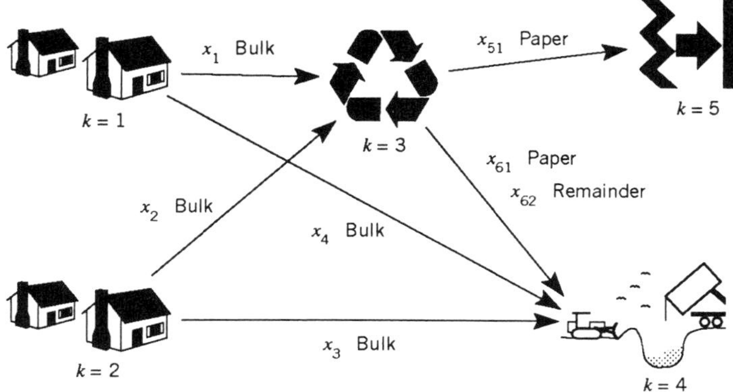

Since no MSW separation is provided at the sources, material shipped over links 1, 2, 3, and 4 must be bulk MSW. The material shipped over link 5 is restricted to recycled paper. Bulk material is shipped over link 6. Since MSW can be separated at the recycling center at node $k = 3$, the material being shipped over link 6 must be identified as being paper and remaining MSW after paper is removed.

The control variables are:

x_1, x_2, x_3, x_4 = bulk mass of MSW being shipped over links 1, 2, 3, and 4, respectively
x_{51} = mass of recycled paper
x_{61} = mass of paper disposed of in the landfill
x_{62} = mass of remaining MSW that is disposed of in the landfill after paper is removed

The unit costs to ship and process MSW over each link are:

$$c_1 = s_1 d_1 = (1)(10) = \$10 \text{ per ton}$$
$$c_2 = s_2 d_2 = (1)(15) = \$15 \text{ per ton}$$
$$c_3 = s_3 d_3 + u_2 = (1)(15) + 40 = \$55 \text{ per ton}$$
$$c_4 = s_4 d_4 + u_4 = (1)(10) + 40 = \$50 \text{ per ton}$$
$$c_5 = s_5 d_5 - p_4 = (1)(10) - 50 = \$-40 \text{ per ton}$$
$$c_6 = s_6 d_6 + u_4 = (1)(10) + 40 = \$50 \text{ per ton}$$

Let y = total network cost; thus, the objective function is

$$\text{minimize } y = 10x_1 + 15x_2 + 55x_3 + 50x_4 - 40x_{51} + 50x_{61} + 50x_{62}$$

Mass balance relationships will be established at nodes $k = 1$, 2, and 3.
At $k = 1$,

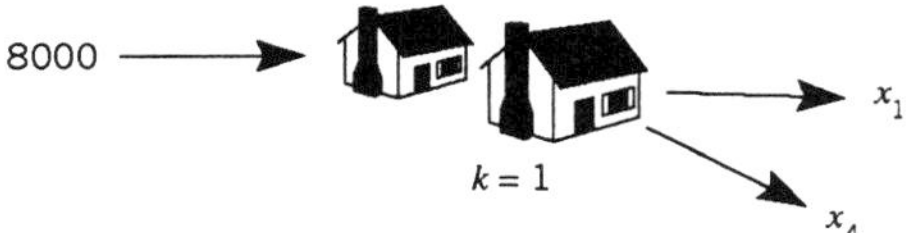

$$\textit{Mass in} = \textit{Mass out}$$
$$8000 = x_1 + x_4 \qquad \text{(Town 1)}$$

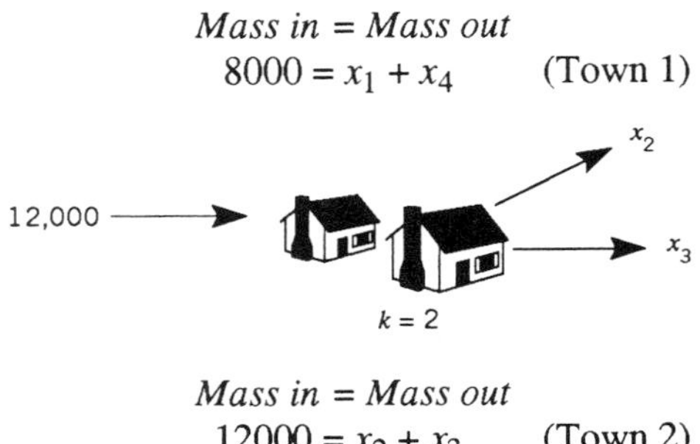

$$\textit{Mass in} = \textit{Mass out}$$
$$12000 = x_2 + x_3 \qquad \text{(Town 2)}$$

At $k = 3$ the mass of material is identified as recycled paper and remaining MSW. Twenty-five and 45 percent of the MSW from towns 1 and 2, respectively, is paper.

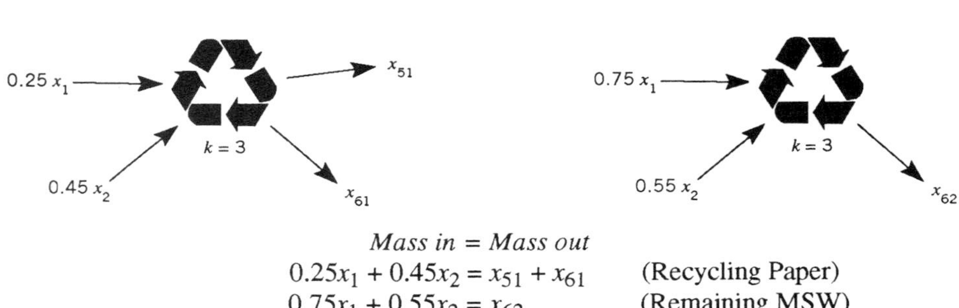

$$\textit{Mass in} = \textit{Mass out}$$
$$0.25x_1 + 0.45x_2 = x_{51} + x_{61} \qquad \text{(Recycling Paper)}$$
$$0.75x_1 + 0.55x_2 = x_{62} \qquad \text{(Remaining MSW)}$$

Capacity Constraints

$$x_1 + x_2 \le 15{,}000 \qquad \text{(Recycling Capacity)}$$
$$x_3 + x_4 + x_{61} + x_{62} \le 15{,}000 \qquad \text{(Landfill Capacity)}$$
$$x_{51} \le 5{,}000 \qquad \text{(Paper demand)}$$

The model is summarized in an array.

x_1	x_2	x_3	x_4	x_{51}	x_{61}	x_{62}		b	
−1	0	0	−1	0	0	0	=	8,000	(Town 1)
0	−1	−1	0	0	0	0	=	12,000	(Town 2)
0.25	0.45	0	0	−1	−1	0	=	0	(Recycling Paper)
0.75	0.55	0	0	0	0	−1	=	0	(Remaining MSW)
−1	−1	0	0	0	0	0	≤	15,000	(Recycling plant Capacity)
0	0	−1	−1	0	−1	−1	≤	15,000	(Landfill Capacity)
0	0	0	0	−1	0	0	≤	5,000	(Paper demand)
10	15	55	50	−40	50	50			Cost

The optimum solution is solved with the simplex method. The minimum cost solution is

$$y^* = \$721{,}000$$

Superimposing these results on the network diagram shows that no MSW from town 1 is recycled. Since capacity constraint on recycling at $k = 5$ is active, it is only possible to ship 11,111 tons of MSW from town 1 to the recycling center. The remaining MSW of 889 tons must be shipped directly to the landfill. The demand for paper is critical.

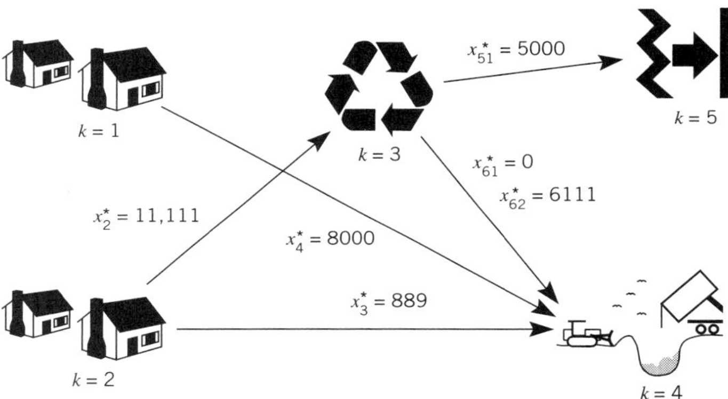

All values in the figure are given in tons per year.

FEEDBACK ANALYSIS OF SOURCE SEPARATION MODEL

In order to study the impact of MSW source separation, a feedback analysis is performed. The same network diagram and unit costs used previously are applied. The primary difference between the two models is that the control variables x_1, x_2, x_3 and x_4 from above are redefined with x_{11}, x_{12}, x_{21}, x_{22}, x_{31}, x_{32}, x_{41} and x_{42} and introduced into a source separation model. The mass balance relationships at nodes k = 1, 2, and 3 are replaced with mass balance relationship based on the redefined control variables. These details are not given here. (See Problem 3.3.6). We will concentrate our discussion on comparing the optimum solutions from the source and no-source separation model.

The minimum cost solution with source separation is

$$y'^* = \$660{,}000$$

The optimum allocation of MSW is

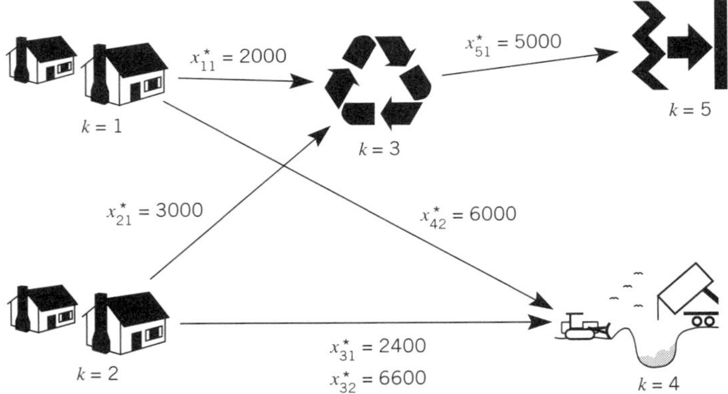

Comparing the solutions with and without MSW source separation shows that a cost savings is associated with source separation of

$$w = y^* - y'^* = \$721{,}000 - \$660{,}000 = \$61{,}000$$

Most importantly, the shipping patterns of the two options is dramatically different. Once again, the demand for paper proves critical as in the no source separation model.

DISCUSSION

As with all systems analyses, critical evaluation is a necessity. For example, the shipping costs for both analyses were assumed to be the same. Careful analysis of the source separation model implies that trucks must make two runs. One run is to pick up recycled paper, and the other is to pick up the remaining MSW. Making two runs may not be reflected in the unit costs in the source separation model. In fact, some studies indicate that it is more cost effective not to separate at the source because it saves time to collect bulk MSW and separate it at a materials recovery facility (MRC) equipped with a conveyor belt system for MSW separation.

Of course, the assumptions that all residents in the source separation model will cooperate and recycle should be challenged. Sensitivity and feedback analyses may prove useful in evaluating the robustness of these two models under different conditions if the actual number of volunteers that will recycle is unknown.

QUESTIONS FOR ANALYSIS

3.3.1* Consider two alternatives:

- Ship MSW to a recycling center for processing.
- Ship MSW to a landfill with no processing.

Find a minimum cost solution assuming source separation.

A town generates 3640 tons of MSW per year, 50% of which is paper. The unit shipping cost is $s = \$1$ per ton-mile. The landfill tipping fee is $u = \$50$ per ton. The market price for recycled paper is \$20 per ton.

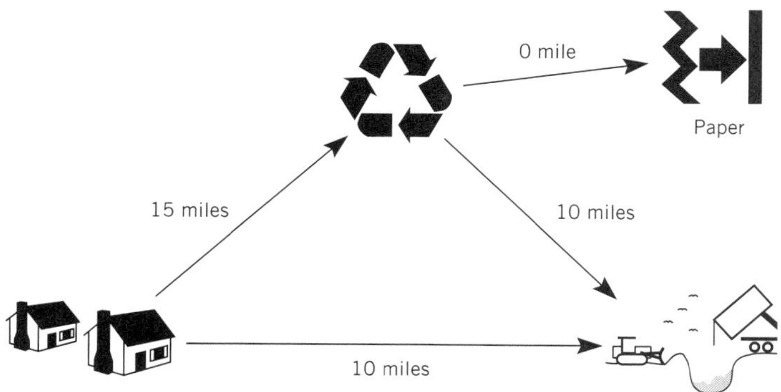

Formulate and solve as an LP model.
Use the following network diagram to define the control variables:

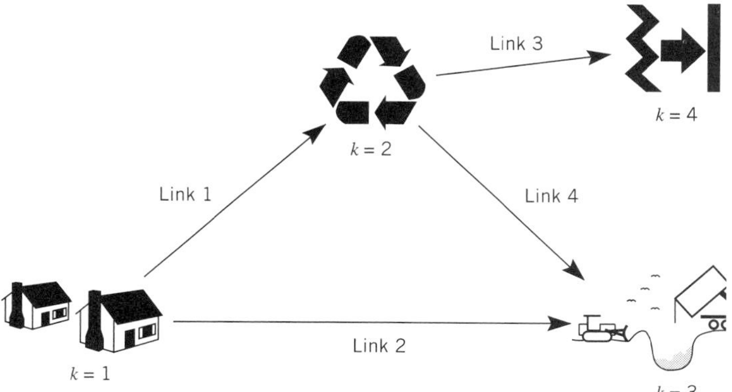

3.3.2* Consider the same two alternatives given in Problem 3.3.1:

- Ship MSW to a recycling center for processing.
- Ship MSW to a landfill with no processing.

Unlike the previous problem, assume that MSW separation will be performed at the recycling center only.

Formulate and solve as an LP model. Use the network diagram given in Problem 3.3.1 to define the control variables and to formulate an LP model.

3.3.3* Consider the same two alternatives given in Problem 3.3.1:

- Ship MSW to a recycling center for processing.
- Ship MSW to a landfill with no processing.

The project objective in this problem is to extend the life of the landfill as much as possible and to do this within a budget of $150,000. Like Problem 3.1.1, assume source separation.

(a) Formulate a mathematical model to minimize the total mass of material deposited in the landfill for the network diagram and the information given in Problem 3.3.1.

(b) Perform a sensitivity analysis and show that there is no feasible solution if the budget is limited to $100,000 or less.

Use the network diagram given in Problem 3.3.1 to define controls variables and to formulate an LP model.

3.3.4* Consider three alternatives:

- Ship MSW to a recycling center for processing.
- Ship MSW to a landfill with no processing.
- Have the residents separate and sell recycable paper directly at the marketplace.

If the residents do not select the third alternative, then it is assumed that they will not separate their MSW for either of the first two alternatives. It is estimated that the residents will directly recycle no more than 300 tons of paper.

Find a minimum cost solution. Formulate and solve as a LP model.

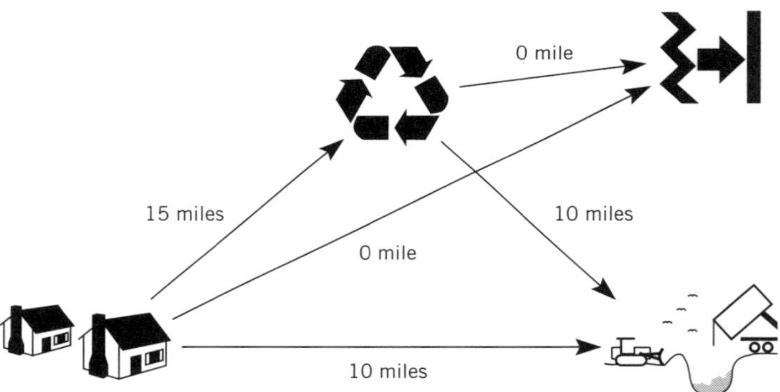

Use the following information to define the control variables:

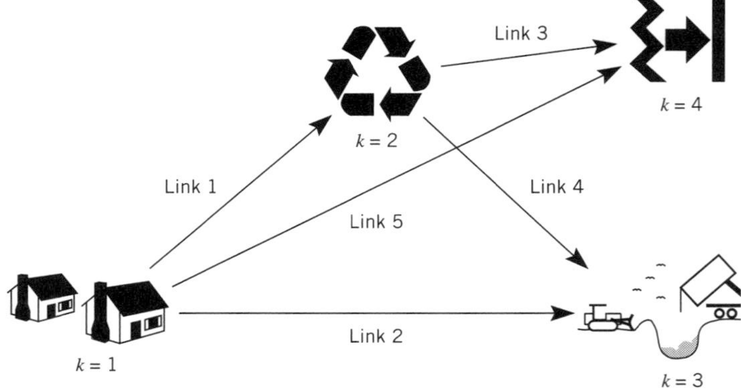

3.3.5* Of a total of 7500 tons of a town's MSW, 1500 tons constitute hazardous material. Currently, this hazardous material is incinerated at a waste-to-energy plant because it is highly combustible and its byproducts of combustion do not endanger the public health. However, this material can be sold as recycled material for p_2 = \$185 per ton. The waste-to-energy facility, which operates at e_3 = 40% thermal efficiency, sells recovered energy at p_3 = \$0.10 per kwh/ton. The unit shipping cost is s = \$1 per ton-mile for all links.

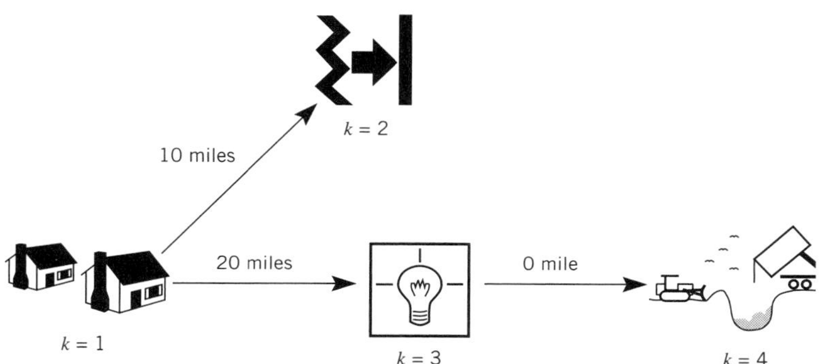

Do you recommend that the hazardous material be incinerated or recycled? Formulate and solve as an LP model.

j	Component	e_j Energy Content (kWh/ton (dry mass))	a_j Ash (% of as collected mass)	m_j Moisture (% of as collected mass)
1	Hazardous material	8380	10	2
2	Remaining MSW*	3376	26	22

*after removing hazardous material

k	Facility	Capacity, b_k (tons per year)	Tipping fee (\$ per ton, u_k)
2	Marketplace for hazardous material	1,000	185 (profit)
3	Waste-to-energy plant	10,000	50
4	Landfill	8,000	0

3.3.6* Confirm the results for the feedback analysis given in Example 3.9. Show the feedback model formulation and solve for its optimum solution.

REFERENCES

Ackerman F., and J. Schall (1988). *WastePlan: A Computer Model for Solid Waste Planning*, Fourth International Conference on Urban Solid Waste Management and Secondary Materials, Philadelphia, PA.

Chapman, R. E., and E. B. Berman (1983). *The Resource Recovery Planning Model, A New Tool for Waste Management*, NBS Report 657, Washington, DC.

Gottinger, H. W. (1988). "Computational Model for Solid Waste Management with Application," *European Journal of Operations Research*, **35** (3), 350–364.

Kaila, J. (1987). *Mathematical Model for Strategy Evaluation of Municipal Solid Waste Management Systems*, Technical Research Centre of Finland, Publications 40, Espoo, Finland.

Karam, J. G., C. St. Cin and J. Tilly (1988). "Economic Evaluation of Waste Minimization Options", *Environmental Progress*, **7**, 192–197.

Mashayehekhi, A. N. (1988). *Long Term Financing of Solid Waste Disposal in New York State: A Dynamic Analysis*, The Nelson Rockefeller Institute of Government, State University of New York.

Murphy, W. R., and G. McKay (1982). *Energy Management*, Butterworths, London.

Ossenbruggen, P. J. and P. C. Ossenbruggen (1992). "SWAP: A Computer Package for Solid Waste Management," *Computers, Environment and Urban Systems*, **16** (2), 83–100.

Reed, S. (1984). "Monitoring and Long Term Planning: Essential Ingredients of Waste Management," *Conservation and Recycling*, **7** (2–4), Pergamon Press Ltd., Oxford, England.

Reuse of Solid Wastes (1982). Conference Proceedings organized by Institute of Civil Engineers in London on November 11–12, 1981, Thetford Press, Thetford, Norfolk, England.

Saad, G. H. (1988). *A Multi-attribute Model for Management of Urban Solid Waste and Secondary Materials*, Fourth International Conference on Urban Solid Waste Management and Secondary Materials, Philadelphia, PA.

Suess, M. J. (1985). *Solid Waste Management, Selected Topics*, World Health Organization, Copenhagen, Denmark.

Tchobanoglous, G., and H. Theisen (1977). *Solid Wastes, Engineering Principles and Management Issues*, McGraw-Hill Book Co., New York.

Chapter 4

Capital Investment Analysis

Thousands of engineers can design bridges, calculate strains and stresses, and draw up specifications for machines, but the great engineer is the man who can tell whether the bridge or machine should be built at all, where it should be built, and when.

Eugene G. Grace (1876–1960)

Most of this chapter is devoted to techniques for assessing capital investments. Remember that a capital investment is a commitment of monetary resources whose purpose is to obtain a future benefit. A land developer, for example, will purchase land and build housing units, shops, or offices in anticipation of future property sales or rentals. Before making a decision to proceed, the developer must weigh the capital cost of investment against future benefits. The decision requires that consideration be given to

- **Time:** Benefits from the investment could take months or years before they are received.
- **Opportunity costs:** Capital spent on this project is unavailable for immediate consumption and for other investments; thus, these opportunities are lost.
- **Economic risk:** Since there is no guarantee of success; money invested could be lost.

These factors will be incorporated into the alternative selection process for capital investment.

4.1 THE PROFIT MOTIVE

In a market economy, the profit motive is a fundamental principle used in decision making. Whereas in earlier chapters, this principle was employed to formulate least cost optimization models, in this and following chapters, it is used for evaluating and selecting alternatives based on capital investment analysis for private and public investment. In Section 1.2, the basic concept of net worth was introduced; it is valuable to repeat it here.

Net worth or net profit, nw, is defined to be the difference between total revenue and total production cost, or

$$nw = w - k$$

or written as a function of the production level x

$$nw(x) = (p - c)x$$

where

c = unit cost to produce an item ($/#)
p = unit price of the item ($/#)
w = total revenue or sales ($)
k = total production cost ($)
x = number of items produced (#)

Let us take a closer look at the total production cost. The total cost of production can be written as a linear combination of fixed and variable costs. Since a firm controls the production level, it is expressed as a function of x,

$$k(x) = k_f + k_v(x)$$

where

k_f = fixed cost ($)
$k_v(x)$ = variable cost function ($)

Fixed costs are overhead costs, such as administrative salaries, property rental, equipment purchases and other expenses that do not vary with output x. Variable costs include those of material, labor and energy associated with items that are directly related to producing x units. The unit cost can also be expressed as a function of x:

$$c = c(x) = \frac{k(x)}{x}$$

A price is a measure of the value of the item and is determined in the marketplace, where a consumer (buyer) and seller bargain until they both agree on an exchange value of the item. The *market price p* is a function of total consumer demand q, or $p(q)$. In the model, $nw = (p - c)x$, and the price p is a constant and average value, reflecting all individual buyer-seller transactions. For our purposes, the price p is assumed known. The function $p(q)$ is discussed in Section 4.5.

Productivity and Incentives

A primary purpose of a capital investment is to improve *productivity*. Productivity is a measure of how efficiently goods and services are produced. For our purposes, it will be measured as either a net worth, $nw = w - k$, or as a unit cost of production c. Productivity increases are achieved by employing various means, including installing new equipment, improving operation and control schemes, and improving worker efficiency through education and other incentives. Their effects are measured with k and c.

Capital investment can give a firm a competitive edge over its competitors. After a firm introduces a cost-saving measure, $c(x)$ decreases and nw increases. This assumes that the market price p is unchanged and that sales x remains constant. The firm has another option: it can attempt to increase its *market share*, its proportion of the total sales of a good or service. After a firm improves its production efficiency, it can pass on a portion of its savings to its customers in the form of lower prices. At the same time, it may increase its net profit by increasing sales x and total revenue px. If its competitors do not reduce their price, the firm should capture a larger share of the market.

In theory, the net effect of increased productivity in a competitive marketplace is that society as a whole will benefit. Under this system, all firms have an incentive to find ways to increase productivity and to increase profits. Once a firm introduces a more productive measure, it may initially attempt to reap larger profits by holding firm on its selling price. However, its competitors, which also have an incentive to maximize profits, will also introduce more productive methods. Some are expected to increase profits by gaining a larger market share by reducing the selling price. In order for the original firm to maintain its profit and market share, it will at some point in time reduce its selling price. In the long term, the consumers benefit by paying lower selling prices. The firms will receive increased profits and consumers will pay lower prices.

Economy of Scale

An *economy of scale* for a system exists when the unit cost $c(x)$ decreases with increases in production x. In Figure 1, a capital investment is shown to have two major parts: there is a downward shift in the unit production cost plus the old system, which has no economy of scale, is replaced by a new system with an economy of scale. Clearly, a producer would be wise to consider both capital investment and the production level x to maximize net profit nw.

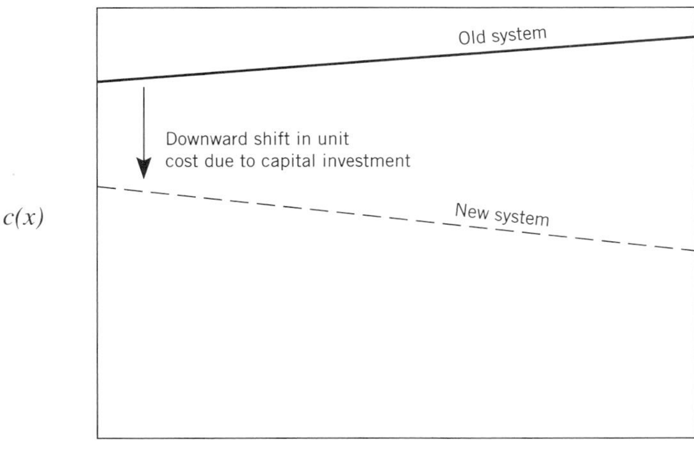

Figure 1 A capital investment and its effect on productivity.

As a consequence of economy of scale, systems are often designed and constructed with an overcapacity in anticipation of future growth in demand x on the system. If the growth is realized, the unit cost will decrease at some future time. Of course, there is the risk that the anticipated growth is not met.

Not all systems have economies of scale. Economic projections for MSW waste-to-energy systems show unit operating costs increasing with increasing plant capacity. In lieu of providing one large system, it is sometimes possible to reduce capital costs through economies of scale associated with the multiple use of smaller units.

EXAMPLE 4.1 *Economy of Scale and Economic Risk*

Suppose, for example, that a one-ton capacity truck makes trips costing \$5. Assume that this cost is independent of the load; thus, a truck filled to the one-ton capacity or to any fraction of capacity will cost the same, \$5.

(a) Evaluate truck usage using the principle of economy of scale.

(b) Discuss the effect of operating the truck at 25% of capacity when there is little evidence of growth in demand.

SOLUTION

(a) Using a unit cost definition gives a different perspective. The unit cost function is simply derived,

$$c(x) = \frac{k(x)}{x} = \frac{\$5}{x} \quad \text{for } x \leq 1 \text{ ton}$$

where x = load (tons). The figure shows that an economy of scale exists. The unit shipping cost $c(x)$ decreases with increased load x.

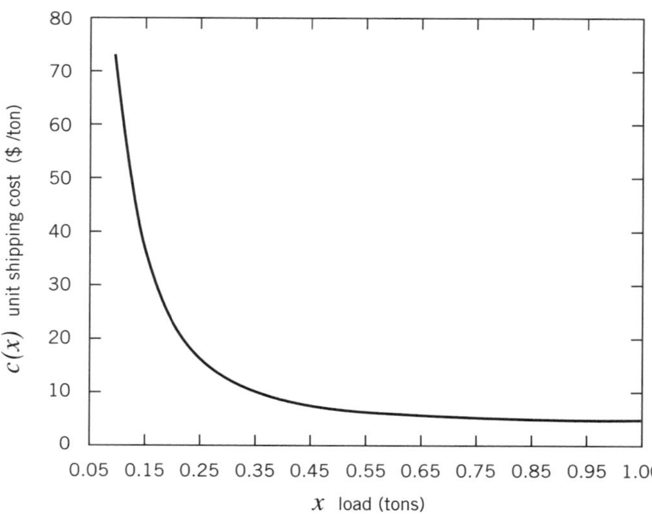

(b) Providing an overcapacity in anticipation of some future demand does not come without economic risk. For example, suppose the demand is $x = 0.25$ ton and a one-ton truck is provided. The unit cost of operation from the graph is $c(0.25) = \$20$ per ton. This is four times the unit cost at capacity, $c(1) = \$5$ per ton. If there is no evidence that demand will not grow rapidly, then a smaller truck operating near capacity may prove to be a better choice.

The Breakeven Point

Let us now turn our attention to alternative selection and the use of sensitivity analysis. Some systems operate more cost effectively under certain conditions than others. This idea is illustrated in Figure 2. In the figure, the total cost functions for alternatives A and B are plotted as $k_A(p)$–p and $k_B(p)$–p.

The *breakeven point* is the point where alternatives have the same value. In Figure 2, the breakeven price for two alternatives A and B is the price where

$$k_A(p') = k_B(p') \rightarrow p' = \$3.60/\text{unit}$$

For the alternative selection where the total cost is minimized, B is an optimum solution for $p < p'$; otherwise, A is an optimum solution for $p > p'$.

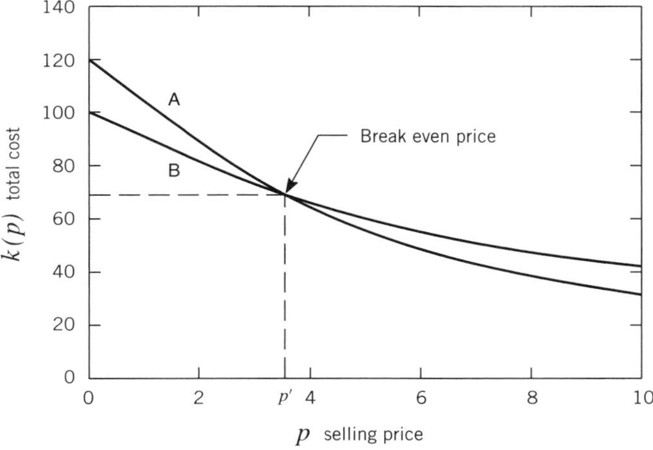

Figure 2 Breakeven price.

The breakeven point is used in a variety of contexts. For example, the breakeven price is where the net worth is zero, $nw = (p - c)x = 0$, or $p' = c$. The following problem uses the breakeven production level or x' for an evaluation.

EXAMPLE 4.2 ***Recycling Beverage Containers***

A survey of the beverage industry showed that the cost of nonreturnable containers was \$0.60 per case and returnable containers, \$3.00 per case. The filling, retailing, and distribution cost for both container types was \$1.20 per case.

Determine whether or not the beverage industry has a profit incentive to use returnable containers.

SOLUTION

The total costs for a beverage packaged in a case of returnable and nonreturnable containers are:

A. $k(x) = 0.60 + 1.20x \rightarrow$ nonreturnable

B. $k(x) = 3 + 1.20x \rightarrow$ returnable

where x = trippage, number of times the case is reused.

Since $x = 1$ for nonreturnable containers, $k_A = k(1) = \$1.80$. It follows that $c_A = k_A/1 = \$1.80$ per case.

The \$3.00 per case cost is considered a fixed cost and the processing cost of \$1.20 per case is considered a variable cost. Its unit cost is

$$c_B(x) = \frac{k_B(x)}{x} = \$1.20 + \frac{3}{x}$$

Since $c_B(x)$ decreases with an increase in x, an economy of scale exists as shown in the following figure. For nonreturnable containers, c_A is constant. The figure also shows a breakeven trippage as $x' = 5$ trips.

According to this analysis, it is economically feasible for the beverage industry to use returnable containers. In fact, the beverage industry will achieve additional profits when $x' > 5$. The additional profits, as a unit savings function, is:

$$w(x) = c_A - c_B(x) = 0.60 - \frac{3}{x}$$

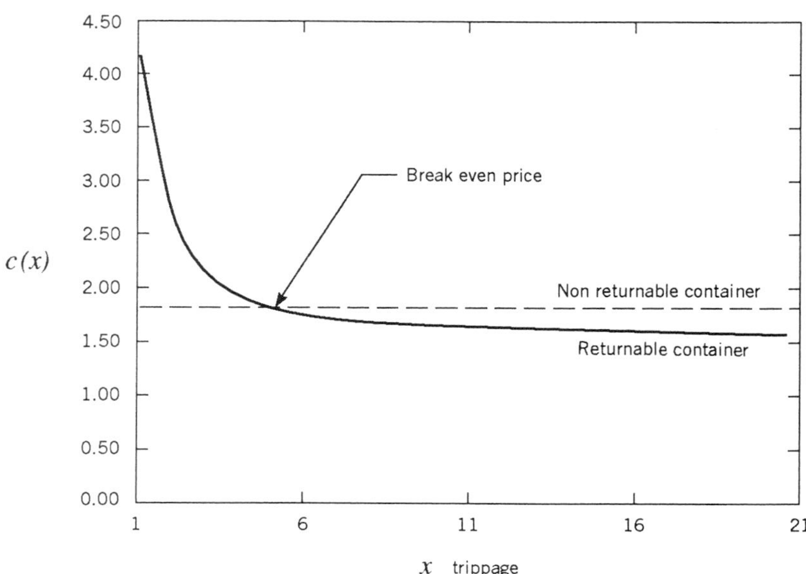

DISCUSSION

This analysis shows that the beverage industry should have an incentive to use returnable containers. Consider some additional facts that have made the use of returnable containers a controversial issue.

Studies revealed that returnable containers were less detrimental to the environment than nonreturnable containers. In addition, society can incur savings by conserving natural resources and by reducing solid waste volume. In spite of these profit incentives and societal benefits, the beverage industry has resisted offering returnable beverage containers.

As a rationale for its resistance, the beverage industry cites consumer survey data illustrating the decline of trippage rates over time for customers with a choice between nonreturnable containers and returnable containers with deposits. Thus, the beverage industry offers the more popular nonreturnable container in order to please the majority of its consumers.

Studies show that consumers find nonreturnable containers to be more convenient than returnable containers because deposit systems tend to be confusing and troublesome. Increasing the amount of deposit on the containers is marginally effective as an inducement for consumers to purchase products in returnable containers.

Finding ways to facilitate the use of returnable containers has proved difficult. Consumers and the beverage industry must cooperate. Some critics claim that the bottling industry can do more. For example, it can inexpensively educate consumers by providing information about deposit return and recycling procedures on the container. A more expensive and fundamental change would be to bottle beverages in standardized, interchangeable containers so that different bottling companies would use them. Product identification would be provided by the label. Since storage and handling by suppliers and consumers would become simpler and easier, we could expect a cheaper price for bottler, transporter, and consumer alike. When changes like these are not made voluntarily, the government steps in and takes action for the overall good of the society. In some states, legislative action has been applied to impose the use of deposits on beverage containers.

EXAMPLE 4.3 *Breakeven Price for Recycled Paper*

Residents have the choice to recycle or incinerate their waste. The town receives \$0.10 per kWh for recovered heat energy from a waste-to-energy plant with a 25% efficiency rating.

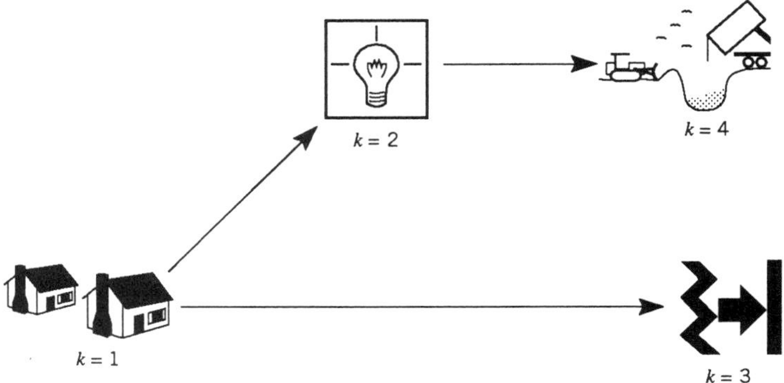

The task is to determine whether or not to

A. Recycle paper, or

B. Burn paper in a waste-to-energy plant.

The town has a population of 10,000 people and generates annually 0.6 ton of MSW per person, or 6000 tons of MSW per year. Forty percent of the solid waste is paper.

The tipping fees and capacities are

	k	p_k, selling price	f_k, unit costs ($ per ton)	b_k, capacity (tons per year)
Waste-to-energy plant	2	$0.10 per kWh	50	10,000
Landfill	4	—	100	10,000
Market for paper	3	p_3	—	10,000

All shipping costs are assumed to be zero. The material properties are

Component	j	h_j, Heat content (dry mass) kWh	m_j, Moisture content (% of as collected mass)	a_j, Ash content (% of dry mass)
Recyclable paper	1	4438	10.2	5.4
Bulk MSW[a]	2	2655	31.1	41.9

[a]Properties excluding recyclable paper.

Perform a sensitivity analysis by varying the selling price of recycled paper from $p_3 = \$0$ to $200 per ton.

SOLUTION

Alternative A:

The revenue from recycling paper, $x = (0.4)(6000) = 2400$ tons of paper per year, is

$$w_{13}(p_3) = p_3 x = 2400\, p_3$$

where

$$0 \le p_3 \le 200$$

Alternative B:

The bulk MSW, the mass of material excluding paper, will be incinerated regardless of whether or not paper is recycled; therefore, it is not entered into the analysis. The cost and profits from incinerating bulk MSW are introduced into the analysis. The revenue from incinerating paper is dependent on its heat content and the selling price of the energy it generates and independent of p_3, or

$$w_{12}(p_3) = (p_2\, h_1' - c)x$$

where

p_2 = selling price of usable heat energy = \$0.10 per kWh
h_1' = usable heat energy equal to the dry heat value of paper times the thermal efficiency of the plant

or

$$h_1' = h_1\,(1 - m_1)\,e_1 = 4438\,(1 - 0.102)\,(0.25) = 996 \text{ kWh per ton}$$

and

c = cost of incinerating the paper and landfill disposal of ash

or

$$c = c_2 + a_1(1 - m_1)c_4$$
$$c = \$50 + 0.054\,(1 - 0.102)(100)$$
$$= \$54.85 \text{ per ton}$$

Thus,

$$w_{12}(p_3) = (\$0.10 \text{ x } 996 - 54.85)\ 2400 = \$107{,}000.$$

The $w_{12}(p_3) - p_3$ and $w_{13}(p_3) - p_3$ functions for the two alternatives are plotted on the following figure. The breakeven price is calculated to be

$$p_3' = \frac{\$107{,}000}{2400} = \$45 \text{ per ton.}$$

This analysis shows that paper should be recycled when its price is greater than p_3'. The recommendation is

A. Recycle paper when $p_3' \geq \$45$ per ton, and

B. Incinerate paper when $p_3' < \$45$ per ton.

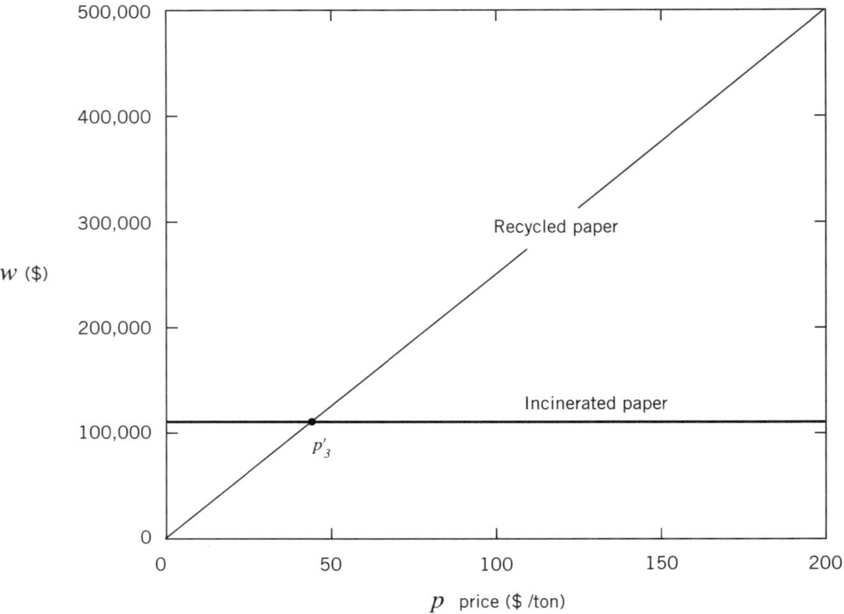

QUESTIONS FOR DISCUSSION

4.1.1 Define the following terms:

breakeven point	market price
capital investment	market share
production cost	productivity
economy of scale	

4.1.2 Describe how, theoretically, the whole society benefits from competition, the profit motive, and increased productivity in the long term.

4.1.3 When is it more profitable to provide a system with an overcapacity? Discuss in terms of economy of scale and economic risk.

4.1.4 The analysis of the beverage industry in Example 4.2 showed that from the point of view of a profit motive, the beverage industry should be offering returnable containers.

(a) In your opinion, what is the primary reason for industry resistance?

(b) Do you favor legislative action that would require beverage containers to be returned and reused? Explain your answer whether or not you favor it.

(c) Recycling of bottles and aluminum cans is one way of allowing the beverage industry to offer nonreturnable containers. Under this scheme, the beverage industry is not involved in the collection of used containers. Consequently, the cost of recycling is borne not by the beverage industry but by local communities, who must collect and sell the used containers. In your opinion, is this system fair? Explain your answer.

QUESTIONS FOR ANALYSIS

Instructions: Problems marked with an asterisk (*) are most easily solved with the aid of a computer spreadsheet program and computer graphics.

4.1.5 The unit production cost function for a system, measured in $M per unit, is

$$c = 100e^{0.1x}$$

where

$$x = \text{number of units produced}$$

(a) Does an economy of scale exist? Why?

(b) If $x = 8$ units are to be produced, would you recommend: (1) the use of one system with a production capacity of $x = 8$ units, (2) the use of eight systems with a capacity of $x = 1$ unit. Use total cost as a measure of effectiveness.

4.1.6* The total production cost is

$$k(x) = \$80 + 3x$$

(a) Determine the unit cost function, $c(x)$.

(b) Plot the $k(x) - x$ and $c(x) - x$ relationships.

(c) In what range of x do you feel it is reasonable to assume that no economy of scale exists?

4.1.7* Each item has a unit selling price of $5 and a production cost of $3 per item. The total fixed cost production is $80.

(a) Determine the net profit function, $nw(x)$.

(b) Plot the $nw(x) - x$ function and identify the break-even profit level x'.

(c) Does an economy of scale exist for this production?

(d) Does an economy of scale guarantee a profit for the firm?

4.1.8 An operating cost function for a system is

$$k = ax^b$$

where

$$x = \text{number of units produced.}$$

(a) What range of b will an economy of scale exist?

(b) The total production level is $x = 8$ units per year. Consider two alternatives where $a =$ \$1M and $b = 1.2$:

A. one system operating at $x = 8$ units.

B. eight systems operating at $x = 1$ unit.

(c) Repeat (b) for $a = \$1M$ and $b = 0.8$.

4.1.9* Two processes with the following production costs are being considered for selection:

A. $k = \$300x^{1.2}$, where $x \leq 15$ production units

B. $k = \$400x^{0.8}$, where $x \leq 15$ production units and

where x = production level

(a) Assume a selling price of $p = \$500$ per item. Derive a net profit function $nw(x)$ for each alternative and compare the two alternatives by plotting their $nw(x) - x$ curves. Determine the production level range where alternatives A and B are recommended.

(b) Identify the breakeven profit points x' for each alternative.

4.1.10* A city of 100,000 people generates no more than 40,000 tons of paper per year. The city owns and operates a material recovery facility where paper is separated and prepared for shipment to a marketplace for sale. The facility costs $500,000 per year to operate, which is independent of the level of paper that it recycles. The city has determined that the selling price of recycled paper in $ per ton declines exponentially as a function of the amount sold, or

$$p = 60e^{-0.00004q}$$

where q = annual amount of recycled paper in tons.

The goal of the city engineers is to determine the optimum amount of paper q^* to be recycled to maximize annual profit nw^* from paper recycled.

(a) Derive an optimization model to maximize net worth in terms of the control variable q.

(b) Plot the $nw(q) - q$ curve and identify the breakeven level(s), or $nw(q') = 0$.

(c) Use differential calculus to determine the optimum solution q^* and $nw(q^*)$.

4.1.11* Determine the effects of shipping distance d and moisture content m on the unit profit function for a waste-to-energy plant that incinerates MSW with an energy content of 3500 kWh per ton dry weight with energy recovery rate of 40%. Assume a selling price of energy of $0.10 per kWh and the shipping cost of $1.00 per ton-mile.

(a) Derive a selling price function, $p(d, m)$.

(b) Plot $p(d,m) - m$ functions for $d = 0, 40, 80$ miles and $m = 0$ to 1 at intervals of 0.1. Determine the breakeven moisture contents for each d.

4.2 TIME AND THE CONCEPT OF DISCOUNTING

In this section, we examine the notion of the time value of money. This basic principle is used for all capital investment analysis methods, including the net present worth, net annual worth, internal rate of return and benefit-cost ratio methods. Regardless of the method employed, the aim is the same — to select an alternative solution that satisfies an investor's goal to maximize profit. In this book, we recommend the net present worth method. However, since the concepts of annual worth and internal rate of return are important, they are used to give deeper meaning to the alternative selection process used by the net present worth method.

Consider a capital investment scenario. An investor commits capital to a project, forfeiting the opportunity to receive immediate benefit or satisfaction from this money in anticipation of receiving future monetary benefits. The evaluation centers on determining whether the future long-term benefits outweigh the investment cost. Since the investor must wait for his or her monetary reward and thus gain satisfaction from spending it, future benefits will be discounted or penalized. The amount of the penalty or discount is assumed to be a function of the time when the future benefits are received.

Long-term projects also include future costs, such as operating, labor and maintenance costs. They are also subject to discounting.

The principle of discounting and the formulas used to transfer payments from one point in time to another used for capital investment analysis are also employed for financial analysis. These formulas are used to calculate both debt payments on loans and future returns from a savings account. We begin the discussion by considering the alternative selection problem.

The Decision Rule for Alternative Selection:

The decision rule for a single alternative solution is:

The Feasibility Test: If the monetary benefits w received from an alternative exceed the total costs k to purchase, operate, and maintain it, then the alternative is considered to be feasible; otherwise, it is considered infeasible. Excluding the time value of money considerations of discounting future benefits and costs, which will be introduced presently, an alternative is feasible when

$$nw = w - k \geq 0 \rightarrow \text{feasible}$$

Otherwise,

$$nw = w - k < 0 \rightarrow \text{infeasible}$$

Selection: If only one alternative is being considered for selection, then it is recommended for selection if it passes the feasibility test. If it is infeasible, then it is rejected and a recommendation of no selection is made. In other words, no action or the *do-nothing alternative* is recommended.

The decision rule for two or more mutually exclusive alternative solutions employs the feasibility test for a single alternative. Only those alternatives passing the feasibility test are considered for selection. The feasible alternative offering the greatest net worth nw is recommended for selection. For example, if alternatives A and B are both feasible, $nw_A \geq 0$ and $nw_B \geq 0$ and $nw_A > nw_B$, then alternative A is recommended. On the other hand, if both alternatives are infeasible, $nw_A < 0$ and $nw_B < 0$, then the do-nothing alternative is recommended. No action is deemed the best choice in this situation.

This decision rule can be applied theoretically to the most mundane problems we face in our daily lives. For example, if we are hungry, we can purchase a meal, consume it, and relieve our hunger. The choice as to whether or not to buy a meal depends on whether we consider the benefit from the meal greater than its cost. The actual value that is placed on the meal is subjective and personal. The quality of the restaurant, the kind of food it serves, its prices, and how hungry we are, as well as other factors, play a part in the decision.

The problem with applying a capital investment method arises from the necessity to make value judgments and assignments. In a capital investment analysis, all benefits and costs must be monetary measures. Consequently, a monetary value must be assigned to the benefit received from eating. The other important issue is time. The time that we ate last and how fast the restaurant can serve us may be critical in a value assignment. Clearly, the restriction that all benefits and costs be monetary measures is solely for convenience and may severely limit the usefulness of the method for some problems. In spite of this difficulty, capital investment methods are employed to analyze various private and public investment projects. Public investment projects are discussed in Section 4.5.

Private Investment

This discussion is limited to private business investments where a capital investment will lead to a future return on this investment. A cash flow diagram is a convenient means to describe the problem. Figure 3 shows that a capital investment $k(0)$ is made in year zero, with an annual return of $nw(t)$ where $t = 1, 2, \ldots, n$ years.

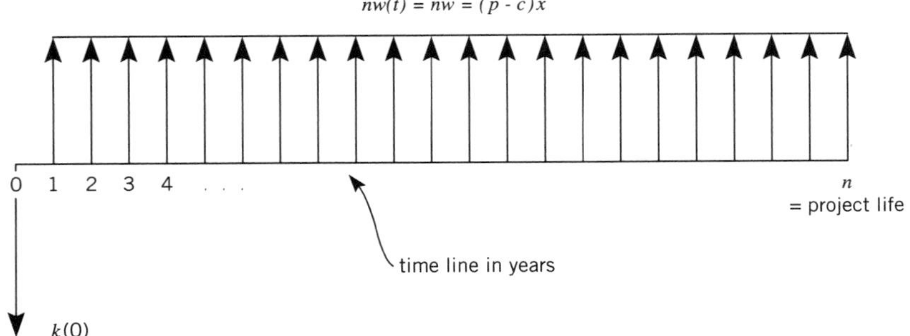

Figure 3 Cash flow for a uniform payment series.

It is always important to identify who is paying and receiving the benefits. In this diagram, the investor pays for the investment and receives the benefits. An outgo, payment, or loss of money is shown with an arrow pointing in the downward direction. The net annual profits $nw(t)$ in year t are shown with upward arrows and calculated as

$$nw(t) = (p - c)x$$

where

$$\begin{aligned} p &= \text{unit selling price} \\ c &= \text{unit production cost} \\ x &= \text{production level} \end{aligned}$$

The cash flow diagram is a time series of constant annual payments of nw. If the annual values of p, c, or x are a function of time, then the functional notation $p(t)$, $c(t)$, and $x(t)$ is used. The annual profit in year t can be written in a more general form as

$$nw(t) = w(t) - k(t)$$

where $w(t) = p(t)x(t)$ and $k(t) = c(t)x(t)$. In Figure 3, it is understood that p, c, and x are independent of time t and all uniform series values over time $t = 1, 2, 3, \ldots, n$ years.

The annual net profit of $nw(t) = nw$ of Figure 3 for any year t is assumed to be received at the beginning of year t. In reality, sales and operation expenses are paid during the time period from year $t - 1$ to t, but the cash flow diagram considers it to be an accumulated value and counted only at time t. In other words, all money exchanges are assumed to be annual lump-sum values.

Other important assumptions can be made about all cash flow problems. The future net worth values of nw are forecasts and so are, of course, subject to forecast uncertainty. There is an economic risk that these forecasts will not be realized. Economic risk and forecast uncertainty are addressed in Section 4.3.

The Discount Factor

The decision to recommend to invest or not to invest is a value judgment and subjective. It can and should be guided by knowledge and by the proper use of data, sound assumptions, and principles. One of the most basic principles used in evaluating and selecting among different capital investment projects is that of discounting future revenues or benefits. Qualitatively, the concept is simple and is illustrated most easily by an example.

Suppose an investor is given two alternatives for investment:

A. $k(0) = \$1M$ with $w(1) = \$1.1M$ benefit is received after year 1

B. $k(0) = \$1M$ with $w(10) = \$1.1M$ is received after 10 years and $w(t) = 0$ for intermediate years or years $t = 1, 2, \ldots, 9$.

Alternative A is recommended because the investor will receive \$1.1M at the end of one year and will not have to wait 10 years for the same return. In fact, the wait is so long that the investor may perceive alternative B as a losing proposition and unworthy of serious consideration.

This is a simple example with an obvious choice. For more complex cash flows, this subjective evaluation procedure may prove unworkable. By introducing the discount factor, the selection process will become an objective selection process. The discount factor is

$$\frac{1}{(1+i)^t}$$

where

i = annual discount rate with $i \geq 0$
t = year at which a cash exchange is made

The discount factor,

$$\frac{1}{(1+i)^t}$$

is called the present worth compound amount factor, *pwcaf*(i, t). Any sum of money $w(t)$ can be multiplied by the discount factor to transform its value evaluated in year t dollars to year zero dollars or

$$pw(t) = pwcaf(i,t)w(t) = \frac{w(t)}{(1+i)^t}$$

Hence, the term, *present worth* of benefits for year t is applied to this product. The present worth pw of a project with $w(t)$ payments where $t = 1, 2, \ldots, n$ years is the sum of the individual present worth values, or

$$pw = \sum_{t=1}^{n} pw(t) = \sum_{t=1} \frac{w(t)}{(1+i)^t}$$

This relationship forms the basis for the net present worth method of alternative selection.

Monetary comparisons in an alternative selection problem are considered unbiased and fair when the same monetary scale is used. That is, we will only compare benefits and costs using the same-year dollars. We prefer to use the net present worth method, and we use year-zero dollars for comparison purposes. Future payments, including future benefits and costs, can be transferred to year-zero dollars using the discount rate formula. Let us return to our example to illustrate the effects of discounting, that is, the use of year-zero dollars, and incorporate them into decision rules for alternative selection.

The present worth from benefits received from A and B in year-zero dollars is

$$pw_A = \frac{w(1)}{1+i}$$

and

$$pw_B = \frac{w(10)}{(1+i)^{10}}$$

where pw = present worth of discounted dollars in year-zero dollars. Suppose the discount rate is $i = 5\%$ per year. The net present worth of A and B are calculated as

$$pw_A = \frac{1.1}{1.05} = \$1.05\text{M}$$

$$pw_B = \frac{1.1}{1.05^{10}} = \$0.68\text{M}$$

Now, that pw_A and pw_B are expressed in year-zero dollars, we can calculate their net present worth values and then perform the feasibility test. The net present worth of the alternatives is defined to be

$$npw = pw - k(0)$$

Thus

$$npw_A = \$1.05\text{M} - 1\text{M} = \$0.05\text{M (profit)}$$
$$npw_B = \$0.68\text{M} - 1\text{M} = \$\text{–}0.32\text{M (loss)}$$

Again, A is the recommended selection. In fact, since $npw_B < 0$, it is considered to be an infeasible alternative and is not recommended for selection.

The assignment of i is critical in the selection process. If $i = 15\%$, for example, both A or B are infeasible and neither investment is recommended; a do-nothing alternative selection is considered the best recommendation. The choice of the discount rate i will, in part, reflect the economic risk. The greater the value of i and thus the smaller the value of pw, the greater the risk.

Note that we often say that future benefits are anticipated returns of investment. In all capital investment analysis problems, future benefits and costs are forecasts. Thus, they are subject to uncertainty. The choice of discount rate should reflect the level of economic risk and uncertainty. The greater the risk and uncertainty about future benefits, the greater the discount rate that is used in the analysis. We discuss this issue in greater detail in Section 4.3.

EXAMPLE 4.4 *A Net Present Worth Calculation*

Suppose that a new piece of equipment, costing \$1M with a usable life of five years, will replace an inefficient production unit. It is estimated that the annual savings will be \$0.25M.

Determine if this is a feasible investment. Assume a discount rate of 10% per year.

SOLUTION

Only one alternative is under consideration; thus, if it is found to be feasible, then the new piece of equipment is considered cost effective and should be recommended for selection. Otherwise, the do-nothing alternative or no-action plan is recommended.

The following cash flow diagram shows that \$1M is spent in year 0 in anticipation of annual future benefits of \$0.25M per year.

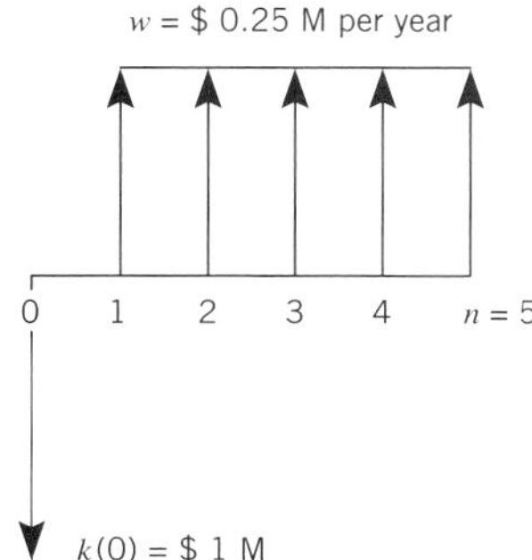

Its annual net worth is

$$nw(t) = w - k(t)$$

where $k(t) = \$1\text{M}$ for $t = 0$, $k(t) = 0$ for $t = 1, 2, \ldots, 5$ years, and $w = 0$ for $t = 0$. Its net present worth for year t is

$$npw(t) = \frac{nw(t)}{(1+i)^t} \quad \text{for } t = 0, 1, 2, \ldots, 5 \text{ years}$$

The net present worth for the entire time frame of $n = 5$ years is

$$npw = \sum_{t=0}^{n} npw(t) = \sum_{t=0}^{n} \frac{nw(t)}{(1+i)^t}$$

A spreadsheet program is a nice way to perform time value of money calculations. The last column of the following table clearly shows the effect of discounting. Since $npw < 0$, the capital investment is infeasible. One may conclude that the investment opportunity is not cost effective and that the new equipment should not be selected for installation.

t	$pwcaf(0.1, t) = \frac{1}{1.1^t}$	$k(t)$	$w(t)$	$nw(t) = w - k(t)$	$npw(t) = \frac{nw(t)}{1.1^t}$
0	1.00	1.00	0.00	–1.00	–1.00
1	0.91	0.00	0.25	0.25	0.23
2	0.83	0.00	0.25	0.25	0.21
3	0.75	0.00	0.25	0.25	0.19
4	0.68	0.00	0.25	0.25	0.17
5	0.62	0.00	0.25	0.25	0.16
				npw =	–0.05M

Amortization of Capital Debt

The present worth compound interest rate formula, $pwcaf(i, t)$, is used for some financial analyses. However, the interest rate i used in these evaluations is not a discount rate but an interest rate for loans or savings given by a bank or other financial institutions. We must not confuse financial interest rates with the discount rate for alternative selection. When confusion might arise, subscripts are used to distinguish between different kinds of financial interest rates. For example, the notation, i, i_f, and i_b, is used to indicate discount rate, the annual rate of return being offered on a savings account, and the annual rate of interest being charged on a bank loan, respectively.

Consider the purchase of a new piece of equipment costing $k(0)$ dollars where the money to pay for it is borrowed. The loan debt will be paid in $n = 5$ years at an annual interest rate of $i = i_b$ = annual interest rate for the loan. *This interest rate is not a discount rate but a rate of interest paid on a loan*! Annual uniform series payments of $k(t) = k$ will be made in fixed installments at the end of each year. The cash flow diagrams of Figure 4 show the debt in year zero and the loan debt repayment scheme as a uniform series of annual payments.

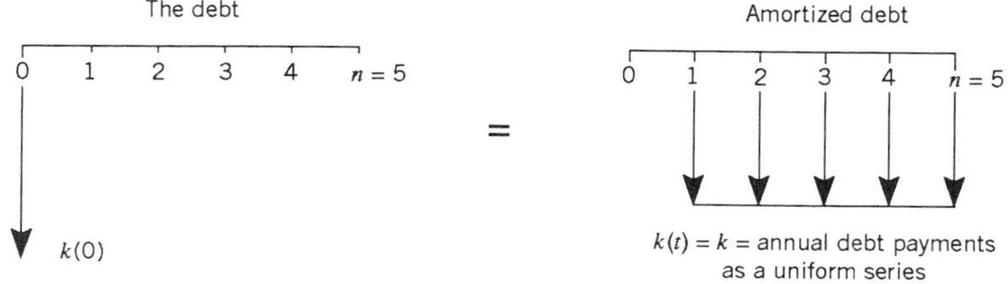

Figure 4 An amortized debt as an equivalent uniform time series of payments.

The two cash flows are considered to be equivalent. The one on the left represents the debtor's obligation to repay the loan in year-zero dollars. The one on the right shows the debt being paid in five equal payments. The $k(t) = k$ payment consists of a portion of $k(0)$ plus an annual interest charge.

The unknown annual debt payments of k may be derived by using the present worth function. Since the two cash flows are considered to be equivalent, the following equality must hold:

$$k(0) = \sum_{t=1}^{n} \frac{k(t)}{(1+i)^t} = k \sum_{t=1}^{n} \frac{1}{(1+i)^t}$$

With the aid of finite series, it may be shown that

$$uspwf(i,n) = \sum_{t=1}^{n} \frac{1}{(1+i)^t} = \frac{(1+i)^n - 1}{i(1+i)^n}$$

where $uspwf(i, n)$ is called the *uniform series present worth factor.*

The inverse of $uspwf(i, n)$ is called the *capital recovery factor,*

$$crf(i,n) = \frac{i(1+i)^n}{(1+i)^n - 1}$$

The annual debt payment for a loan of interest rate i_b of n-year duration is

$$k = k(0)crf(i_b, n) = k(0)\frac{i_b(1+i_b)^n}{(1+i_b)^n - 1}$$

The important point of this example is that the single-debt payment $k(0)$ and the $n = 5$ uniform payments of k are considered to be equivalent cash flow payments. From the lender's point of view, he or she is equally pleased with receiving a single payment of $k(0)$ dollars or a series of five payments of k dollars each. As a $n = 5$ time payment series, the lender receives interest payments for the opportunity costs associated with the time the money is not in his or her hand. Conceptually, the lender receives the same satisfaction from possessing $k(0)$ or receiving five payments of k each. Furthermore, the lender receives interest only for the period of time the money is in the borrower's possession.

TABLE 1 Time Value of Money Formulas as Transfer Functions

Notation:

k = uniform annual series of k dollars for years, 1, 2, 3, . . ., n

$k(0)$ = value of k dollars in year 0

$k(n)$ = value of k dollars in year n

t = year where $t = 1, 2, 3, \ldots, n$ years

i = annual interest rate

Given	Find	Cash Flow Diagram	Transfer Function	Payment Factor
		1. Future Worth of a Single Payment		
$k(0)$	$k(n)$	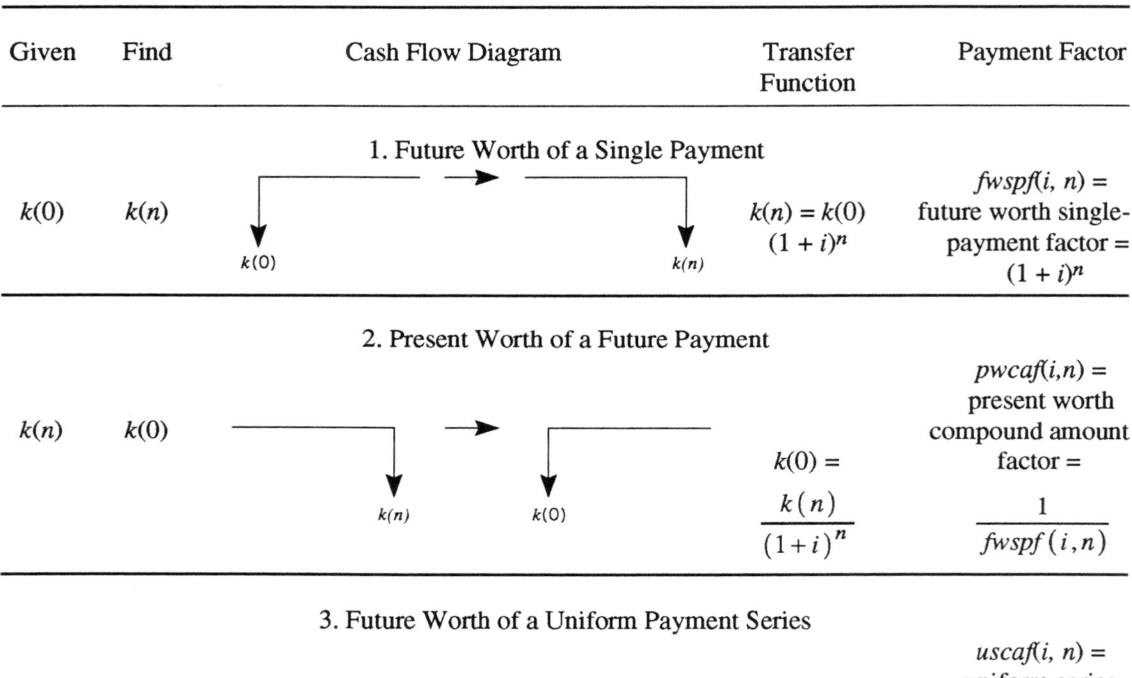	$k(n) = k(0)(1 + i)^n$	$fwspf(i, n)$ = future worth single-payment factor = $(1 + i)^n$
		2. Present Worth of a Future Payment		
$k(n)$	$k(0)$		$k(0) = \dfrac{k(n)}{(1+i)^n}$	$pwcaf(i,n)$ = present worth compound amount factor = $\dfrac{1}{fwspf(i,n)}$
		3. Future Worth of a Uniform Payment Series		
k	$k(n)$	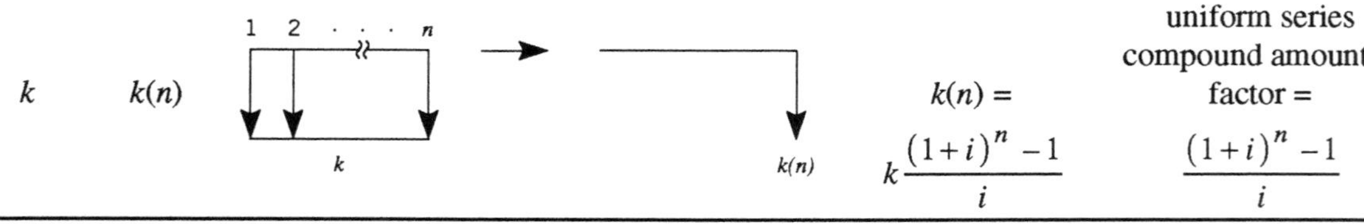	$k(n) = k\dfrac{(1+i)^n - 1}{i}$	$uscaf(i, n)$ = uniform series compound amount factor = $\dfrac{(1+i)^n - 1}{i}$
		4. Annual Worth of a Future Payment		
$k(n)$	k	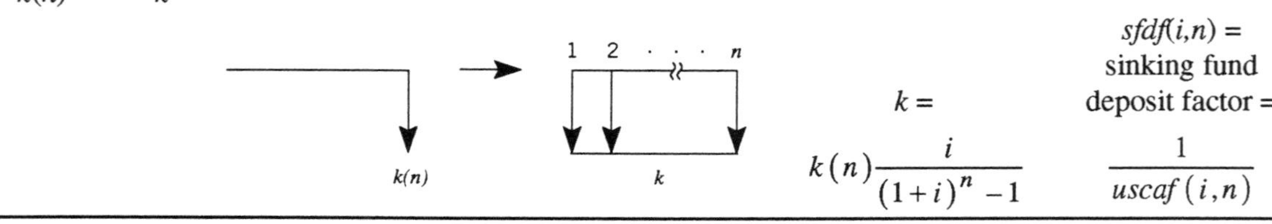	$k = k(n)\dfrac{i}{(1+i)^n - 1}$	$sfdf(i,n)$ = sinking fund deposit factor = $\dfrac{1}{uscaf(i,n)}$

5. Future Worth of a Single Payment expressed as a Uniform Payment Series

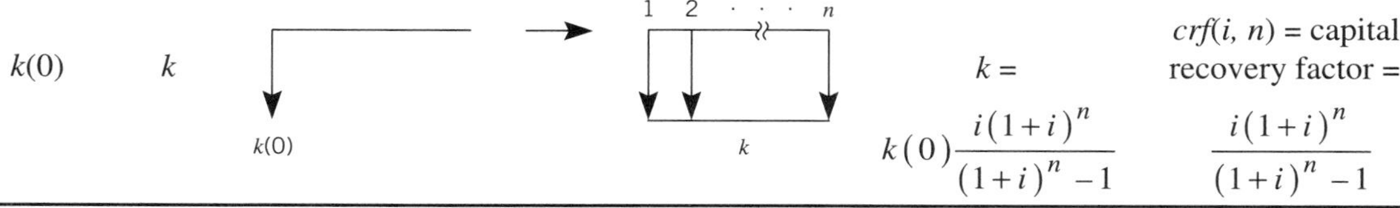

$k(0)$ k

$k = k(0)\dfrac{i(1+i)^n}{(1+i)^n - 1}$

$crf(i, n)$ = capital recovery factor = $\dfrac{i(1+i)^n}{(1+i)^n - 1}$

6. Present Worth of a Uniform Payment Series

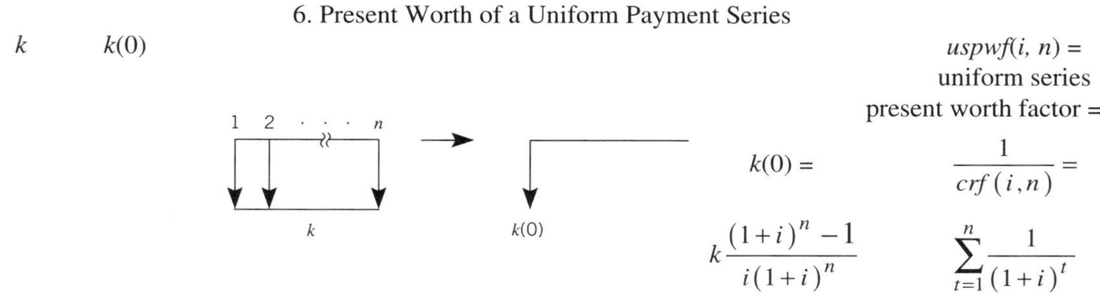

k $k(0)$

$k(0) = k\dfrac{(1+i)^n - 1}{i(1+i)^n}$

$uspwf(i, n)$ = uniform series present worth factor = $\dfrac{1}{crf(i,n)} = \displaystyle\sum_{t=1}^{n} \frac{1}{(1+i)^t}$

Typical Capital Investment Calculations

Table 1 shows six payment plans where $k(0)$, k, or $k(n)$ are shown as present and future worth transfers of one another. All formulas are based on the principle of compounding and the discount factor, $1/(1 + i)^t$. Note that cases 1 and 2, cases 3 and 4, and cases 5 and 6 are inverses of one another. These formulas are used for typical capital investment and financial analysis as illustrated in Tables 2 and 3.

Examples 1 and 2 of Table 2 show application of the transfer functions given in Table 1 for capital investment problems using the net present worth method. Here, we take the position of an investor; thus, payments and returns on investment are shown in the cash flow diagrams as losses (down arrows) and profits (up arrows), respectively. The notation also reflects this point of view.

Example 3 employs the annual net worth method. The illustration shows the cash flow for stage construction as an equivalent uniform series. In order to calculate the *anw* as a uniform series, the present worth is first calculated and then transferred to a uniform series. This example is intended to show that present and annual worth are transforms of each other. Consequently, using either *npw* or *anw* values for alternative selection will lead to the selection of the same alternative. The feasibility test is $anw \geq 0$, using the annual worth method.

In all three examples, the interest rate i is assumed to be a discount rate.

TABLE 2 Typical Capital Investment Calculations

Notation:

anw = annual net worth as a uniform payment series ($ per year)

npw = net present worth in year 0

$k(t)$ = capital investment made in year t

$w(t)$ = total worth as a benefit, revenue, or savings received in year t

$nw(t)$ = net worth as the difference between gross revenue minus cost received in year t

n = project life in years

i = discount rate given as an annual interest rate

Present Worth Calculations

EXAMPLE 1. A CAPITAL INVESTMENT

Find the net present worth, npw, for a capital investment $k(0)$ receiving an anticipated net revenue $nw(t) = nw$ as a uniform series for $t = 1, 2, 3, \ldots, n$.

Cash Flow Diagram:

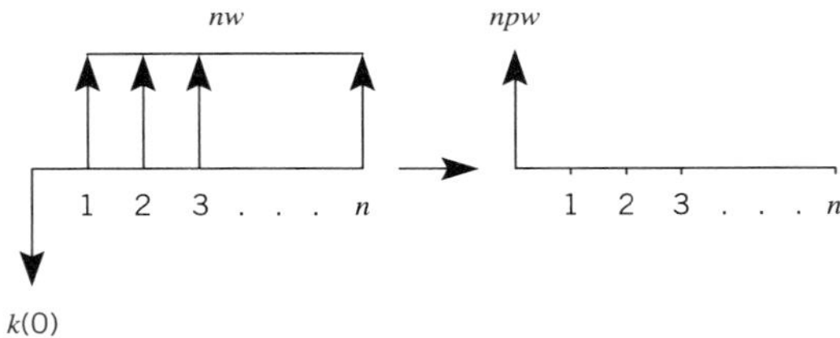

Formula: $npw = nw\ uspwf(i, n) - k(0)$

EXAMPLE 2. A CAPITAL INVESTMENT USING STAGE CONSTRUCTION

Find the net present worth, npw, for Example 1 where capital investments are made in years 0 and m, $k(0)$ and $k(m)$, respectively.

Cash Flow Diagram:

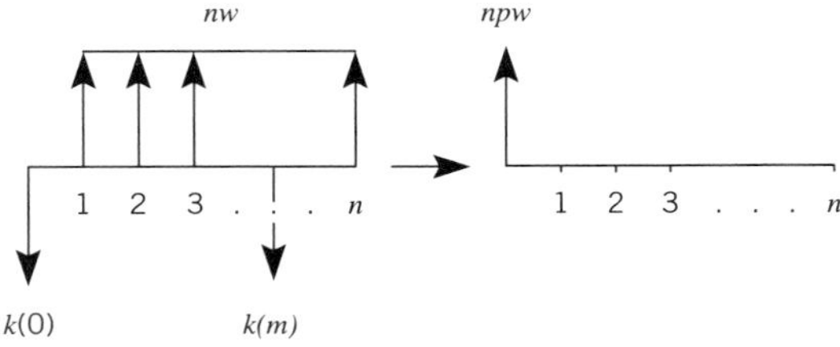

Formula: $npw = nw\ uspwf(i, n) - k(0) - k(m)\ pwcaf(i, m)$

Annual Worth Calculations

EXAMPLE 3. EXAMPLE 2 AS AN ANNUAL WORTH

Find *anw* for the Stage Construction Problem.

Cash Flow Diagram:

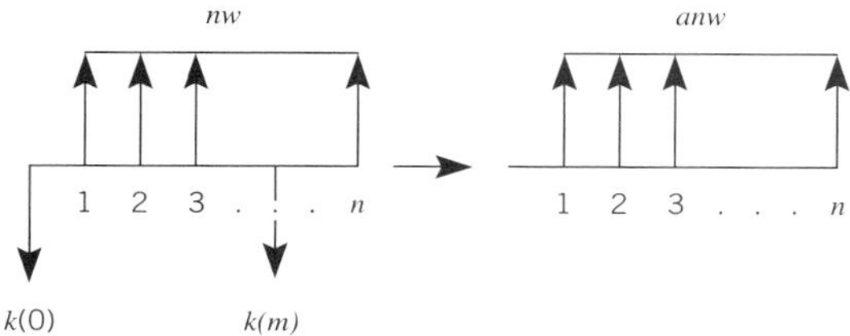

Formula: $anw = npw \;\; crf(i, n)$ where *npw* is given in Example 2.

Typical Financial Analyses Calculations

Table 3 shows the formulas of Table 1 used for typical financial analyses problems. The notation and cash flow diagrams take the point of view of an investor for the savings plan examples and of a borrower for the debt amortization example.

Consider a bank or some other financial institution that offers a savings plan, Example 1 of Table 3, where the future benefits are calculated as compounded interest payments with an annual savings interest rate i_f. The future worth of a single payment or deposit made in year 0 or $k(0)$ is

$$w(n) = k(0)(1 + i_f)^n$$

where $fwspf(i_f, n) = (1 + i_f)^n$ is called a future worth single payment factor. Compounded means that the principle of $k(0)$ plus interest on the principle of $ik(0)$ is reinvested at the end of the first year. The total value of $k(0)$ after 1 year is $k(1) = k(0)(1 + i)$. The process is repeated for each subsequent year. For example, after two years, $k(2) = k(1)(1 + i) = k(0)(1 + i)^2$.

The future benefit using simple interest is

$$w(n) = k(0)(1 + ni).$$

The investor favors the compounded interest rate payments to simple rate payments because the return $w(n)$ for compounded interest payments is greater given the same simple savings interest rate i.

A sinking fund, Example 2 of Table 3, is an account where an individual places fixed annual payments $k(t) = k$ in an account earning compound savings interest rate i_f for a period of n years. On withdrawal, the individual receives $w(n)$ as shown in case 3 of Table 1. The future worth of a uniform series payment of annual payments k for n years is calculated as

$$w(n) = k\frac{\left(1+i_f\right)^n - 1}{i_f}$$

where

$$uscaf(i_f, n) = \frac{\left(1+i_f\right)^n - 1}{i_f}$$

is called the uniform series compound amount factor.

A capital recovery fund, Example 3 of Table 3, is an account where an individual places a fixed sum of money $k(0)$ into an interest-bearing account and withdraws fixed annual sum payments w for n years. The annual benefits are calculated for a savings interest rate i_f as

$$\frac{i_f\,(1+i_f)^n}{(1+i_f)^n - 1}$$

for $t = 1, 2, 3, \ldots, n$ years.

The same transfer function formula is used for the debt amortization problem given as Example 4. The only difference lies in the interpretation of results. Since the borrower repays a loan at interest rate i_b, the annual payments are given as k, or

$$k = k(0)\frac{i_b\,(1+i_b)^n}{(1+i_b)^n - 1}$$

for $t = 1, 2, 3, \ldots, n$ years, and where i_b is the interest rate charged on the loan of $k(0)$ dollars.

TABLE 3 Time Value of Money Formulas Used For Typical Financial Analyses Problems

EXAMPLE 1. A SAVINGS ACCOUNT

Determine the future worth, $w(n)$, of a single deposit, $k(0)$, made in year 0, placed in a compound interest account for n years where i_f = savings interest rate.

Cash Flow Diagram:

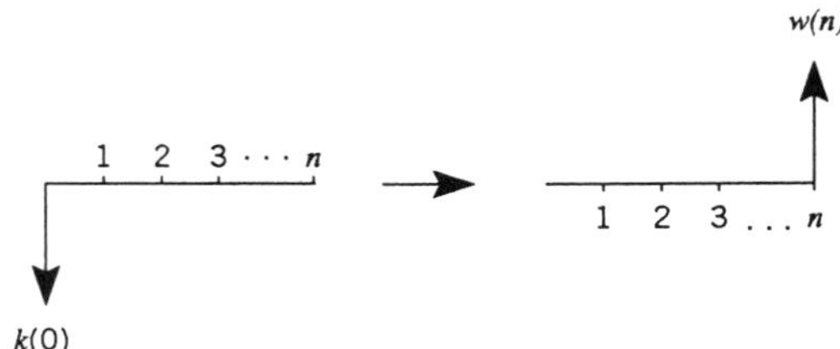

Formula: $w(n) = k(0)\,fwspf(i_f, n)$

Note that the deposit is shown with an arrow pointing downward, indicating a loss of revenue k in year 0, and future payment is shown as an up arrow, indicating a gain in revenue or wealth in year n. Similar notation is used for the other examples.

EXAMPLE 2. A SINKING FUND

Determine the future worth, $w(n)$, of a uniform annual series of deposits of k dollars placed in a compound interest account for years 1, 2, 3, . . . , and n where i_f = savings interest rate.

Cash Flow Diagram:

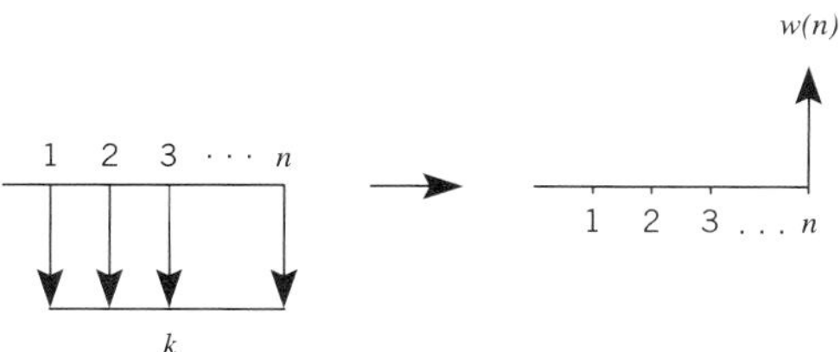

Formula: $w(n) = k\ uscaf(i_f, n)$

EXAMPLE 3. A CAPITAL RECOVERY FUND

Determine the annual future worth of n equal payments, w, returned on the single deposit $k(0)$ where i_f = savings interest rate.

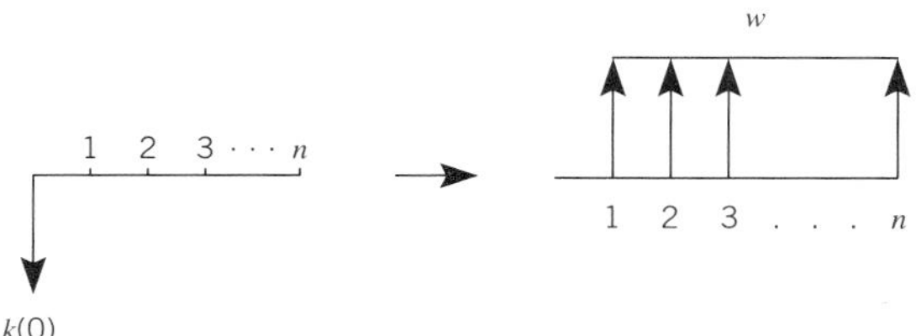

Formula: $w = k(0)\ crf(i_f, n)$

EXAMPLE 4. DEBT AMORTIZATION

Determine the repayment of a debt $k(0)$ as a uniform series of annual payments k made over n years where i_b = loan interest rate.

Cash Flow Diagram:

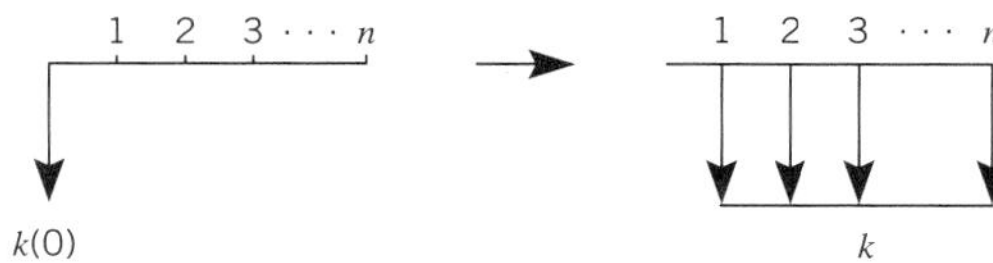

Formula: $k = k(0)\ crf(i_b, n)$

EXAMPLE 4.5 *An Annual Net Worth Calculation*

Reconsider Example 4.4 once again. This time, the selection will be based on the net annual worth *anw* method.

SOLUTION

The annual net worth is calculated as

$$anw = nw(t) = w(t) - k$$

where k = amortized debt expressed as uniform time series of constant yearly payments. The cash flow diagram shows both w and k as a uniform series.

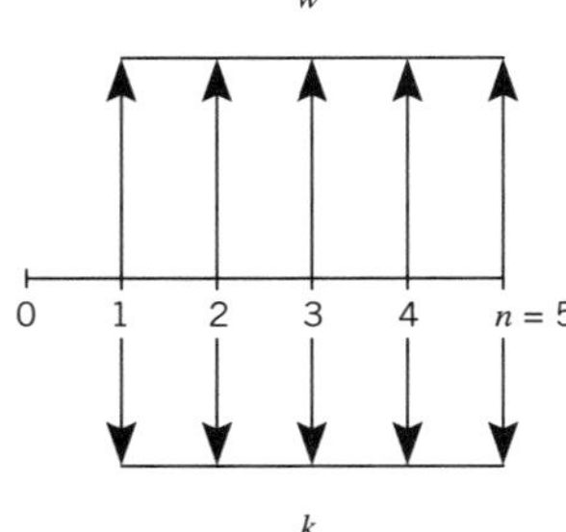

Using the capital recovery factor relationship with a discount rate $i = 10\%$ per year, we find that the annual payments are

$$k = k(0)\ crf(i, n) = \$1\text{M}\left[\frac{0.1(1.1)^5}{(1.1)^5 - 1}\right] = \$0.26\text{M per year}$$

The annual net worth is

$$anw = \$0.25\text{M} - \$0.26\text{M} = -\$0.01\text{M per year} < 0$$

Since there is annual net loss, once again the investment plan is not recommended. Given this result, it is more profitable to "do nothing."

QUESTIONS FOR DISCUSSION

4.2.1 Define the following terms:

amortized debt	feasible alternative
compound interest	present worth

discount factor
simple interest
do-nothing alternative

4.2.2 List three important assumptions in performing a capital investment analysis for evaluating the feasibility of a multiple-year investment project.

QUESTIONS FOR ANALYSIS

Instructions: While transfer functions can be used to solve capital investment analysis problems, it is strongly recommended that spreadsheet programs be used to solve these problems. The spreadsheet program gives additional insights and makes it possible to perform sensitivity analyses most easily. Problems where a spreadsheet program is recommended are indicated with an asterisk (*).

4.2.3 Compare the return in 10 years from a savings account that earns simple interest of $i_f =$ 10% per year, with a savings account that earns compound interest of $i_f = 10\%$ per year. Let $k(0) = \$1000$.

4.2.4 An investor is considering investing $k(0) = \$50{,}000$ into one of the following projects with project life of $n = 1$ year:

A. A bank is offering a savings plan earning an annual interest of $i_f = 5\%$ per year. The bank will return to the investor the principal $k(0)$ plus an interest payment on $k(0)$ after one year.

B. An industry offering $w(1) = \$54{,}000$ after one year.

(a) Draw a cash flow diagram for each alternative investment.

(b) Recommend the best alternative using the present worth method. Assume a discount rate of $i = 8\%$ per year.

(c) Show that the recommendation made in (b) is the same one if the calculation is performed using the annual worth method. Again, use $i = 8\%$ per year.

4.2.5* The "rule of 72" states that the time to double one's money, compounded annually, is equal to $72/i$, where i = the annual interest rate expressed as a percent (%). For example, if $i = 8\%$, then it takes $t = 72/8 = 9$ years to double one's money.

Write a spreadsheet program to calculate $w(9)$ to show that the rule of 72 is valid. Let $k(0) = \$100$.

4.2.6* Consider a uniform time series of \$1000 per year payments and assume a discount rate of 9% per year.

(a) Write a spreadsheet program to calculate pw(t), where t = 0, 1, . . . , 10 years. Find *npw* as the sum,

$$\sum_{t=0}^{10} pw(t)$$

(b) Check the answer from (a) by using the *uspwf* equation to find *npw*.

4.2.7 (a) Give the governing equation for calculating the net present worth *npw* for a stage construction project depicted in the cash flow diagram. Assume a discount rate of i is known.

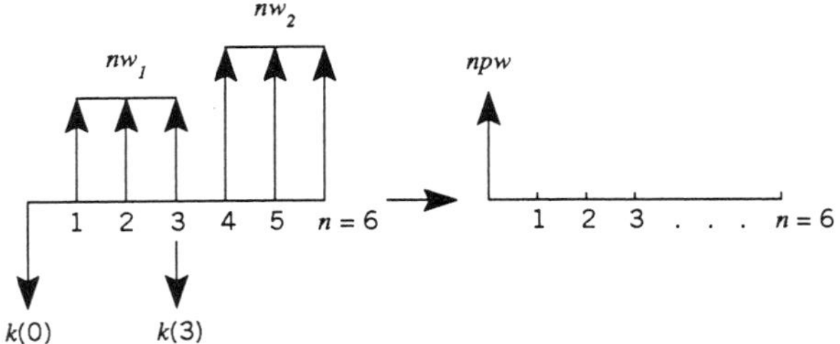

(b) Give the governing equation for calculating the annual net worth anw for this stage construction project.

4.2.8 The revenue from an investment of $k(0) = \$15,000$ is received in year 3. No monetary benefits are received in years 1 and 2. The rate of return on investment, $i' = 16\%$ per year. What is the net profit $nw(3)$ in year 3 dollars if $k(0)$ is borrowed and repaid in year 3 with a bank loan of $i_b = 10\%$ per year? In other words, how much money will an investor have in his or her hand after paying the loan?

4.2.9 A recycling center is expected to process 1500 tons of MSW per year. Its tipping fee in \$ per ton will be based on the projected annual labor and maintenance costs of \$60,000 per year plus the amortization cost of a construction loan of \$200,000. Estimate a tipping fee to cover these costs. Assume that the life of a loan with 15% annual interest rate is 25 years.

4.2.10 Estimate the net unit profit p in \$ per kWh for a waste-to-energy plant that recovers and sells recovered electrical energy totalling 30,000,000 kWh per year. The annual labor and maintenance costs are \$500,000 for a plant costing \$20M. The plant sells power to its customers at \$0.10 per kWh and repays a 30-year construction loan debt at an annual interest rate of 5%.

4.2.11 A 5-year loan was made at an interest rate of $i = 10\%$ per year. The borrower agreed to repay the loan with five equal end-of-year payments of \$1000. Two on-schedule payments have been made. How much will the borrower have to pay if she decided to pay the outstanding portion of the loan in year 3?

4.2.12 Draw a cash flow diagram describing the following stage construction project with life $n = 20$ years.

Year t	Construction cost $k(t)$ in \$M	Predicted annual net worth, $nw(t)$ in \$M
0	\$10	\$0
1	20	0
2	30	0
3, 4, . . .	0	4M

Use the net present worth method to determine if the project is feasible. Assign a discount rate of $i = 8\%$ per year.

4.2.13 Draw a cash flow diagram describing the costs to construct and operate a facility with a design life n = 20 years.

Year t	Construction cost, $k(t)$ in \$M	Annual operating costs $k(t)$ in \$M
0	\$10	\$0
1	20	0
2	10	0
3, 4, . . .	0	5

There are 10,000 people who will be served by the facility. Each person will pay 20 annual tax payments. Calculate the annual tax assuming the facility was originally financed by a government bond with $i_b = 8\%$ per year.

4.3 OPPORTUNITY COSTS AND ECONOMIC RISK

The assignment of the discount rate i is a most critical assignment in a capital investment analysis using the net present and annual worth methods. The discount rate should reflect the opportunity cost of the investment as well as its economic risk. Since this assignment is such an important issue this entire section is devoted to it.

Guidelines for Assigning a Discount Rate

Consider a capital investment problem where a person seeks to maximize profit by investing in a project with a life of n = 1 year as shown in Figure 5. The investor has the following choices; he or she can select the project or the do-nothing alternative.

Since the information regarding the do-nothing alternative is introduced into the analysis of the project, a cash flow diagram for the do-nothing alternative is not needed.

The capital investment return of $w(1)$ is not guaranteed. The actual return can be more or less than $w(1)$, but the investor must wait for one year to know the exact amount that will be received. In fact, the possibility that $w(1) = \$0$ and the investor loses all or part of $k(0)$ is not ruled out. Since the decision of whether to invest in the project or to do nothing is made in year zero, $w(1)$ is merely a *forecast* and is only an *anticipated future benefit*.

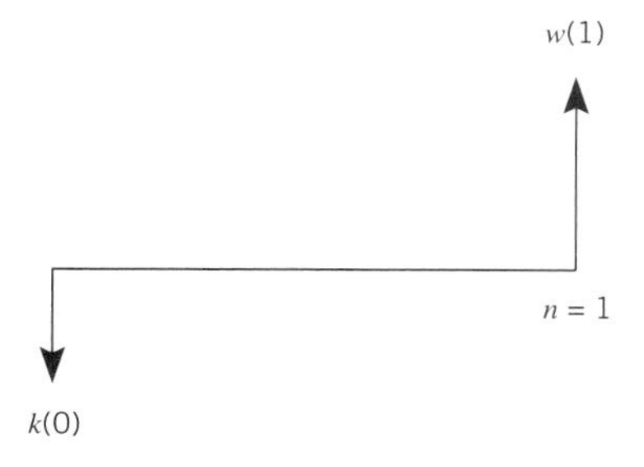

Figure 5 Cash flow diagram for benefit $w(1)$ and capital investment $k(0)$.

The crux of the problem is the economic risk associated with uncertainty of receiving $w(1)$ dollars after one year. The amount demanded for $w(1)$ is an individual decision and dependent on how much worth the investor places on $w(1)$ and the possible loss of $k(0)$. The magnitude of $w(1)$ will depend in part on the investor's *opportunity cost*, that is, the monetary value the investor places on the inability to use $k(0)$ dollars for a one year period. The final selection decision is an individual determination, but guidelines can be established to help the investor make a good decision. Let us consider other facts and assumptions.

The investor has no obligation to select the project alternative. If conditions are unsatisfactory, the investor will select the "do-nothing" alternative. The term means that the investor takes no action toward the capital investment project alternative, but it does not mean that the investor will not use his or her $k(0)$ dollars in some other way. We assume that the investor will act in a financially conservative and risk averse manner and place the money into a savings account earning interest of i_f per year. Since savings accounts are insured against bank failure, we consider the do-nothing alternative to be a risk-free investment. After one year, the investor is guaranteed to receive the principle plus interest:

$$w_f(1) = k(0)(1 + i_f).$$

In selecting the project alternative, the investor will expect a return in year one equal to or greater than the return from the do-nothing alternative:

$$w(1) \geq w_f(1).$$

The *anticipated rate of return interest rate* i' for the capital investment is calculated as

$$i' = \frac{w(1) - k(0)}{k(0)}.$$

If $i' \geq i_f$, then the investor can expect that his or her goal to maximize profit will be realized. For this simple cash flow example, comparing the interest rates i' and i_f is equivalent to comparing $w(1)$ and $w_f(1)$.

The anticipated rate of return interest rate i' can also calculated as a *internal rate of return interest rate* i'. It is defined to be a breakeven discount rate that satisfies the equality,

$$npw(i') = 0.$$

For this example, the net present worth is calculated as

$$npw(i) = \frac{w(1)}{1+i} - k(0).$$

The internal rate of return i' is

$$npw(i') = \frac{w(1)}{1+i'} - k(0) = 0,$$

or

$$i' = \frac{w(1) - k(0)}{k(0)}.$$

This is the same relationship obtained by calculating it as an anticipated rate of return interest rate.

The internal rates of return can be used for alternative selection. The values of i' are calculated for each alternative under consideration. The alternative offering the maximum return is selected. The basic principles of the internal rate of return method for alternative selection have been used in this simple cash flow example when i' and i_f are compared and the capital investment project is recommended for selection only when $i' \geq i_f$. Otherwise, when $i' < i_f$, the do-nothing alternative is recommended. In spite of the appearance of simplicity, the internal rate of return method often leads to analysis difficulties and ambiguous results for complex cash flows (see Example 4.6).

When the net present worth method is used, the investor must assign a discount rate i. Remember that i is considered to be a penalty factor associated with waiting for some future return as well as the uncertainty associated with the project. In the context of this discussion, we give the discount rate i an added meaning,

i = *acceptable* rate of return on capital investment.

If the investor assigns a discount rate satisfying the following inequality, $i' \geq i$, then it follows from the preceding discussion that $npw \geq 0$. The capital investment project is deemed to be a feasible alternative. This is most easily seen by substituting $w(1) = k(0)(1 + i')$ into the npw equation,

$$npw = \frac{w(1)}{1+i} - k(0).$$

This equation reduces to

$$npw = \frac{k(0)(1+i')}{1+i} - k(0) = k(0)\,\frac{i'-i}{1+i}.$$

Clearly, the feasibility condition, $npw \geq 0$, is satisfied only when $i' \geq i$.

Before recommending the capital investment project for selection, the investor must decide on an acceptable rate of return on investment and assign it as a discount rate i. The logical choice of a financially conservative investor is to assign a discount rate that satisfies the inequality, $i \geq i_f$, where i is a lower bound acceptable rate of return. An assignment of i that satisfies the inequality gives some assurance that the project will give a return on investment for the project that is better than the do-nothing alternative.

Let us alter the one-year capital investment project into the situation where the investor will borrow the capital investment money $k(0)$ and will pay it back with a one-year loan secured from the bank at an annual interest rate ib where $i_b > i_f$. The investor has an obligation to pay

$$k_b(1) = k(0)(1 + i_b).$$

The net worth in year one dollars is $nw(1) = w(1) - k_b(1)$. For this scenario, the financially conservative investor will want a return of investment of least i_b; therefore, he or she will assign a discount rate satisfying the inequality, $i \geq i_b$. In this case, the lower bound acceptable rate of return for this example is i_b. Let us summarize these observations into a guideline, which can be applied to the selection of multiple alternatives with more complex cash flows than illustrated here.

A recommended guideline for a financially conservative capital investment analysis for choosing i is:

1. Determine i_f, and if applicable, determine i_b.
2. Initially assign i = maximum(i_f, i_b).
3. If the forecast of long-term net benefits is highly speculative, adjust i upward. The assignment of i is an individual determination; therefore, no rule can be offered as to how much this value should be increased.

The following numerical example shows the importance of assigning i and the pitfalls associated with not carefully assigning it in alternative selection.

EXAMPLE 4.6 *The Effect of the Discount Rate on Alternative Selection*

Consider two \$10,000 alternatives A and B where a bank loan of 12% per year can be obtained for the purchase of new equipment. For A, the new equipment is forecast to bring a profit of \$11,500 after one year.

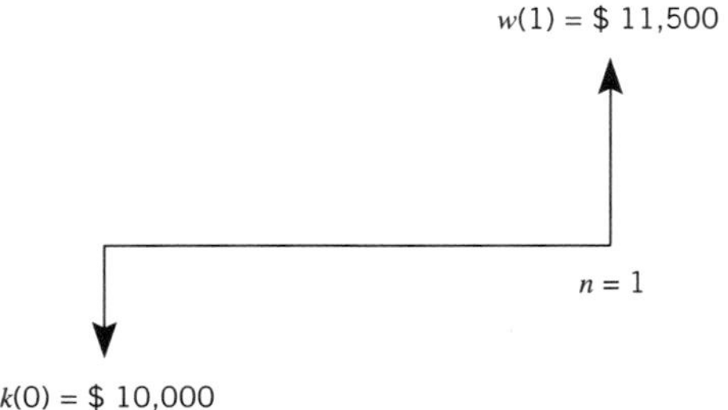

For B, the new equipment is forecast to bring a profit of \$12,000 after two years.

Assume that the bank rate on an insured saving account of 8% per year can be obtained.

(a) Use the present net worth method to aid in this selection process. Show that it is better to recommend the selection of A for the following discount rate assignments:

A. $i = i_f = 8\%$ per year, the lower bound, risk free interest rate case.

B. $i = i_b = 12\%$ per year, the bank loan rate.

(b) Show that the internal rate for alternative A is $i' = 15\%$ per year.

(c) Assign a discount rate $i = 8\%$ per year for the evaluation of alternative A. Compare the *npw*, as calculated in (a) and (b) to the *npw* of the anticipated net worth of A after a loan repayment rate of $i_b = 12\%$ per year is made. The assigned discount rate violates the financially conservative guideline for assigning the discount rate, $i \geq i_b \geq i_f$. Show that the *npw* at $i = 8\%$ per year tends to distort the true worth of alternative A.

SOLUTION:

(a) The present net worth for alternatives A and B are calculated as:

A. $$npw = \frac{w(1)}{(1+i)} - k(0)$$

and

B. $$npw = \frac{w(2)}{(1+i)^2} - k(0)$$

The following table shows the importance of the discount rate assignment on the npw values for the two alternatives. The higher the rate of i, the less attractive alternatives A and B are.

	npw for i =	
Alternative	8% per year	12% per year
A	\$648	\$268
B	\$288	\$-434

For all given discount rates of i, $npw_A > npw_B$; therefore, A is always preferred to B. In fact, when $i = 12\%$ per year, alternative B is an infeasible alternative. The extra year wait for receiving profit for B proves decisive.

(b) The net present worth for alternative A from (a) is evaluated over a range of discount rates, $0 \leq i \leq 25\%$ per year as

$$npw = \frac{w(1)}{(1+i)} - k(0).$$

The results of the analysis are shown in the following plot:

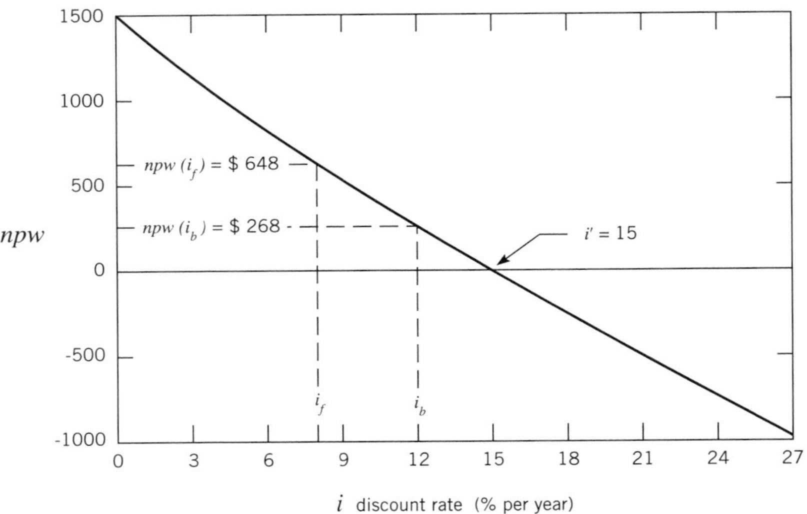

The value of internal rate of return is $i' = 15\%$ per year. The *npw* values for $i = i_f$ and i_b are also shown.

(c) The net present worth for alternative A at $i = 8\%$ per year is $npw = \$648$. The net profit after repaying the loan in year one dollars is

$$nw'(1) = w(1) - k(0)(1 + i_b) = \$11{,}500 - \$10{,}000(1.12) = \$300$$

At a discount rate of $i = 8\%$ per year, its net present worth is

$$npw' = \frac{nw'(1)}{(1+i)} = \frac{\$300}{1.08} = \$278.$$

There is a disturbing difference between $npw = \$648$ and $npw' = \$278$. The formulas *npw* and *npw'* are suppose to be measures of the same thing; therefore, the equality of $npw = npw'$ is expected. Let us investigate these equations more carefully. Expand and write *npw'* as

$$npw' = \frac{nw'(1)}{(1+i)} = \frac{k(0)(i' - i_b)}{(1+i)}$$

Contrast this formula with the *npw* formula from (b). Rewrite it as a function of i'.

$$npw = \frac{w(1)}{(1+i)} - k(0) = \frac{k(0)(1+i')}{1+i} - k(0),$$

or

$$npw = \frac{k(0)(i' - i)}{(1 + i)}$$

The *npw* function does not include i_b; therefore, there is no way that an assignment of $i = 8\%$ per year can reflect the obligation that a loan of $i_b = 12\%$ must be paid! At discount rate $i = 8\%$ per year, we claim $npw = \$648$ distorts the true worth of the investment.

We contend that the $i = i_b$ is a more acceptable assignment because the obligation to repay the loan is included in the analysis. With $i = i_b = 12\%$ per year, $npw = \$268$ and $npw \approx npw'$.

Whereas the investor may be free to choose any discount rate that he or she desires, $i = i_f$ cannot be recommended for a financially conservative, risk averse analysis, because the *npw* formula gives an overly optimistic view of the value of alternative A.

DISCUSSION

Certainly, a financially conservative investor will demand a selection procedure that will give a more conservative recommendation. Increasing i is a conservative step because it always makes the do-nothing alternative a more likely recommendation. We conclude that the *minimum* acceptable rate of return is i_b and that the discount rate should satisfy the assignment as $i = i_b$!

In reality, remember that there is no guarantee that the forecasted benefit of $w(1) = \$11{,}500$ will be received. The economic risk associated with this investment has not been addressed. If there is little uncertainty associated with receiving the return of $w(1)$, then the assignment of $i = i_b$ may be appropriate. The assignment of the discount rate is an individual decision; therefore, for greater uncertainty the investor may choose to use a larger discount rate for a minimum rate of return on investment where $i \geq i_b$.

Consider the *npw-i* diagram and the assignment of different discount rates of i. For an assigned discount range of $12\% \leq i \leq 15\%$ per year, $npw \geq 0$ and alternative A is recommended for selection. If the investor demands a return $i > 15\%$ per year, alternative A is infeasible alternative and the do nothing alternative is recommended. In other words, the investor should avoid economic risk by placing \$10,000 in a savings account and be guaranteed a return of $w(1) = k(0)(1 + i_f) = \$10{,}800$.

We repeat, the approach of assigning high interest rates to speculative investment schemes does not assure the anticipated rate of return i' will be received. The assignment only makes it more difficult to justify the selection of alternatives with this method of selection. Higher discount rates tend to make the selection of the do-nothing alternative more likely.

The Present Worth Method for Alternative Selection

The key results from above can be placed into a selection strategy for the net present worth and annual worth methods for alternative selection. Even though these methods employ the same concepts, the net present worth method is preferred for complex cash flows; therefore, it is presented first.

1. Establish a discount rate i, where i represents a minimum acceptable rate of return on investment $i \geq i_b \geq i_f$. The assignment of discount rate i is often dictated by the interest rate of a loan i_b and by considerations of forecast uncertainty, as shown in Example 4.6. Loan rates reflect various market forces; therefore, they are often referred to as being a *market interest rate*. The overall state of the national economy, the availability of money for investment, and the credit rating of the person or company seeking a loan are important factors when one applies for a loan.
2. Use year-zero prices to forecast all future year benefits $w(t)$ and costs $k(t)$ for $t = 1, 2, \ldots, n$ years. For example, the net profit for year t for alternative j is calculated as

$$nw(t) = (p - c)x$$

where

p = unit selling price of the product in year zero prices
c = unit production cost in year zero prices
x = forecast of annual sales for year t

Of course, the unit prices and costs and annual sales can be a function of time and introduced into $nw(t)$ as $p(t)$, $c(t)$, and $x(t)$.

Some may question using year-zero prices because prices and costs are subject to price inflation. If price inflation is an important consideration, the year-zero prices and costs, p, c, and k can be adjusted. Price inflation is introduced in the Section 4.4 of this chapter.

3. Calculate the npw value and determine the feasibility of each alternative. The net present worth for alternative j is the sum of the discounted future net profits $nw(t)$ for $t = 0, 1, 2, 3, \ldots, n$ years, or

$$npw_j = \sum_{t=0}^{n} \frac{nw(t)}{(1+i)^t}$$

where $nw(t) = w(t) - k(t)$ with $w(t) = (p - c)x(t)$ and p and c are in year-zero unit prices and unit costs.

This form of npw is advocated in this text because it is most easily interpreted and can be easily programmed on a spreadsheet for complex cash flows. If the future cash flow is a uniform time series, then it may be convenient to use the transfer functions given in Table 1.

In order to fairly compare alternatives, each alternative should be adjusted so that they all have the same project lives of n years each. If alternative j, for instance, has a shorter project life than its competition, say $n_j < n$ years, then it does not earn benefits during the years $t = n_{j+1}, n_{j+2}, \ldots, n$ and it may put it at a disadvantage during these years. It is reasonable to assume that after the useful project life of n_j years, the production unit may be replaced with a new unit and be productive for another n_j years.

When discount rate i and n_j are relatively large values, discounting of far future benefits of $nw(t)$ will make the contribution to npw_j insignificant; thus, in this case the equal project life adjustment is unnecessary.

Use the capital investment method rules for selection. If $npw_j \geq 0$, then the plan is considered feasible and selected for further consideration; otherwise, it is designated as infeasible and a do-nothing alternative.

4. Repeat steps 2 through 3 for each alternative. Select the maximum feasible npw_j among the competing alternatives. If all alternatives are infeasible the do-nothing alternative is selected.

The Annual Net Worth Method of Alternative Selection

The net present worth and annual net worth methods of selection employ the same principles. The same steps described for the net present worth method apply to the annual net worth method. The major difference is that anw_j values are used in place npw_j values. Examples 2 and 3 of Table 2 show that *anw* and *npw* calculations are transforms of one another.

The annual net worth is calculated as a uniform time series. When $p(t) = p$, $c(t) = c$ and $x(t) = x$ are all constants in time, the annual worth method is simply calculated as

$$anw = (p - c)x - k$$

where the capital cost $k(0)$ is calculated as a uniform time series or amortized debt

$$k = k(0)\ crf(i,\ n)$$

When p, c, or x is a function of time t, the calculation becomes somewhat awkward. Take, once again, the stage construction problem of Table 2, Examples 2 and 3. Since the *anw* is treated as a uniform series, the *npw* is calculated first and then the *anw* is calculated as the product of the capital recovery factor times the *npw*.

Since the *anw* gives the investor an estimate of the average annual income over the project life n, it is often used in conjunction with the net present worth method. For example, since *npw* is a lump sum, it may be useful to transform it into *anw* in order to evaluate the alternative from a different point of view.

EXAMPLE 4.7 *The Effects of Discount Rate and Benefit Distribution*

The effects of discount rate and benefit distribution have an important bearing on alternative selection. Note that the sum total of benefits are the same in each case, \$150.

A. A uniform series of equal benefits

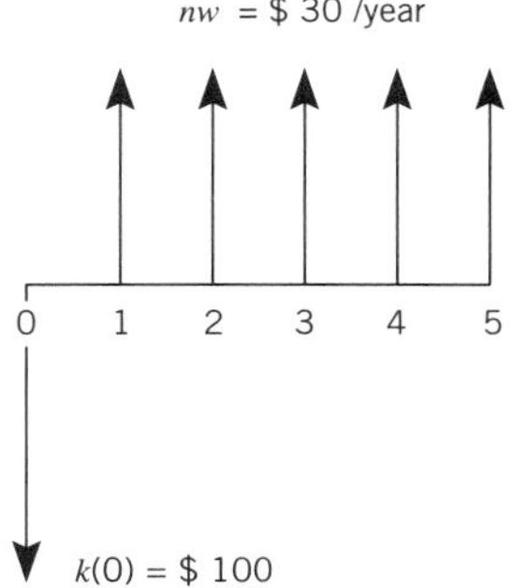

B. A single lump-sum benefit

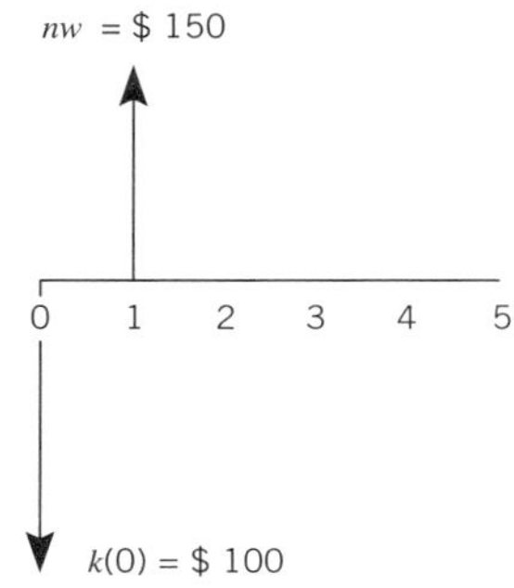

C. Linearly increasing series of benefits, $nw(t)$

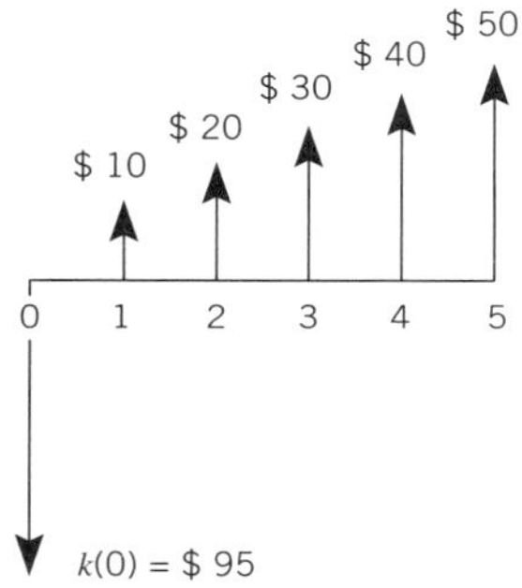

(a) Perform a sensitivity analysis to determine the effect of discount rate i on these alternatives. Use the net present worth method.

(b) Compare the internal rate of return method with the net present worth method of alternative selection.

SOLUTION

(a) The net present worth for each plan is calculated as

$$npw = \sum_{t=0}^{5} \frac{nw(t)}{(1+i)^t}$$

where $nw(t)$ = net annual revenue in year t. The design life n is 5 years. The results of the calculations are summarized in the following graph and table.

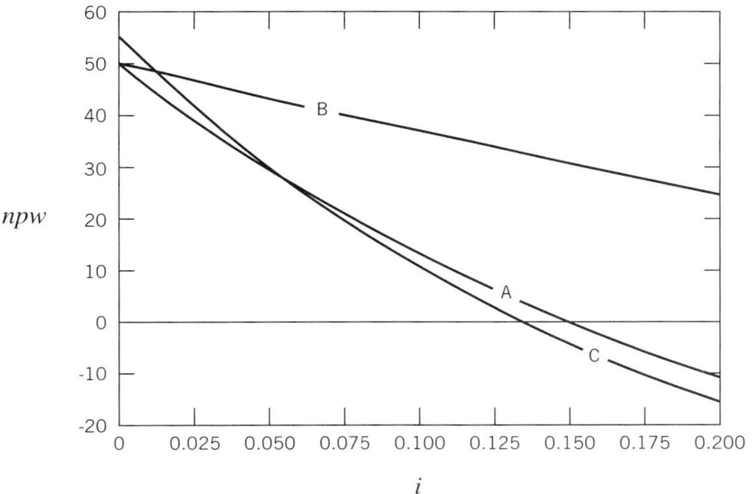

Net Present Worth for an Annual Discount Rate i

Plan	0.0%	2.5%	5.0%	7.5%	10.0%	12.5%	15.0%	17.5%	20.0%
A	\$50.00	\$39.37	\$29.88	\$21.38	\$13.72	\$6.82	\$0.56	(\$5.11)	(\$10.28)
B	\$50.00	\$46.34	\$42.86	\$39.53	\$36.36	\$33.33	\$30.43	\$27.66	\$25.00
C	\$55.00	\$42.08	\$30.66	\$20.54	\$11.53	\$3.48	(\$3.73)	(\$10.20)	(\$16.03)

Figures in parentheses ( ) are negative numbers, that is, $npw < 0$.

Without discounting, that is, $i = 0\%$, plan C is the most attractive alternative. There is no penalty assessed to waiting for a benefit. In other words, in plan A, the same value of benefit placed on benefits received in year 1 is the same in year 5, $nw(1) = nw(5) = \$30$. Obviously, a zero discount rate is of academic interest because an opportunity cost and economic risk are excluded from this evaluation.

The results show that as the interest rate increases, the less attractive all alternatives become. Overall for $i > 0\%$, plan B is the most attractive. It remains a feasible solution for all interest rates shown. In comparison, plans A and C become infeasible, $npw < 0$, at interest rates starting about $i = 15.0\%$ and 14.00%, respectively.

The combination of a high interest rate i and a long wait for benefits reduces the discount factor,

$$\frac{1}{(1+i)^{t'}},$$

thus making future benefits less valuable, and the do-nothing alternative a more likely result. This effect is most dramatically illustrated by comparing plans A and C. In the range of discount rates from 3% to 6.5%, plan A is preferred to plan C. For a range of i between 6.50% and 14% per year, plans A and C are both feasible alternatives but plan A is preferred. At higher interest rates, the large monetary benefits received for plan C in years 4 and 5 tend to be negated.

(b) The internal rates of return satisfy the condition, $npw(i') = 0$. For alternatives A and C, they are most easily determined from the $npw - i$ plot shown above. By inspection,

$$i'_A = 15\% \text{ per year}$$
$$i'_C = 14\% \text{ per year}$$

For alternative A, the value is most easily determined as

$$i'_B = \frac{w(1) - k(0)}{k(0)} = 50\% \text{ per year}$$

Since $i'_B > i'_A > i'_C$ alternative B is the best selection. This result is consistent with the *npw* method of selection above for some discount rate assignments.

If we compare *npw* and the internal rate of return method of selection for A and B, we find some inconsistency. From (a), we found that the selection of A and B depends on the assignment of the discount rate i.

- $0\% \le i \le 2\%$ per year, $npw_A \ge npw_B \rightarrow$ recommend A, and
- $2\% < i$, $npw_A \ge npw_B \rightarrow$ recommend B,

Using the internal rate of return method, plan B is always recommended.

The value of i' from the internal rate of return method is an overall measure of effectiveness of an alternative. It is independent of the assigned discount rate i and factors associated with economic risk. For example, a minimum acceptable rate of return of investment is neither considered nor is a discount rate assigned. The internal rate of return is considered to be an indirect measure of effectiveness. The net present worth method of selection is preferred in this book because a value of npw does not suffer from this criticism.

QUESTIONS FOR DISCUSSION

4.3.1 Define the following terms:

acceptable rate of return	market interest rate
internal rate of return	risk free rate of return

4.3.2 Does opportunity cost vary with personal wealth? In other words, do you agree with the statement that the rich are less risk averse than the poor? Explain your opinion.

4.3.3 Cite an example where an internal rate of return method recommendation is inconsistent with a net present worth method recommendation.

4.3.4 The questions deal with guidelines for selecting a discount rate for alternative selection.

(a) In the discussion of selecting discount rate i, two interpretations of the meaning of discount rate are given. What are they?

(b) What is the governing equation for determining the internal rate of return interest rate i'?

QUESTIONS FOR ANALYSIS

4.3.5 (a) Suppose that a capital investment project is to be financed with a loan at $i_b = 9\%$ per year. The bank also offers a risk-free (savings plan) interest rate $i_f = 6\%$ per year. Furthermore, suppose that the internal rate of return is estimated to be $i' = 8\%$ per year. What minimum discount rate i would you assign for evaluating the project? Why? In other words, why did you choose i_b, i_f or i' for i.

(b) Reconsider the recommendation given in (b). Give your answer from (d), what is your recommendation?

4.3.6 Determine the internal rate of return interest i' for an investment with $k(0) = \$10{,}000$ and $nw = \$7000$ for $t = 1, 2$ years.

4.3.7* (a) Determine net present worth npw for $i = 3, 6, 9, 12$, and 15% per year.

$k(0)$ ($)	nw ($ per year)	n (years)
$2M	$0.25M	20

(b) Estimate the internal rate of return rate i' for this investment.

(c) Determine the annual net worth anw for alternative given in (a) and for $i = 3, 6, 9, 12$, and 15% per year.

4.3.8* Consider the following alternatives:

	$k(0)$ ($)	nw ($ per year)	n (years)
A	$1.2M	$0.3M	6
B	$0.6M	$0.3M	3

Assume a discount rate of $i = 5\%$ per year.

(a) Draw cash flow diagrams for each alternative. Make an appropriate assumption so that alternatives A and B have the same design lives of $n = 6$ years each.

(b) Use the net present worth method for alternative selection to determine the better choice. Write a spreadsheet program.

(c) Confirm the recommendation made in (b) by using the transfer function equations.

4.3.9* Consider the following alternatives:

	stage t (years)	$k(t)$ ($)	c ($ per unit)	p ($ per unit)	x (units per year)
A	$0 < t \leq 10$	$k(0) = \$1.0M$	$35	$50	4000
B	$0 < t \leq 5$	$k(0) = \$0.5M$	$35	$50	2000
	$5 < t \leq 10$	$k(5) = \$0.5M$	$35	$50	6000

Assume a discount rate i = 6% per year.

(a) Draw cash flow diagrams for each alternative.

(b) Use a spreadsheet to calculate the net present worth of each alternative. Recommend the better alternative.

(c) Confirm the recommendation made in (b) by using the transfer function equations of Table 1.

4.4 PRICE INFLATION

Inflation is the persistent upward trend in the general price level. It results in a decline in consumer purchasing power. Price inflation is caused by many factors, including increased demand for a good or service while the labor supply and industrial capacity is fully utilized, increased wage rates without comparable increase in productivity, and a shortage in materials. In this section, we will discuss the effects of price inflation on capital investment analysis.

Total and Unit Prices

For capital investment analyses, it is important to distinguish between long-term and capital costs. In this context, we have used $k(t)$ to represent a capital investment made in year t and most simply, the product $k = cx$ to represent the annual production costs. But in certain financial analyses, it is desirable to combine these costs into a composite figure.

The total production cost, as an annual cost in dollars per year, has been defined to be

$$k = k_f + k_v(x)$$

Define the fixed cost to be the sum

$$k_f = k' + k'' = k(0)\ crf(i, n) + k''$$

where

k' = the amortized debt on the capital investment $k(0)$ calculated as a uniform series of constant annual payments

k'' = fixed annual costs for salaries, property rental, etc.

Since $c = k/x$, the unit cost includes capital and other fixed costs plus the variable costs of production. The net revenue, expressed as the unit value $p - c$, and, of course, p also reflect this unit cost. Clearly, profits are realized only when $p > c$.

Compound Inflation Rate

Price inflation in year t is assumed to be compounded and equal to

$$p(t) = p(1 + f)^t$$

where

p = price in year zero dollars

f = annual average rate of price inflation

The price in this context is considered a generic variable. Similar formulas for $k(t)$ and $c(t)$, for example, can be written as $k(t) = k(1 + f)^t$ and $c(t) = c(1 + f)^t$, where k and c are the total and unit costs in year zero dollars, respectively.

The primary reason for using year zero estimates for p, c, and k is to reduce, as much as possible, forecast uncertainty. Since current market conditions for a product are available and can be studied, forecasts to some future year t can be made with the compound inflation rate formula. For obvious reasons, more confidence can be placed in year zero price forecasts than in future year forecasts in year t dollars.

The Real Interest Rate

Let us incorporate the compound inflation rate formula into the present worth formula. By definition, the present worth of a $p(t)$ item is

$$pw(t) = \frac{p(t)}{(1+i)^t} = p\left(\frac{1+f}{1+i}\right)^t$$

where i = the discount rate. Now, let us simplify the formula by substituting

$$1 + r = \frac{1+i}{1+f}$$

into $pw(t)$ and reducing it to

$$pw(t) = \frac{p}{(1+r)^t}$$

By rearranging $1 + r$, we calculate r to be a function of i and f, or

$$\mathrm{r} = \frac{i-f}{1+f}$$

The *real rate of return interest rate*, or real interest rate for short, is defined to be the difference between the market interest and inflation rates, or

$$r' = i - f$$

It is a measure of the true gain from investment. Since $r \neq r'$, the inflation rate r is called a corrected real rate of interest of r', or

$$r = \frac{i-f}{1+f} = \frac{r'}{1+f}$$

If f is relatively small, then r' and r are approximately the same value. The corrected real interest rate r is used in this text.

Prices That Keep Pace with Inflation

As previously illustrated, a high market interest rate i decreases the value of future benefits; thus, it has the affect of making the do-nothing alternative a more likely choice. However, if the effect of inflation is passed on to the consumers in the form of higher prices, its effect on the producer is mitigated to some extent. When this assumption holds, prices are said to *keep pace with inflation*.

Suppose that the annual net worth in year t is calculated as

$$nw(t) = (p - c)x$$

Further suppose that the production costs c rise with the average annual rate of inflation f. It may be reasonable to assume that these additional costs will be passed on to producers' consumers at the same rate f in the form of a higher price p. The real interest r may be used in lieu of i in the net present worth calculation. Since the same inflation rate f applies to $p(t)$, $c(t)$, $k(t)$ the net present worth formula is written as

$$npw = \sum_{t=0}^{n} \frac{nw(t)}{(1+r)^t}$$

Since $r < i$, a recommendation for alternative selection is a more likely outcome.

If the assumption that prices keep pace with inflation cannot be justified, then i must be used in the net present worth calculation. For example, individuals contribute part of their earnings to a retirement fund. The money is invested and then, upon retirement, the individual removes fixed annual sums of money from the fund for life or some fixed number of years n. If these annual benefits $w(t)$ do not keep pace with inflation, and the net present worth of the fund for n payments is

$$npw = \sum_{t=0}^{n} \frac{w(t)}{(1+i)^t}$$

where assigned the discount rate i is used. Certainly, a retiree would like to have payments keep pace with inflation.

Price Indices

Suppose a capital investment analysis is to be conducted. In keeping with the guidelines for conducting the analysis, zero year prices of p are used to forecast future year prices $p(t)$. If the 1980 price is known, then the 1988 price can be estimated to be

$$p(1988) = p(1980)(1 + f)^8$$

This approach is discouraged because the correction is only as good as estimate of f. A much superior approach is to use a price index.

The consumer price index (CPI), producer price index (PPI), building cost index (BCI), and construction cost index (CCI) give an overall indication of price changes over time in the United States.

- The CPI is a measure of the average change in prices paid by urban consumers of the 17 largest metropolitan areas of the United States for a fixed market basket of goods and services.

- The PPI is a measure of the average price change of domestic producers in all stages of processing. It is based on a sample of about 3100 commodities and about 75,000 quotations per month in the manufacturing, agriculture, forestry, fishing, mining, gas and electricity, and public utility sectors of the economy.
- The BCI is based on a 20-city average wage of skilled labor, carpenters, bricklayers, and structural steelworkers, and average material cost for fixed quantities of structural steel, cement, and lumber.
- The CCI uses the same assumptions as the BCI, except that it is based on the cost of common labor instead of skilled labor.

The price adjustment from year 1980 to year 1988 is simply

$$p(1988) = \frac{\text{PI}(1988)}{\text{PI}(1980)} p(1980)$$

where PI is a price index. The average price for capital equipment, construction, consumer prices, and processed fuels have changed by the following amounts:

- capital equipment:

$$p(1988) = \frac{\text{PPI}(1988)}{\text{PPI}(1980)} p(1980) = \frac{114.3}{85.8} p(1980) = 1.33\, p(1980)$$

- construction:

$$p(1988) = \frac{\text{BCI}(1988)}{\text{BCI}(1980)} p(1980) = \frac{2598}{1941} p(1980) = 1.34\, p(1980)$$

and

$$p(1988) = \frac{\text{CCI}(1988)}{\text{CCI}(1980)} p(1980) = \frac{4519}{3237} p(1980) = 1.40\, p(1980)$$

- consumer prices:

$$p(1988) = \frac{\text{CPI}(1988)}{\text{CPI}(1980)} p(1980) = \frac{118.3}{82.4} p(1980) = 1.44\, p(1980)$$

- processed fuels:

$$p(1988) = \frac{\text{PPI}(1988)}{\text{PPI}(1980)} p(1980) = \frac{71.2}{85.0} p(1980) = 0.70\, p(1980)$$

With exception to processed fuels, price increases were about the same.

Controlling Inflation

In order to control price inflation, the Federal Reserve Bank adjusts its *discount rate*, that is, the amount of interest the government charges commercial banks to borrow money. The federal discount rate is different from the market interest rate i used in the discount factor formula, $(1 + i)^{-t}$.

After receiving a federal loan at the federal discount rate, commercial banks lend this money to their customers for capital investment. The actual interest rate i charged by the commercial bank to its borrowers reflects profit, bank service charges, the risk of the investment, the availability of bank money, the credit rating of the borrower, tax laws, the amount requested, and other local and national factors. Moreover, the market interest rate i reflects the federal discount rate. As a rule of thumb, the commercial bank's best customers, large corporations with good credit ratings, pay interest rates that are about 3% higher than the federal discount rate that is charged to it.

In periods of high inflation, the Federal Reserve Bank will increase its discount rate, thus making money more expensive for businesses and consumers to borrow. Since the market interest rate i charged to commercial banks will be higher, investment capital becomes more expensive, and projects will be more difficult to justify. Recall that high market interest rates i, when used as a discount rate, tend to make a recommendation based on net present worth calculations favor the do-nothing alternative. In addition, higher prices and higher interest rates for personal loans lessen consumer demand. Over time, the inventory supply of goods will increase. Businesses will eventually reduce prices. Thus, the federal government uses its discount to control capital investment and price inflation.

EXAMPLE 4.8 *The Effect of Price Inflation*

An investor may invest the $50,000 in a company promising to return a sum of $56,000 (profits plus investment capital) after one year. There is no guarantee against loss for this investment.

The investor has the option of depositing the $50,000 in a bank offering an interest rate of $i_f = 5\%$ per year. Bank deposits of up to $100,000 are fully guaranteed against loss.

The investor is concerned about the investment risk associated with the capital investment alternative and decides to apply the net present method. The investor investigates the state of the economy and estimates the annual inflation rate to be $f = 3\%$ per year. In addition, the investor has experience with companies with similar product lines, success histories, and bank credit ratings. The investor estimates that the company could receive a capital investment loan at a rate of $i_b = 8\%$ per year. Given the added assumption that the company products will keep pace with inflation, estimate the net present worth for the alternative.

Should capital investment alternative be recommended?

SOLUTION

The cash flow diagram for capital investment is:

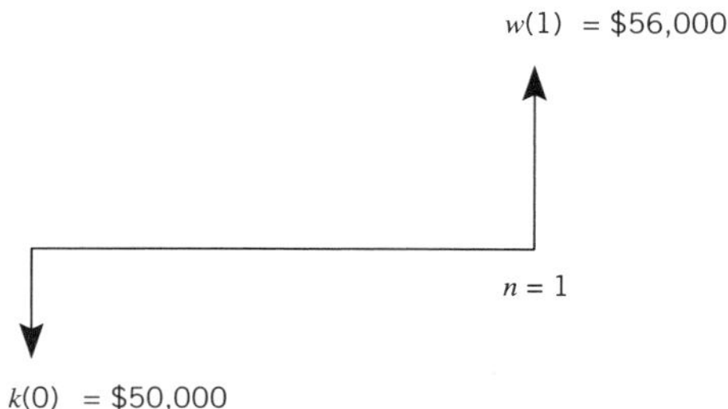

Assume an acceptable rate of return interest rate of i_b; thus, the discount rate is assigned $i = i_b = 8\%$ per year for the net present worth analysis. Since $i_b > i_f$, the analysis is considered to be conservative.

Assuming that prices keep pace with inflation, the corrected real rate of return where the discount rate is assumed to be $i = i_b = 8\%$ per year is

$$r = \frac{i - f}{1 + f} = \frac{0.08 - 0.03}{1 + 0.03} = 0.049.$$

The net present worth for the alternative is

$$npw = \frac{56,000}{1.049} - 50,000 = \$3,407 > 0$$

The alternative is recommended.

EXAMPLE 4.9 *The Effects of Inflation*

An individual is considering investing $k(0) = \$250,000$ into a recycling facility that will serve a region that generates $x = 6000$ tons of MSW per year. Owing to population growth, it is anticipated that a maximum $g = 500$ tons of additional MSW will be generated each year. The investor forecasts that the market price of recyclable materials will remain stable at $p = \$50$ per ton. Forty percent of the MSW is recyclable. The cost to process the waste is forecast to be $c = \$12$ per ton of MSW. The investor can secure a loan at an annual interest rate of $i_b = 8\%$.

The investor anticipates annual rate of inflation of $f = 5\%$ on waste-processing cost. The investor is unsure that

- market price for recyclable materials, while stable, will grow at $f = 5\%$; or that
- the MSW growth g will be realized.

Perform a sensitivity analysis to determine the effects of f and g on the npw values where their ranges are $0\% \leq f \leq 5\%$ per year and $0 \leq g \leq 500$ tons of MSW per year.

SOLUTION

The net present worth is calculated with discount rate $i = i_b$ as

$$npw = \sum \frac{nw(t)}{1.08^t}$$

where the annual net worth is

$$nw(t) = (p - c)x(t) - k(t)$$

with

$$x(t) = (0.4)(6000 + gt)$$
$$c(t) = \$12(1 + f)^t$$
$$p(t) = \$50(1 + f)^t$$

The cash flow diagram shows $p(t)$, $c(t)$ and $x(t)$ all to be function of time t, the growth rate g, and price inflation rate f.

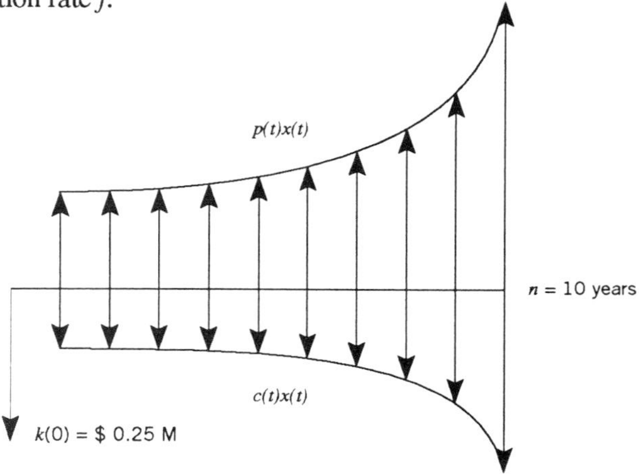

The results are most easily analyzed with the aid of a spreadsheet program. The following is a sample output for $f = 0.05$ and $g = 500$ per year.

t	$k(t)$ ($M)	$p(t)$ ($/#)	$c(t)$ ($/#)	$x(t)$ (#)	$nw(t)$ ($M)	$pw(t)$ ($M)
0	0.25	50.00	12.00	0	0	−0.25
1	0	52.50	12.60	2600	0.10	0.10
2	0	55.13	13.23	2800	0.12	0.10
3	0	57.88	13.89	3000	0.13	0.10
4	0	60.78	14.59	3200	0.15	0.11
5	0	63.81	15.32	3400	0.16	0.11
6	0	67.00	16.08	3600	0.18	0.12
7	0	70.36	16.89	3800	0.20	0.12
8	0	73.87	17.73	4000	0.22	0.12
9	0	77.57	18.62	4200	0.25	0.12
10	0	81.44	19.55	4400	0.27	0.13

The following table shows the effect of g and f on npw.

	npw for	
f (inflation rate)	g (annual growth)	
	0	500
0.00	\$0.36M	\$0.61M
0.05	\$0.53M	\$0.88M

The investment is feasible for all combinations of g and f and worthy of selection consideration.

QUESTIONS FOR DISCUSSION

4.4.1 Definitions:

federal discount rate	price inflation
price index	real interest rate

4.4.2 (a) Explain why in periods of high interest rates and high price inflation individuals are reluctant to invest in new capital projects.

(b) According to the theory presented in this section, when prices keep pace with inflation the real interest rates of r' for $i = 15\%$ and $f = 12\%$ per year and for $i = 4\%$ and $f = 1\%$ per year are the same, or

$$r' = 3\% \text{ for each case}$$

Compare r, not r', for $i = 15\%$ and $f = 12\%$ per year and for $i = 4\%$ and $f = 1\%$ per year. Do you feel that the net present worth method of alternative selection is consistent with the observation made in (a). Explain.

4.4.3 Why are year-zero prices used as forecasts in capital investment analysis?

QUESTIONS FOR ANALYSIS

4.4.4* Compare the effects of average inflation rates of 3% and 6% per year on the sales price of \$100 on a time series plot for time ranging from 0 to 25 years.

4.4.5* An individual receives 10 equal \$1000 end-of-year payments. Inflation will reduce the purchasing power of each future year payment. Define purchasing power loss in year t as

$$l(t) = \text{difference between inflated cost of goods in year } t \text{ minus the amount of money received in year } t.$$

Assume an inflation rate of 5% per year. Plot purchasing power loss over time.

4.4.6* The annual production and sales are forecast to increase at an annual rate of 10% per year as given by the compounded growth rate expression, $g(t) = 10(1.1)^t$.

Calculate the net present value of a \$1000 capital investment. The selling price and operating costs are \$10 per item and \$100 per year, respectively. The project life is 20 years and the discount rate is 8% per year.

4.4.7* Consider the following alternative.

$k(0)$ (capital cost)	$nw(t)$ (net worth)	n (years)
\$0.2M	\$0.05M	5

(a) Determine the internal rate of return interest rate i' assuming prices do not keep pace with inflation.

(b) Determine the internal rate of return interest rate i' when prices are assumed to keep pace with inflation. $f = 10\%$ per year.

4.4.8* The design life of a project is 25 years. A fleet of garbage trucks is replaced every 5 years. Assume the fleet costs \$20M and after 5 years it has no salvage value. That is, the selling price of a 5-year-old truck is zero. The rate of inflation is 4% per year and the 5-year 7.5% loan may be secured. Plot the amortized debt for the 25 year period assuming a fleet is purchased in year 0.

4.4.9* Consider the following alternative:

$k(0)$ (\$)	$p - c$ (\$ per year)	n (years)
\$2M	\$0.01M	10

Assume that the production grows at an annual rate that is governed by the compound interest formula

$$x(t) = 10(1 + g)^t \text{ units per year}$$

The discount rate i is 4% per year. Determine the minimum growth rate g', where the alternative is a feasible alternative for selection using the net present worth method.

4.4.10 Compare the PPI for finished goods, CPI, BCI for all items, and CCI by estimating the average annual inflation rate f for each price index for the 10-year period between 1979 and 1989. Use the compound rate equation to estimate f. The price indices are given in the Appendix.

4.4.11 Consider the following alternatives:

	t(years)	$nw(t)$	$k(t)$
A	$1 \le t \le 5$	\$1.5M	$k(0)$ = \$6M
	$6 \le t \le 10$	\$2.5M	$k(5)$ = \$6M
B	$1 \le t \le 2$	\$5.0M	$k(0)$ = \$3M

$$f = 2\% \text{ per year}$$
$$i_f = 4\% \text{ per year}$$
$$i_b = 7\% \text{ per year}$$

(a) Assign a minimum acceptable rate of return discount rate assuming that no bank loan is needed.

(b) Assume prices keep pace with inflation. Calculate r.

(c) Use the net present worth method to recommend the better alternative.

4.5 THE PUBLIC SECTOR

The same basic analytical procedures that are used to analyze private investments in Sections 4.1 to 4.4 are applied to public projects. The major difference between public and private projects is that the government has a primary obligation to maximize social services, not maximize net profit. The profit motive is not always applicable to the public sector. Let us explore this conceptual difference and then look at related modifications to the net present worth method.

User Benefits

Analytically, the capital investment methods of evaluating private and public projects appear to be the same. The npw formula for the net worth present method is written as before:

$$npw = \sum_{t=0}^{n} \frac{nw(t)}{(1+i)^t}$$

The method of analysis is modified in two fundamental ways:

- The social benefit is measured as user savings, not a revenue from the sale of goods and services.
- The market interest rate i reflects social concerns, not market forces.

The same modifications apply to the annual worth method. Here, the discussion focuses on the net present worth method.

Some public services have no clearly identifiable products to sell. Thus, the net profit is sometimes impractical and impossible to calculate. Consider the net worth in year t

$$nw(t) = w(t) - k(t)$$

where

$w(t) = px$, the annual profit
$k(t)$ = capital investment plus annual production costs in year t

The difficulty lies with evaluating the total revenue term, $w(t) = px$. Protecting public health through environmental protection is an example where the benefit is not sold as a commodity or service. On the other hand, the $k(t)$ costs are often the same for both public and private projects.

For public projects, the measure of effectiveness $w(t)$ is called the user benefit for year t. It is a measure of *social benefit* or service, often evaluated as an annual user savings. Since the capital investment methods are restricted to monetary measures of effectiveness, $w(t)$ must be assigned a monetary value. This assignment can be difficult even when the public pays for the service. Take public transportation as an example.

From a historical perspective, when some private transportation firms could no longer operate at a profit, they were taken over by the government and operated for the common public good. Many publicly owned urban bus and rail transport services are now operated by nonprofit government agencies and are given subsidies to keep fares low in an effort to maintain ridership. These subsidies are justified on nonmonetary grounds including minimizing traffic congestion and automobile air pollution emissions which, in turn, protects the urban environment and public health. The same arguments for providing subsidies can be used for capital investment. Thus, maximizing the level of service of an urban public transit system as a social benefit $w(t)$ may be a more worthy project objective than maximizing transit revenue.

Supply and Demand Functions

The principle of *market equilibrium* is used as a theoretical basis for measuring user benefit $w(t)$. Producer supply and consumer demand functions play a fundamental role in this principle. Commonly, the consumer demand is assumed to be an exponential function,

$$p = \alpha q^{-\varepsilon}$$

The demand function is always assumed to have a negative slope as shown in Figure 6.

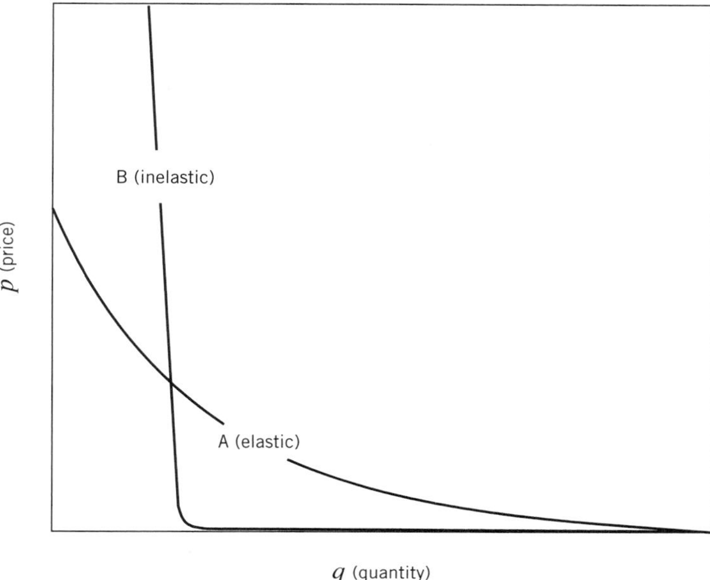

Figure 6 Consumer demand curves.

Demand elasticity ε' is defined as the ratio of the percentage change in consumer demand to the percentage change in price

$$\varepsilon' = \frac{\%\,\Delta q}{\%\,\Delta p} = -\frac{p}{q}\frac{dq}{dp}$$

For $p = \alpha q^{-\varepsilon}$, demand elasticity is simply $\varepsilon' = 1/\varepsilon$, where $\varepsilon > 0$. Thus, the magnitude of ε controls the slope of the function. For $0 < \varepsilon < 1$, the demand is considered elastic because a relatively small change in p will affect the consumer demand q, as illustrated by A in Figure 6. Luxury items are characterized as being elastic. Curve B has an inelastic demand, $\varepsilon > 1$, and is characteristic of items necessary for survival, such as drinking water. Figure 6 shows the demand for B to be relatively constant for a wide range of p.

The supply curve, Figure 7, shows the relationship between the selling price p and the total output or supply q. It is called a production function and is often assumed to have an exponential function

$$p = \delta q^{\gamma}$$

where $\gamma > 0$. As the demand q increases, the price p is assumed to increase.

Market Equilibrium

In a competitive market, the selection of a feasible plan implies that

- the investment will lead to production efficiency, and
- this efficiency will lower consumer prices

As previously discussed in Section 4.1, the profit motive is assumed to be the driving force to improve production efficiency by capital investment. In the long term, society as a whole is expected to benefit. The same principles are assumed to apply to public investments. The supply shift of Figure 7 is assumed to be the result of capital investment and production efficiency. With improved production efficiency, the same number of items q can be produced or the same number of people q can be served at lower cost. This is indicated as a supply shift.

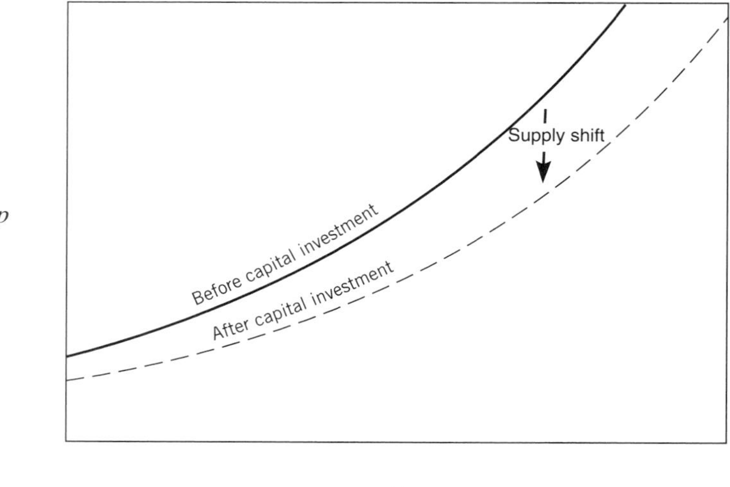

Figure 7 Supply curve.

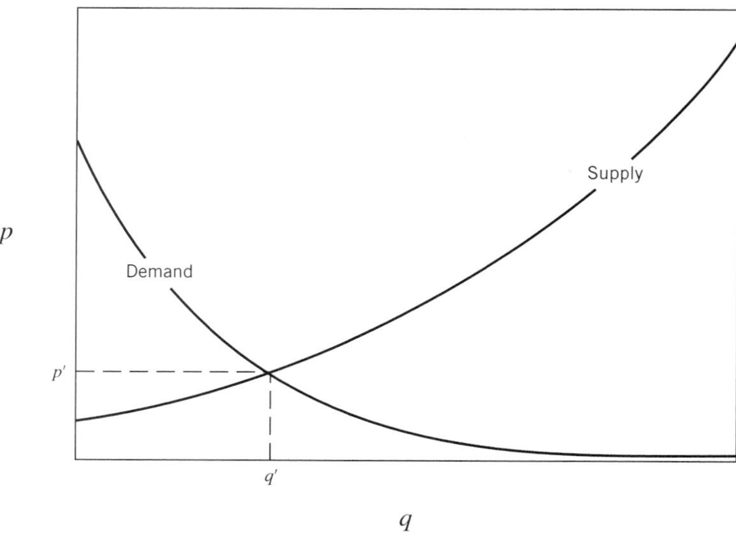

Figure 8 Market equilibrium.

The second point that production efficiencies lead to lower consumer prices will be illustrated with the concept of market equilibrium. Market equilibrium, as shown in Figure 8, is a point where the supply and demand curves intersect. The intersection point is described by

q' = total consumer demand for the good or service
p' = the average price for these goods or services

The values of q' and p' can be determined analytically as

$$p = \alpha q^{-\varepsilon} = \delta q^{\gamma}$$

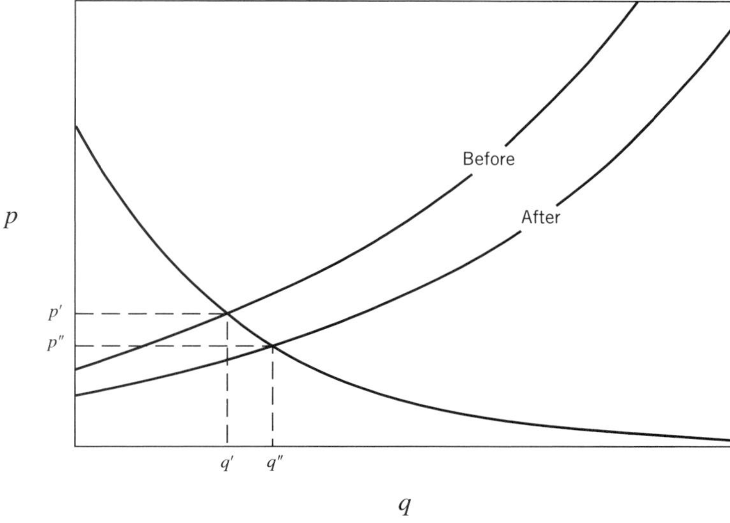

Figure 9 The effect of new production efficiency on market equilibrium.

The effect of the supply shift from new production efficiencies on market equilibrium is shown in Figure 9. Note the effect on consumers:

- Consumer prices are lowered, $p' > p''$.
- More consumers are served, $q' < q''$.

User Benefits as a Social Savings

At and below the market equilibrium price p' as shown in Figure 10, there are q' consumers willing to pay a price of p equal to or greater than the market equilibrium price p', or $p \geq p'$. The difference, $p - p'$, is a measure of individual consumer satisfaction. The total consumer satisfaction or consumer surplus is calculated to be the area shown in the figure.

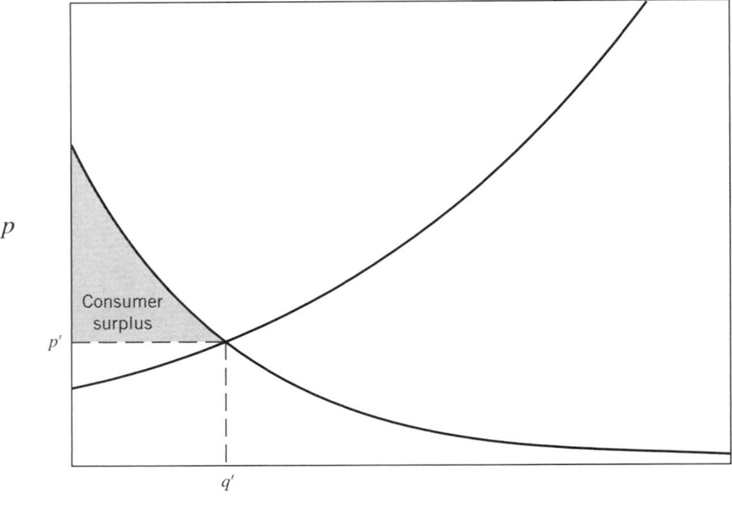

Figure 10 Consumer surplus.

After the capital improvement, there are more consumers, $q'' > q'$, paying less money, $p' < p''$. In other words, there is a change in consumer surplus, shown as the area in Figure 11. The *direct user benefit* used for public projects is defined as

$$w(t) = \frac{1}{2}(p' - p'')\,(q' + q'')$$

It is calculated as the area of a trapezoid.

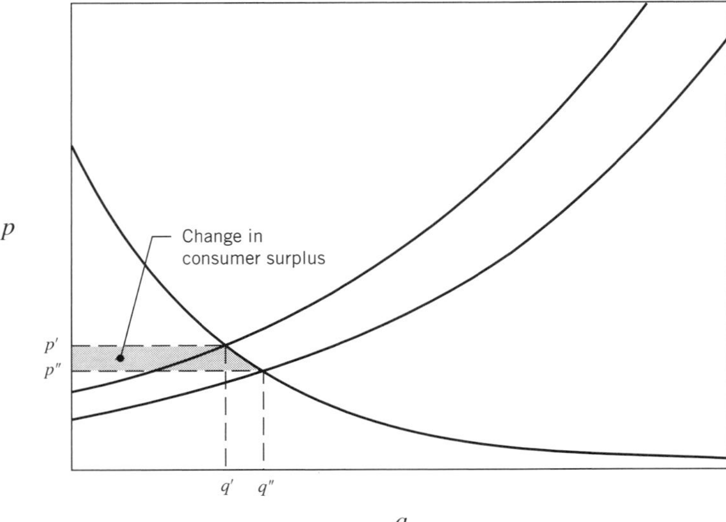

Figure 11 User savings as a change in consumer surplus.

The Least Cost Method of Alternative Selection

The notion of user benefits as user savings is most suitably applied to public MSW collection and processing, public transport, water projects, and other public utilities where user charges are made. However, some public services are offered at no charge and cannot be represented by supply-demand principles. In some cases, it is possible to justify the assumption that all alternatives offer the same user benefits $w(t)$. When comparing competing alternatives with capital investment methods, this factor cancels out, and there is no need to calculate $w(t)$. The measure of effectiveness reduces to an annual cost measure only:

$$nw(t) = k(t)$$

For a net present worth analysis, a least cost alternative solution decision rule is used.

The least cost method vividly demonstrates a fundamental difference between public and private alternative selection processes. For private projects, only feasible alternatives, $npw > 0$, are considered for selection. In fact, if all alternatives are infeasible, the do-nothing alternative is selected. For public projects, since the benefits are not measured in a least cost analysis, no feasibility check is made.

The Discount Rate

The assignment of the discount rate i for private and public investments usually differ. Private investments are influenced by competition in the marketplace, taxes, and other market forces. In most instances, the discount rate i can be safely assumed to be equal to the bank interest charged to the private firm for securing a loan for a capital investment. Since the bank is seeking a return of its investment, it will evaluate the credit history of the firm, the likelihood that the proposed investment will be successful, as well as other factors. The overall state of the economy and the availability of money for investment affect interest rates. If the investment is considered too risky, the firm could be denied a loan. If a loan is made, the interest rate i is assumed to reflect the level of investment risk.

On the other hand, the government borrows money and uses tax dollars to pay for projects. The discount interest i for public projects may be less than the market interest rate paid on borrowed money for private investment. Since lower interest rates favor long-term benefits, the use of lower rates for i is justified as being an investment for future generations. Individuals generally feel justified in using lower discount rates for projects dealing with public health and the environment, such as cleanup of hazardous waste sites and other past environmental abuses.

Some economists argue that i should be at least equal to the interest rate offered by risk-free government bonds. Clearly, the actual assignment of i is a complex issue involving many economic and social factors. Often government legislates the interest rate to be used in capital investment calculations. There have been instances when i is less than the risk-free interest rate.

EXAMPLE 4.10 ***Selecting an MSW Collection Fleet***

A total of 20 tons of MSW must be collected each day. A new fleet of trucks must be purchased. Two alternatives are being considered:

A. Trucks without garbage compactors

B. Trucks with garbage compactors

	Truck Capacity				
	ρ, Garbage density (ton per m^3)	v, Volume (m^3)	Capital Cost $k(0)$ (\$/truck)	Operating Cost c (\$/ton)	n (years)
A	0.2	15	\$15,000	\$35	5
B	0.35	15	\$20,000	\$25	5

The town currently operates as a nonprofit service with a truck that is described by alternative A. The residents pay for this service in form of a property tax.

Currently, the town operates trucks without compactors with characteristics described by A. It is assumed that there are 250 working days per year. The population remains the same for all future years. The town may receive a loan for $i_b = 8\%$ per year for a period of $n = 5$ years. Inflation is estimated to be $f = 4\%$ per year.

Which truck should be purchased?

(a) Use a least cost approach and net present worth method.

(b) Use the direct user benefit method and the net present worth method.

SOLUTION

(a) First, let us identify the client. Clearly, since the residents pay for the service, they are the client. Since this service is offered as nonprofit service, all increases in service cost caused by price inflation will be paid by the clients. Since prices are assumed to keep pace with inflation, the real interest rate r can be used in the analysis where discount rate is $i = i_b = 8\%$ per year. The inflation rate is forecast to be $f = 4\%$ per year:

$$r = \frac{i-f}{1+f} = \frac{0.08-0.04}{1.04} = 0.04$$

In this least cost analysis approach, there is no do-nothing alternative. New garbage trucks must be purchased. The question remains, Should it be trucks with or without compactors?

Using the least cost method, the consumer benefits are assumed the same for both A and B; therefore, $w(t)$ need not be quantified. By this method, the focus is on costs.

The number of trucks required per day are:

$$\text{A. } n = \frac{20}{\rho v} = \frac{20}{(0.2)(15)} = 6.67 \rightarrow 7/\text{day}$$

$$\text{B. } n = \frac{20}{(0.35)(15)} = 3.8 \rightarrow 4/\text{day}$$

The capital costs of investments are

A. $k(0)$ = 7 x 15,000 = \$105,000 (trucks only)

B. $k(0)$ = 4 x 20,000 = \$80,000 (trucks with compactors)

The yearly operating costs are

A. k = \$35/ton x 20 ton/day x 250 days/year = \$175,000/year

B. k = \$25/ton x 20 ton/day x 250 days/year = \$125,000/year

The cash flow diagrams for the two alternatives are as follows:

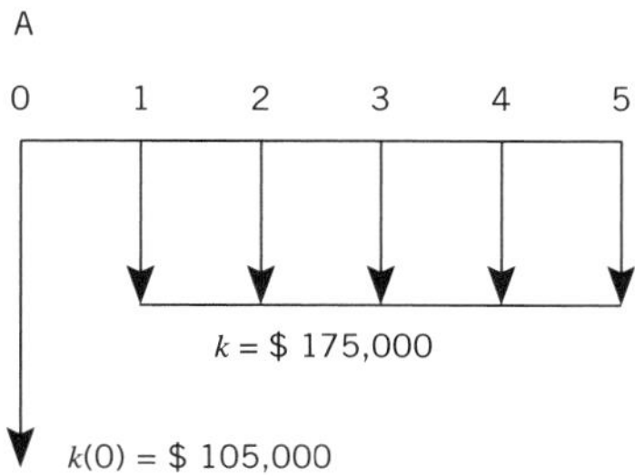

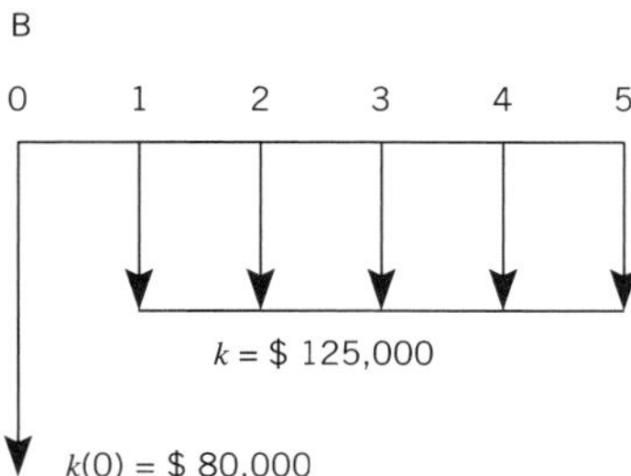

By inspection, the truck with the compactor, alternative B, is the better choice. For completeness, the net present worth is calculated using the *uspwf* factor

$$npw = k\frac{(1+r)^n - 1}{r(1+r)^n} + k(0) = 4.45\ k + k(0)$$

A. npw = 4.45(175,000) + 105,000 = \$883,750

B. npw = 4.45(125,000) + 80,000 = \$636,250

Since $npw_A > npw_B$, truck type B is recommended.

(b) The use benefit approach requires us to calculate the change in consumer surplus,

$$w(t) = \frac{1}{2}(p' - p'')(q' + q'')$$

and

$$k(0) = \text{the capital cost to provide } w(t)$$

where

$$\begin{aligned} p' &= \text{unit price for MSW pickup with current truck fleet} \\ &= c_A = \$35 \text{ per ton} \\ p'' &= \text{unit price for MSW pickup with replacement truck fleet} \\ &= c_B = \$25 \text{ per ton} \\ q' &= \text{amount of MSW pickup with current fleet} \\ q'' &= \text{amount of MSW pickup with replacement fleet} \end{aligned}$$

with

$$q' = q'' = 20 \text{ tons per day x } 250 \text{ days per year} = 50{,}000 \text{ tons per year}$$

because there is no population growth.

For alternative A, the net present worth is

$$npw = \$0$$

because A offers no improvement in service or $w(t) = 0$.

For alternative B, the change in unit price is $p' = p'' = c_A - c_B = \$35 - 25 = \$10$ per ton; thus

$$w = w(t) = \frac{1}{2}(10)(5000 + 5000) = \$50{,}000 \text{ per year}$$

$$k(0) = k_B(0) - k_A(0) = \$80{,}000 - \$105{,}000 = -\$25{,}000$$

or a savings of \$25,000.

The cost flow diagram for B is

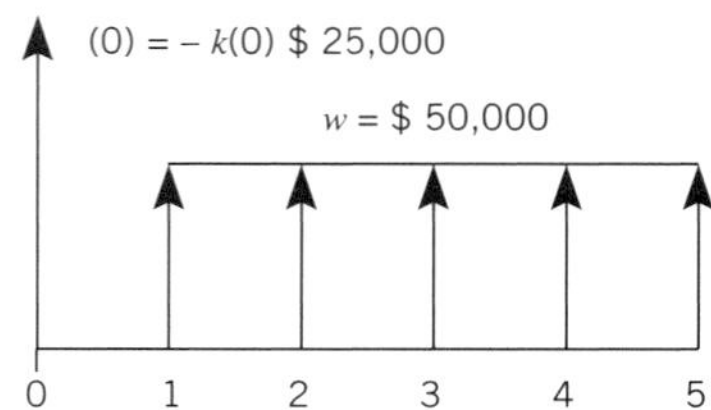

The net present worth is

$$npw = w\,\frac{(1+r)^{n-1}}{r(1+r)^{n}} + w(0) = 4.45w + w(0)$$

$$npw = 4.45(50{,}000) + 25{,}000 = \$247{,}500$$

Since $npw_A = 0$, $npw_B > 0$, and $npw_B > npw_A$, truck B is recommended as a replacement.

DISCUSSION

In spite of their differences in philosophy, the two approaches should be expected to lead to the same recommendation. They give the same net present worth of savings,

Least cost: $npw_A - npw_B = \$883{,}750 - \$636{,}250 = \$247{,}500$
User savings: $npw = \$247{,}500$.

Even though they give the same recommendation, the user savings technique is preferred because it forces one to address and evaluate client needs most directly. In addition, this method permits the feasibility of the solution to be evaluated.

QUESTIONS FOR DISCUSSION

4.5.1. Define the following terms:

consumer surplus
direct user benefit
market equilibrium
social benefit
user benefit

4.5.2 Suppose a law states that a net present worth method for alternative selection is to be applied to a public project and that the assigned discount rate is at least equal to the market interest rate. It is reasoned that using a discount rate at least equal to the market interest rate yields benefits and costs from public projects that are the same as those received from private investments.

These requirements are applied to two project proposals. The results from a net present worth analysis are as follows:

A. A drinking water supply system must be repaired to protect public health. An *npw* analysis shows a plan to be infeasible at a discount rate of 10%, the market rate. The repair plan has a breakeven discount rate of $i' = 5\%$ per year.

B. A public transit system must be repaired to improve service. An *npw* analysis shows a plan to be infeasible at a discount rate of 10%, the market rate. The repair plan has a breakeven discount rate of $i' = 5\%$ per year.

(a) For the public good, would you advocate a change in the law and use a discount rate of 5% for the drinking water supply? Explain your reasoning.

(b) To improve service, would you advocate a change in the law and use a discount rate of 5% for the public transit system? Explain your reasoning.

4.5.3 What is a fundamental difference between the least cost and the direct user benefit methods?

QUESTIONS FOR ANALYSIS

4.5.4 Show that the demand elasticty is $\varepsilon' = 1/\varepsilon$ for the consumer demand function, $p = \alpha q^{-\varepsilon}$.

4.5.5 (a) A demand function is perfectly inelastic. What is its demand elasticity ε' and demand function for this curve (line)?

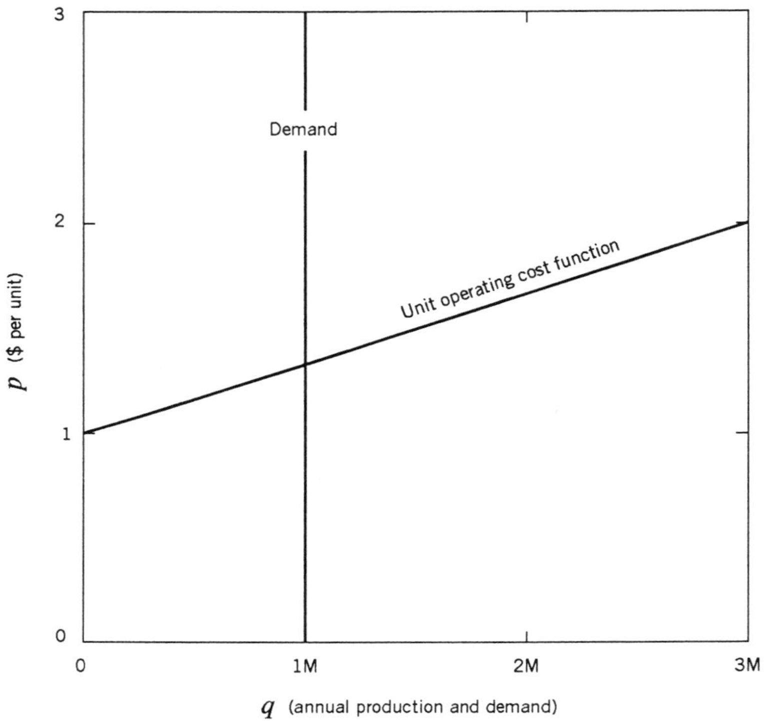

(b) The slope of the unit operating cost line is 1/3. Derive a supply function and determine the market equilibrium price p' and q'.

(c) What is the annual revenue received by the producer?

(d) *Private Investment:* A capital investment will cut the variable cost by 50%. Estimate the annual net present worth, *nw*, that would be used in a capital investment analysis for a private investor.

(e) *Public Investment:* A capital investment will cut the variable cost by 50%. Estimate the annual net present worth, *nw*, that would be used in a capital investment analysis for a public investment.

4.5.6 The demand function is

$$p = 30 - 0.03q$$

and the supply function is

$$p = 20 - 0.009q$$

(a) Determine the market equilibrium price p' and consumer demand q'. Use a graphical solution.

(b) Assume that new production efficiencies can reduce the total cost by 25%. Determine the new market equilibrium price and consumer demand.

(c) Calculate the direct user benefit.

4.5.7 A town is collecting MSW 250 days per year at a rate of 40 tons per day. It must replace its truck fleet. It is considering the following options:

	Truck Capacity				
	ρ, Garbage density (ton per m^3)	v, Volume (m^3)	$k(0)$ (\$/truck)	c (\$/ton)	n (years)
A	0.25	18	\$35,000	\$35	6
B	0.35	12	\$20,000	\$45	3

A town may secure a loan at 5% interest per year. Assume that the collection system is a nonprofit operation.

(a) Use the least cost *npw* method of analysis to determine the better alternative.

(b) Use the direct user benefit *npw* method to determine the better alternative.

4.5.8 The fuel consumption in gallons per vehicle-mile is as follows:

$$\text{Passenger car: } f = 0.03 + \frac{0.7}{v}$$

$$\text{Tractor-trailer truck: } f = 0.17 + \frac{2.43}{v}$$

where v is measured in miles per hour.

The travel time for all vehicle types between two cities, located 50 miles apart, is 3 hours. The traffic volume is 100,000 vehicles per day. Twenty percent of this volume is truck traffic.

A \$750,000 road improvement plan is contemplated. It is estimated that the improved road will generate a 10% per year increase in traffic with an average speed for all vehicle types of 30 miles per hour. For simplicity, assume that there is a step change in traffic flow when the improved road is open to the public but no other traffic growth in the future. The project life is 10 years and the discount rate is assumed to be 5% per year. The cost of gasoline is \$1.50 per gallon.

Use the net present worth method and the principles of direct user benefit to determine the feasibility of the plan.

4.5.9 Assume that the demand for a product is a function of its selling price, $q = 100 - p$, and the unit cost to produce it is \$5 per unit.

(a) In a competitive market, the selling price is \$6 per item. If plant capacity is 100 units, determine the number of items a firm should produce to maximize its net profit.

(b) Assume a monopoly. Determine the quantity q^* and the selling price p^* that a firm should produce to maximize its net profit. What is the selling price at maximum profit? A monopoly controls both the production level q and price p.

REFERENCES

Au T., and T. P. Au (1992). *Engineering Economics for Capital Investment Analysis*, Prentice Hall, Englewood Cliffs, NJ.

Bridgewater, A. V., and K. Lidgren, ed. (1981). *Household Waste Management in Europe: Economics and Techniques*, Van Nostrand Reinhold Company, New York.

——— (1981). "Returnable and Nonreturnable Containers and Household Waste,by J. Fisher," *Household Waste Management in Europe: Economics and Techniques*, Van Nostrand Reinhold Company, New York, 40–59.

Hendrickson, C., and T. Au (1989). *Project Management for Construction*, Prentice Hall, Englewood Cliffs, NJ.

Chapter 5

Risk

It is never possible to predict a physical occurrence with unlimited precision.

Max Planck (1858–1947)

A systems analyst is required to answer questions covering a diverse and broad range of topics. Optimization and capital investment analyses, though useful, do not adequately cover all issues; therefore, other methods of analysis are required. In this chapter, we use the fundamental principles of probability to address questions of risk and uncertainty.

5.1 A PROBABILISTIC CONCEPT

No matter what issue is under investigation, a systems analyst must deal with forecasting and forecasting uncertainty. To forecast simply means to estimate in advance. Unfortunately, a successful forecast can be validated only after an event has taken place. For example, if a facility is designed for earthquake loading, it may be years before an earthquake occurs and the success of the design can be evaluated. The best a designer can do is attempt to predict the earthquake's actual magnitude and to design accordingly. Clearly, uncertainty is associated with this estimate. There is a chance that the facility may not be subject to an earthquake over its entire lifetime. Theoretically, a designer could gamble on the chance that no earthquake will occur. The "up front" or capital costs of design and construction are minimized, but if an earthquake occurs, the social and monetary costs from the loss of human life and property may be great. The uncertainty of not knowing the future can indeed be costly.

Two different kinds of risk must be confronted. First, there are economic or investment risks where an investor may not realize anticipated profits and may possibly lose his or her investment capital. Second, there are risks associated with technology and acts of nature where the environment and public are damaged or harmed, causing monetary losses, personal discomfort, and even death. In this chapter, we are primarily concerned with this second type of risk.

While society generally is risk averse, some people voluntarily choose to engage in harmful and dangerous activities — mountaineering and smoking, for example. Exposure to air pollution and earthquake, on the other hand, are involuntary risks. Since engineering systems that mitigate the hazards associated with natural risks can be constructed, both natural risks and the risks derived from exposure to the toxic byproducts from our industrialized society are classified as technological. In this book we are concerned with these risks primarily for the purpose of decision making.

Questions of design, reliability, and risk are most adequately analyzed as *random processes* using principles of probability theory. The results from these methods of analysis can be introduced into the decision-making process.

Consider an earthquake design problem. Here, we treat the occurrence of an earthquake as a chance event, or more explicitly, we consider the outcome

$$S = \text{an earthquake occurs, or the event } S = \{\text{earthquake}\}$$

The brackets { } indicate an event. The complement of this event is that no earthquake will occur, or

$$S' = \{\text{no earthquake}\}$$

The random events S' and S are mutually exclusive events that span the entire sample space Z, or

$$Z = \{S, S'\}$$

Mutually exclusive events have the property that if one event occurs, say S, then S', or the intersection of S and S', cannot be observed at the same time.

The probability of an earthquake is defined to be

$$P(S) = \theta$$

Since the sum of the probabilities of two mutually exclusive events must be one,

$$P(S') = 1 - P(S) = 1 - \theta$$

If

$$P(S) = \theta = \frac{1}{25}, \text{ then } P(S') = 1 - \frac{1}{25} = \frac{24}{25}$$

According to the laws of probability 1, 2, and 4 given in the Appendix A, the probability of an event must always lie between zero and one, so that $0 \leq \theta \leq 1$ and $0 \leq 1 - \theta \leq 1$ must hold. If an event is certain, a probability of one is assigned, that is, $\theta = 1$. Then it follows that its complement is an impossible event (probability equal to zero).

Expectation

Let the random variable Y for the sample space Z be defined as

$$Y = \begin{cases} 0 \text{ if } S' \text{ occurs} \\ 1 \text{ if } S \text{ occurs} \end{cases}$$

and

$$P(Y = 1) = \theta \text{ and } P(Y = 0) = 1 - \theta$$

or as the probability model,

$$P(Y = y) = f(y) = \theta^{y} (1 - \theta) \text{ for } y = 0, 1$$

The expected value of Y is defined to be

$$E(Y) = \sum yf(y)$$

For the case where Y defines an earthquake–no earthquake process,

$$E(Y) = (1)(\theta) + (0)(1 - \theta) = \theta = \frac{1}{25}$$

This approach is general and applies to a wide variety of problems. When dealing with hazardous byproducts from industry, the dire consequence is the result of exposure to the hazard as well as the effect of this exposure on the individual. The consequences can be acute or chronic. Human exposure to toxic air pollutants from incineration and groundwater pollutants from a failed landfill usually cause chronic effects. The sample space $Z = \{S, S'\}$ can refer to these consequences as {failure, success}, {death, life}, {cancer, no cancer}, {sick, healthy}, and other mutually exclusive outcomes.

Risk

Risk is the expected loss or damage associated with the occurrence of a harmful event, or $R = E(H)$. If the earthquake results in property damage totaling \$2M, let the random variable H be defined as

$$H = \{h = \$2\text{M if } S, \text{ and } h = \$0 \text{ if } S'\}.$$

The expected earthquake damage cost is

$$E(H) = (\$2\text{M})(\theta) + (0)(1 - \theta) = \$2\text{M} \; \frac{1}{25} = \$80{,}000$$

The literature also defines risk as the probability that humans and other living species are exposed to a hazard, or θ. Unfortunately, using the same term to describe an expected value and probability can lead to ambiguity. In this text, risk R is defined as an expected monetary cost,

$$R = E(H) = h\theta$$

where

$h =$ the monetary consequence from a given state of nature or technology
$\theta =$ risk probability

This definition is preferred because it includes the probability of exposure θ as well as the consequence of exposure H.

Since $R = E(H)$ is a monetary value, it can be added to *real monetary costs* and used in decision making. A real cost transaction is one in which money is exchanged to buy goods and services. The expected damage cost in the earthquake example, R = \$80,000, is considered to be a *social cost.*

The tipping fee is a real monetary cost because it covers the debt charge on construction, and the charges for collecting, treating and disposing of waste. Suppose for a facility with an annnual capacity of 5000 tons per year the tipping fee is \$30 per ton. Let us adjust the tipping fee to account for earthquake damage of \$80,000 per year. It is assumed that each year an expected cost of \$80,000 must be paid in addition to real costs. For decision-making purposes, the tipping fee for a facility operating at capacity is calculated to be

$$u = \$30 + \frac{80,000}{5000} = \$46 \text{ per ton}$$

This cost can be introduced into mathematical models and used in decision making. In most instances, it will be treated like other model parameter estimates.

Technological Risk

An environmental hazard is defined to be the exposure to toxic chemicals and dangerous substances that lead to sickness, injury, or death. Exposure includes dermal contact, inhalation, and ingestion. Federal regulators use a lifetime risk of one chance in 1 million (10^{-6}) as a goal for establishing standards. Since lifetime risk is a risk probability, it is given as $\theta = 10^{-6}$.

Time is an important consideration in many analyses. In this text, the unit of time is typically expressed in units of one year. Thus,

$$\theta(t) = \text{probability of failure in year } t$$

where

$$t = 0, 1, 2, \ldots$$

The average life span of an individual is assumed to be on the order of 70 years. The goal or annual risk probability is

$$\theta(t) = \frac{\theta}{70} = \frac{10^{-6}}{70} = 1.43 x 10^{-8}$$

Estimates of $\theta(t)$ are calculated as an annual frequency of death, or

$$\hat{\theta}(t) \quad \frac{\text{number of deaths in year } t}{\text{exposed population}}$$

In order to appreciate the goal of $\theta(t) = 1.43 \times 10^{-8}$, consider statistics for the estimates given in Table 1.

TABLE 1

Hazard Causing Death	$\theta(t)$, Annual Risk Estimates
Black lung disease	8×10^{-3}‡
Cigarette smoking, one pack per day	3.6×10^{-3}§
Cancer	2.8×10^{-3}§
Coal mining accidents	1.3×10^{-3}‡
Mountaineering	500×10^{-6}§
Motor vehicles	240×10^{-6}§
	220×10^{-6}†,‡
Air pollution, eastern U.S.	220×10^{-6}§
Falls	77×10^{-6}†
Home accidents	110×10^{-6}§;
	12×10^{-6}†,‡
Football	40×10^{-6}‡
Police killed in the line of duty	22×10^{-6}§
Air travel, based on one transcontinental trip per year	2×10^{-6}‡

Sources: † EPA (1985b).

‡ 1978 statistics from Glickman and Gough (1990).

§ Wilson and Crouch (1987). Authors state that risks are calculated in various ways.

The Value of Life

Risk and risk probability have straightforward definitions. The difficulty arises when we apply these definitions to design and decision-making problems. Consider two fundamental questions, "What is safe enough?" and "What is an acceptable risk?" Even when a seemingly clear objective is defined, regulatory agencies sometimes have to make tradeoffs. The establishment of a drinking water standard for surface water illustrates the difficulty associated with assigning a limit.

Chlorination is a proven disinfection method for treating surface waters. However, chlorine combines with natural substances found in water to produce trihalomethanes (THMs). Animal studies show that THMs produce kidney tumors in laboratory rats and carcinomas in mice. Epidemiological studies suggest that they cause bladder tumors in humans. The Primary Drinking Water Standard for THMs is estimated to have a lifetime cancer risk of $\theta = 5 \times 10^{-4}$. In order to achieve the goal of 10^{-6}, a lower chlorine dosage is required; however, this dose is ineffective as a disinfectant of bacteria. Furthermore, the scientific evidence is insufficient to prove causality between THMs at relatively low doses and human cancer. The EPA standard is a compromise between risks associated with inadequate disinfection and cancer.

The effect of risk can be studied using the notion of expected value, $R(t) = h\theta(t)$ and $R = h\theta$. This analysis leads to the question, "What is the value of life?" or h.

Three basic calculation methods can be used to answer this question.

1. The human capital approach, also called the "foregone earnings" approach, values a life based on the premature death of the individual. This method is criticized because more weight is given to an accidental death of a young worker than to the death of an older worker from cancer or other chronic disease. Since it is based on wage earnings, it sets low values on the lives of the poor, handicapped, and elderly.
2. The willingness to pay approach uses questionnaires to ask individuals how much they would pay to avoid the risk of death.
3. The cost-effectiveness approach is another alternative that indirectly estimates the value of life. It is equal to the dollars spent on equipment, evacuation procedures, and other safety devices per number of lives saved.

The National Highway and Traffic Safety Administration, using the willingness to pay approach, values life at $360,000 per fatality. In contrast, calculations based on Nuclear Regulatory Commission design code standards to safeguard a population from nuclear power plant failure value a human life from $5M to $10M. A survey of federal regulatory agencies using the cost-effectiveness approach finds that they value each life from $50,000 to $12.1M.

These inconsistencies can be explained partially by the public's perception of risk. Catastrophic risk is deemed less acceptable than individual risk. Americans accept and tolerate 50,000 annual deaths on its highways, but a plane crash taking 100 lives commands much media attention. Since government responds to public pressure, some agencies devote more funds to safety programs than to other programs.

As noted earlier, risks are also classified as voluntary and involuntary. Since consumption of water is needed to sustain life, drinking water can be considered an involuntary risk. On the other hand, smoking is considered a voluntary risk. Consequently, the goal for involuntary lifetime risk of 10^{-6} has been established. The probabilities associated with voluntary risk are typically greater than those associated with involuntary risk. Generally, studies of various activities show voluntary risk to be 1000 times greater than involuntary risk.

EXAMPLE 5.1 *A Risk Assessment*

Residents of a neighborhood of 5000 people are concerned about the potential health risk due to air pollution emissions from the incinerator that has been proposed for their neighborhood. They call for a risk assessment. The following assumptions are made:

$$\theta = \text{risk probability over the life of the project} = 10^{-5}$$

where

$$n = \text{project life} = 25 \text{ years}$$
$$h = \$5\text{M per death}$$

(a) What is the expected number of lives lost for the project life and annually?

(b) What are the lifetime and annual risks as a social cost?

SOLUTION

(a) Let the random variable Y represent the consequence associated with

$$S' = \{\text{life}\}$$
$$S = \{\text{death}\}$$
$$Y = \{y = 1 \text{ if } S, \text{ and } y = 0 \text{ if } S'\}$$

The annual risk probability is

$$\theta(t) = \frac{10^{-5}}{25} = 4 x 10^{-7}$$

The expected values of Y and $Y(t)$, the losses of life over the project life and annually, are

$$E(Y) = (1)(\theta) + (0)(1 - \theta) = \theta = 10^{-5}$$

$$E[Y(t)] = \theta(t) = \frac{10^{-5}}{25} = 4 \text{ x } 10^{-7} \text{ per year}$$

Since every individual is at risk, the lifetime number of expected deaths for the neighborhood is defined to be the random variable,

$$T = xY$$

where

$$x = \text{population of the neighborhood} = 5000$$

The expected number of lives lost over the project life is

$$E(T) = E(5000Y) = 5000\ E(Y) = 5000 \text{ x } 10^{-5} = 0.05 \text{ death}$$

Similarly, as an annual measure,

$$E[T(t)] = (5000)(4 \text{ x } 10^{-7}) = 2 \text{ x } 10^{-3} \text{ death}$$

It can be shown with elementary probability theory that $E(aY) = aE(Y)$ where a = constant. (See Appendix A.)

(b) Let h = unit cost of death = \$5M per death. The lifetime risk, as an expected unit cost, is

$$R = h\theta = \$5\text{M x } 10^{-5} = \$50 \text{ per person}$$

and the annual risk is

$$R(t) = h\theta(t) = \$5\text{M x } 4 \text{ x } 10^{-7} = \$2.00$$

Let

$$H = \text{total social cost to the neighborhood} = Rx$$

and

$$H(t) = \text{annual social cost to the neighborhood} = R(t)x$$

The neighborhood cost, as a lifetime risk, is

$$H = Rx = (\$50)(5000) = \$250{,}000$$

As an annual cost, it is

$$H(t) = R(t)x = (\$2)(5000) = \$10{,}000$$

Note that capital letters are reserved for random variables, Y and Z, and expected values, R, $R(t)$, H, and $H(t)$, for example. For our purposes, they generally refer to the social costs associated with the physical and personal damage from system failure. Real costs are indicated with lower case letters. The total annual cost to the neighborhood, for example, is $k(t) + H(t)$, the sum of real and social costs for year t.

EXAMPLE 5.2 *Effect of Uncertainty on Decision Making*

Two construction plans for a landfill are being considered:

A. Landfill with a groundwater protection system

B. Landfill constructed with a liner system to protect groundwater

The annual amortized costs are

$$k_A(t) = \$200{,}000$$

$$k_B(t) = \$250{,}000$$

The annual risks of failure of these systems are estimated to be

$$\theta_A(t) = 10^{-2}$$

$$\theta_B(t) = 10^{-4}$$

The cleanup cost if the system fails and the groundwater is contaminated is estimated to be $h = \$6\text{M}$.

Since $k_A(t) < k_B(t)$, construction alternative A is preferred. Consider the social cost in the decision.

SOLUTION

The annual social costs for the two alternatives are

$$R_A(t) = h\theta_A(t) = \$6M\ 10^{-2} = \$60{,}000$$

$$R_B(t) = h\theta_B(t) = \$6M\ 10^{-4} = \$600$$

The total costs per year are

$$k_A(t) + R_A(t) = \$200{,}000 + \$60{,}000 = \$260{,}000$$

$$k_B(t) + R_B(t) = \$250{,}000 + \$600 = \$250{,}600$$

The minimum cost solution is sought; therefore, alternative B is recommended.

This result indicates that the social costs are significant and play a large role in decision making.

DISCUSSION

The decision rule for alternative selection is easy to apply. Decisions based on the assignment of risk probabilities, cleanup costs, and monetary values of life sometimes go beyond questions of forecast uncertainty and the applicability of probability modeling. No simple answer is offered, nor can one be offered with systems analysis. Society as a whole and its political process must deal with these decisions. The most probability modeling can offer is better insight and understanding of the problem and the alternative solutions.

QUESTIONS FOR DISCUSSION

5.1.1 Define the following terms:

forecast	risk
mutual exclusive events	real monetary cost
random process	social cost

5.1.2 (a) What are the differences between economic and technological risks?

(b) In your opinion, what are the greatest environmental and/or public health risks associated with landfill disposal and incineration?

(c) Which risk from (b) do you personally fear the most and why?

(d) Are the risks from (b) voluntary or involuntary risks?

5.1.3 (a) List an advantage and disadvantage of the human capital, willingness to pay, and cost-effectiveness methods in estimating the value of human life.

(b) In your opinion, which is the best method of estimating the value of life. Why?

(c) In your opinion, which is the poorest method of estimating the value of life. Why?

5.1.4 In Example 5.2, real and social costs are added together to establish a decision rule for selection. If you were representing your town, would you challenge the analysis? Explain your reason for your opinion.

QUESTIONS FOR ANALYSIS

5.1.5 (a) Estimate the annual risk probabilities $\theta(t)$ for the statistics given for 1973 or 1974 below. Assume a population of 210M and continuous exposure to each hazard.

Hazard Causing Death	Total Deaths per Year
All causes	1,973,003
Heart disease	757,075
Cancer*	351,055
All accidents	105,000
Motor vehicles*	46,200
Work accidents	13,400
Fires, burns	6,500
Firearms, sporting	2,400
Railroads	1,989
Pleasure boating	1,446
Lightning	124

(b) Compare your estimates for hazards marked with an asterisk (*) with the appropriate frequencies given in Table 1.

(c) In your opinion, are the method and assumptions used in (a) to estimate $\hat{\theta}(t)$ for pleasure boating and firearms sporting risk probabilities representative of the true danger associated with these activities? Why? Suggest an alternative estimate of $\hat{\theta}(t)$.

5.1.6 An investigation of watershed data reveals that the largest flood in the last 50 years caused property damage of \$60M and drowned 5 people. Assuming a value of life of \$4M per person, forecast the expected annual flood damage cost including both real and social costs for the 1- in 50-year flood.

5.1.7 Total building construction and earthquake damage costs are

A. Designed without earthquake protection: $k(0)$ = \$15M and h = \$5M

B. Designed with earthquake protection: $k(0)$ = \$18M and h = \$5M

The estimated frequency of earthquake occurrence is one chance in every 10 years. While it is impossible to control the occurrence of an earthquake, it is possible to protect the building against damage. The probability that the building will be exposed to an earthquake exceeding the earthquake design load in any year is 1 in 1000. Determine annual savings associated $w(t)$ with building designed for earthquake loading. In other words, find the difference between the annual real and social costs for the two designs.

Use a discount rate of i = 10% per year and a design life of 30 years.

5.2 RISK ANALYSIS

By law, proposals for constructing large-scale projects are subject to public review. Thus, the decisions on whether or not to construct are often based on and strongly influenced by public opinion, political lobbying, journalistic reporting and interpretation, and the strength of arguments and conviction of advocates and protestors. The selection process in practical application may sometimes appear to be disorderly and imprecise. Mathematical modeling plays an important role in developing our understanding of a system and aids in the decision-making process. A model has limitations, and if its application is overextended, the results may be questionable. Clearly, a final decision incorporates more information than just a technological evaluation.

In this section, we discuss some controversies associated with the use of models and how science and politics are merged to perform risk analyses and to regulate technology.

Alternatives with Dissimilar Attributes

Source reduction is a very appealing approach to reducing waste volume. Consider two very different strategies for implementing this plan. The first alternative involves capital investment to introduce a new manufacturing process. The second alternative is to educate the public in the practice of purchasing products packaged in returnable containers. The question is which plan should be recommended.

In the first case, the performance of the manufacturing process can be modeled using accepted engineering and mathematical modeling principles. In the second case, the response to the initiative can be measured by a survey. In both cases, the amount of material to be thrown away as waste is estimated and compared and preference is given to the alternative with the least residual.

Some argue against this kind of study because the alternatives and the analysis methods are too dissimilar to be of practical use. In essence, they say, human response data and machine data should not be compared. Advocates for the study feel that the information is valuable, but they would modify the selection process. Even though the two groups use the same measuring scale, both consider it difficult or even impossible to make a fair comparison. Another issue is consistency. Studies show that conservation and recycling efforts lose their appeal with time and that many people return to old habits. In contrast, a manufacturing process is expected to give repeatable results.

Those who favor the study recommend the establishment of a decision-making rule that compares "like" items using "like" methods of analysis. For example, the optimization process formulated as an LP model offers a fair comparison of different solid waste management alternatives. The technologies being compared offer similar services, and their costs can be estimated with reasonable confidence. This is not the case in the source reduction study as presented. Advocates for the study believe that a more subjective, adaptive selection process is warranted. Possibly, after the study results are completed, expert opinion could be obtained, issues debated, and a consensus reached.

Risk Analysis: A Merger of Science and Politics

When proposals that might endanger public health are presented, society relies most heavily on scientific evidence, regulation policy, and local, state, and federal law. Let us investigate the products of MSW combustion, regulations for design and operation of an incinerator, and ways that are being employed to reduce its environmental risk. The major products of combustion are placed in two categories:

Flue Gases and Fly Ash	Bottom Ash
Combustion byproducts HCl, CO, SO_2, NO_x, CH_4, and other hydrocarbons	Unburnt organic matter including dioxin and other byproducts of combustion
Organic chlorine byproducts Dioxins, furans, PCE, PAH	Metals: lead, cadmium, zinc
Fly ash Metals: lead, cadmium, zinc, mercury	

Chlorine is a common component in many residual products, including salt in food residues and certain plastics. It is converted to hydrochloric acid (HCl) and organic chlorine byproducts during combustion. HCl and other acids form a corrosive environment, which reduces the useful life of an incinerator, and adds to the acid rain problem, namely, the release of sulfur dioxide (SO_2) and nitrogen oxides (NO_x) into the atmosphere.

Dioxins (polychlorinated dibenzo-p-dioxin, PCDD), furans (polychlorinated dibenzofurans, (PCDF), polycyclic aromatic hydrocarbons (PAHs), and tetrachloroethane (PCE) are the chlorinated hydrocarbons of most concern. One of the most toxic, teratogenic, and carcinogenic chemicals known is the dioxin, 2,3,7,8 TCDD (tetrachlorodibenzo-p-dioxin). For male guinea pigs the LD_{50} (the lethal dose where 50% of the population die) is 0.6 mg/kg body mass. Dioxins and furans bioaccumulate in fish, animals, and humans. They have been found in cow and mother's milk. PCE has induced tumors in animals and caused liver, kidney, and central nervous system damage. PAH is a carcinogen. Since measured concentrations of dioxin and furan have been found in soil and air and in cows and mother's milk, investigations have focused on them.

Metals are not destroyed by the combustion process but are oxidized and vaporized. With the exception of mercury, which occurs in a gaseous phase, metals attach themselves to dust particles in fly ash or settle into the ash. Heavy metals are considered a health concern because they accumulate in the human body. Cadmium, a carcinogen, causes kidney and brain damage. Lead causes kidney damage and anemia. Children exposed to lead can develop learning disability. Mercury causes damage to the brain, nervous system, and kidneys. Zinc irritates the eyes and skin.

Since 1970, 12 major U.S. statutes have been enacted regulating toxic substances in the environment. They include the Clean Air Act (1970 plus amendments and 1990), the Federal Water Pollution Control Act (1970 plus amendments), the Hazardous Materials Transportation Act (1978), and the Resource Conservation and Recovery Act (1984). Federal, state, and local regulations have been enacted to ensure that adverse impacts are minimized. Permission to construct and operate an incinerator requires, for instance, that environmental impact studies be prepared and that air, surface, groundwater, and wetland permits be obtained. A basic understanding of the regulatory process and the establishment of standards are essential, for these requirements affect the design and operation of a plant.

The Process

Since these laws and regulations are complex, it is difficult to generalize. The Environmental Protection Agency (EPA), the Occupational Safety and Health Administration (OSHA) and the Food and Drug Administration (FDA) base their regulatory actions and decisions on risk analysis studies. These studies comprise two distinct phases: risk assessment and risk management. Their relationship is depicted in Figure 1.

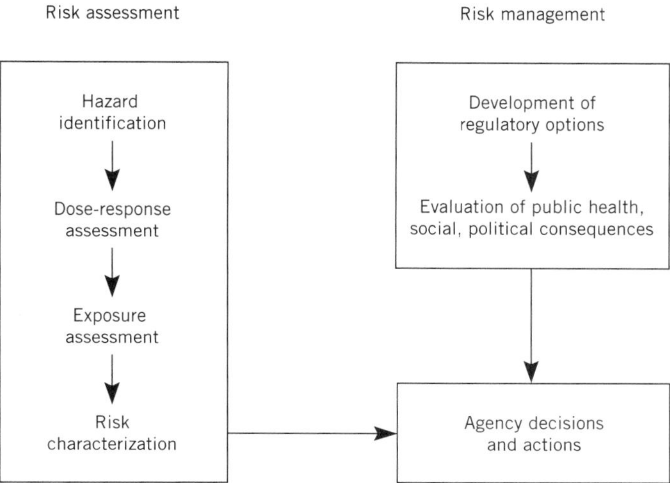

Figure 1 The risk analysis model.

Risk assessment consists of the following steps:

- Hazard identification: Does a chemical cause an adverse effect?
- Dose-response assessment: What is the relationship between the dose of a chemical and its effect on humans?
- Exposure assessment: How many people are affected, and what is the degree of exposure to the chemical?
- Risk characterization: What chemical dosage causes an adverse effect on the public?

Animal and epidemiological studies are the primary methods used in hazard identification. Animals, typically mice and rats, are exposed to high doses of a chemical, and their responses are recorded. These data are used to develop a dose-response relationship. Epidemiological studies are typically conducted on workers who have experienced high exposure to a chemical. These data and models are extrapolated to predict a human response and are combined with assessment information to form the basis for risk characterization.

In the risk management phase, regulatory decision making is based on these scientific results and legislative requirements. Government regulations reflect on societal demands, scientific evidence, economic tradeoffs, and society's perception of risk. Regulations are generally based on one of three methods: (1) risk probability only, (2) risk balancing, which includes the economic costs and benefits of evaluation associated with the regulation, and (3) technological control, which emphasizes the use of the best available technology to reduce adverse effects.

Dose-Response Assessment

Animal toxicity studies are conducted to estimate health risks to humans. The principle of extrapolation from animals to humans has been accepted by the scientific community and is employed by regulatory agencies. The dose-response model is a principal tool used for risk assessment. The acceptable dosage level or virtual safe dose (VSD) depends on whether or not the toxic material is classified as a carcinogen.

In order to better appreciate the challenge that faces decision makers, consider the two dose-response functions given in Figure 2. TD_{50} is defined as the dose d which produces a toxic effect in 50% of the exposed population. When the effect causes death, it is called a lethal dose and is designated as LD. The TD_{50} for chemicals A and B is about the same—about $TD_{50} = 50$ units. However, when the exposure dosage $d < TD_{50}$, chemical A is more lethal than B. In fact, chemical B has a threshold value of about 20 units, below which no adverse affect is observed at all.

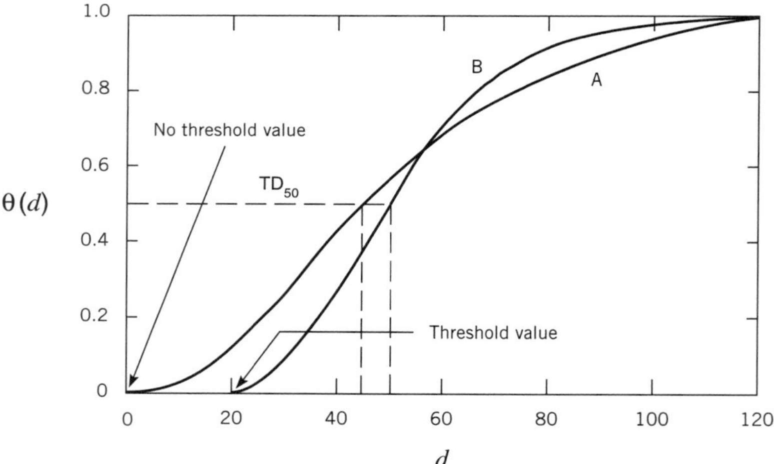

Figure 2 Hypothetical dose-response curves for chemicals A and B.

The challenge is to determine the safe exposure to potentially toxic materials or the VSD of a chemical. The VSD is specified for lifetime risk probabilities, which are of small magnitude—on the order of $\theta = 10^{-4}$ to 10^{-6}. In order to conduct reasonably economic animal studies and to be able to observe a response, animals are exposed to heavy dosages of chemicals relative to their weight. In order to develop a dose-response function $\theta(d)$, data over a wide dose range are needed to be able to fit a statistically significant model.

A major concern is dealing with low doses and their effect on model calibration and forecasting. Consider the hypothetical dose-response curves in Figure 2. Responses at dose levels of TD_{50} or in its vicinity give very little information at low probabilities of 10^{-4} to 10^{-6}. For example, the TD_{50} for chemical B gives no indication that it has a threshold value of 20 units. Thus, for a VSD of 20 units or less this chemical, theoretically, can be considered risk free. The heart of the matter is that with the kind of data and mathematical tools available to us, it is impossible to make statements with a reasonable degree of confidence for low-dose concentrations.

Now, consider the problem of determining whether a threshold value exists for chemical B. In Figure 2, $\theta(0) = \theta(20) = 0$, indicating that no background risk exists. However, when dealing with experimental data, some animals may show a response when not exposed to the chemical. The DDT data presented in Table 2 demonstrate the point. Since, $\hat{\theta}(0) > 0$, the data imply that a background risk exists. Clearly, choosing a model at low dosages poses a dilemma.

TABLE 2 Animal Response to DDT Exposure

Dose, d	Response, $r(d)$	Number of Animals, $n(d)$	Lifetime Risk Estimate, $\hat{\theta}(d) = \frac{r(d)}{n(d)}$
0	4	111	0.036
2	4	105	0.030
10	11	124	0.089
50	13	104	0.125
250	60	90	0.667

Source: VanRyzin (1980).

Dose-Response Functions

Regulatory agencies have used animal response data to fit a variety of dose-response functions. They include

Linear model: $\theta(d) = \beta d$

One-hit model: $\theta(d) = 1 - \exp(-\beta d)$

Multistage model: $\theta(d) = 1 - \exp(-\beta_1 d - \beta_2 d^2 - \ldots - \beta_k d^k)$

Weibull model: $\theta(d) = 1 - \exp(-\beta d^{\gamma})$

The multistage and Weibull models are extensions of the one-hit model.

A hit means that a cell is exposed to the toxic chemical. The probability that a cell will become cancerous is hypothesized to be proportional to the dose d. For the multistage model all powers of d, including k, are positive integers. The risk probability for the Weibull model is assumed to be proportional to d^{γ} where γ is a shape parameter.

The problem of low-dose extrapolation is still unresolved. Clear-cut answers to establish a VSD remains not only a scientific problem, but also one that faces society and its political leaders and policymakers.

The model that appears to give the most conservative risk probability estimates at low dosages is generally the linear model. See Figure 3 for an example. The Environmental Protection Agency uses a linear model for low-dose extrapolation:

$$\theta(d) = \beta d$$

where

$$\begin{aligned} \theta(d) &= \text{excess risk probability} \\ \beta &= \text{unit risk} \\ d &= \text{intake dosage} \end{aligned}$$

The term *excess risk probability* refers to the probability that the individual is exposed to some hazard greater than some background or threshold level found in a healthy environment. If concentration of a hazardous chemical, for instance, is less than the threshold concentration, there is no danger to the individual. Carcinogens are assumed to have a threshold level equal to zero. Thus, $\theta(d)$ are considered to be conservative estimates at low dosage.

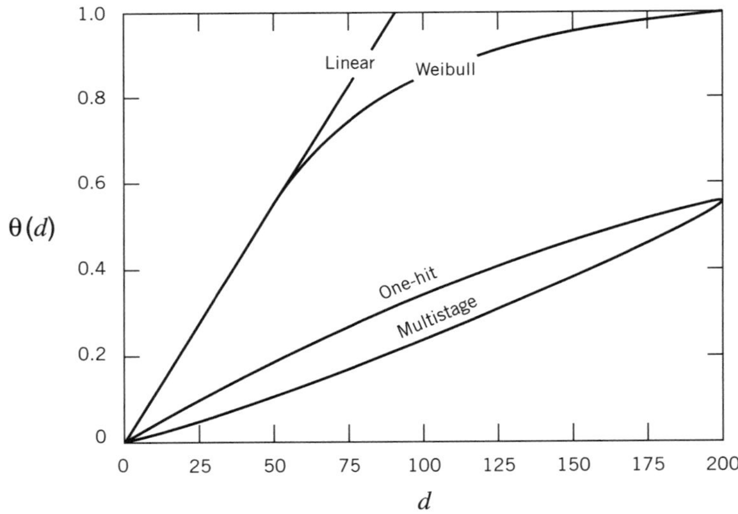

Figure 3 Dose-response functions of DDT.

A unit risk is the slope of an animal dose-response curve β where dose is measured in units of mg per day per kg of animal body weight. It is defined to be the excess cancer risk associated with a lifetime exposure to 1 mg/kg/day of the chemical. Table 3 presents a sampling of available EPA information for residual chemicals from MSW incineration.

Table 3 Unit Risk

Slopes of Dose-Response Relationships at 95% Upper Limit Potency

Chemical	Chemical Formula	β (mg/day-kg body weight)$^{-1}$
Benzene	C_6H_6	2.9×10^{-2}
Cadmium	Cd	6.1
2,3,7,8 TCDD	$C_{12}H_4Cl_4O_2$	1.56×10^5
PCE	C_2Cl_2	5.1×10^{-2}
Vinyl chloride	CH_2CHCl	1.75×10^{-2}

Source: EPA (1985a).

In order to use this model for human populations, risk probabilities are calculated by substituting human weight and inhalation intake rates for animals. The intake dosage is defined as

$$d = \frac{\rho r}{w}$$

where

ρ = chemical concentration
w = average body weight
r = daily intake

Standard values of w and r are given in Table 4.

Table 4

Standard Weight and Daily Intake Rates

	Adults	Children
w, kg	70	10
r, daily water intake, l/day (l = liter)	2	1
r, daily amount of air breathed, m^3/day	20	5

Source: EPA (1986).

Clearly, this makes the calculation simple, but it does not address fundamental biological differences between humans and animals—not to mention questions associated with human lifestyle and the compounding effects of being exposed to many different chemicals over a lifetime.

Establishing Regulatory Standards

For noncarcinogens, a chemical is considered safe if the dosage is less than a threshold dosage, approximated by the no-observable-effect level (NOEL). For chemical B in Figure 2, NOEL = 20. The acceptable daily intake (ADI), in mg/kg/day, is equal to the NOEL divided by a safety factor (*sf*),

$$\text{ADI} = \frac{\text{NOEL}}{sf}$$

Safety factors of 10, 100, 1000, and 10,000 have been used. The choice of safety factor reflects, among other things, the uncertanty about chemical toxicity.

The determination of an ADI for carcinogens and noncarcinogens differs in fundamental ways. For carcinogens, genetic damage of even a single cell of DNA is assumed to cause a series of changes resulting in the production of tumor. Some scientists postulate that a single molecule of a toxic chemical can cause cell damage. Consequently, no threshold dose is assumed to exist, as shown in Figure 2 for chemical A. The VSD is determined for a given risk probability θ. Assuming a linear dose-response function, it is calculated as

$$\text{VSD} = \frac{\theta}{\beta}$$

The ADI is found by assigning a safety factor sf as discussed for noncarcinogens.

$$\text{ADI} = \frac{\text{VSD}}{sf} = \frac{\theta}{\beta sf}$$

The maximum contamination level (MCL) is calculated as

$$\text{MCL} = \frac{w\text{ADI}}{r} = \frac{w\theta}{r\beta sf}$$

EXAMPLE 5.3 ***The Risk Probability of 2,3,7,8 TCDD***

The background air concentration level of 2,3,7,8 TCDD is $\rho = 3.4 \times 10^{-9}$ mg/m^3. This level is reported to be representative of areas in the rural and urban levels of Japan, Germany (West), Sweden, the United States, and the Netherlands.

Estimate the risk probability for an adult.

SOLUTION

$$\theta(d) = \beta d = \beta \frac{\rho r}{w}$$

where standard values of r and w for an adult are used. The value of β is obtained from Table 3.

$$\theta(3.4 \times 10^{-9}) = (1.56 \times 10^{5})(3.4 \times 10^{-9})(20)(\frac{1}{70}) = 1.52 \times 10^{-4}$$

$$(\text{unitless}) = \left(\frac{\text{kg day}}{\text{mg}}\right)\left(\frac{\text{m}^3}{\text{day}}\right)\left(\frac{\text{mg}}{\text{m}^3}\right)\left(\frac{1}{\text{kg}}\right)$$

Now that we have a better appreciation of the challenges associated with risk analysis, consider the design and operation of a modern incinerator.

EXAMPLE 5.4 *The Controversy Surrounding MSW Incineration*

In this example we discuss the design and operation of incinerators, their control, air emissions studies, and society's reaction to the construction of modern incinerators.

DESIGN AND OPERATION

Typical incinerator systems (see pages 15 and 16 in Chapter 1) can be divided into three basic subsystems:

- Furnace
- Boiler
- Flue gas purification

Solid waste is stored in a bin and then transported by crane into the furnace, where the waste is burned in an atmosphere of excess air. The combustion process is rapid: the residence time of flue gases in the entire system is only about 10 seconds. The hot gases rise and are passed through a boiler where their thermal energy can be transferred to cooler water for the production of hot water or steam from which the thermal energy can be recovered. Heavier dust particles drop to a conveyor and are transported to an ash bin for storage and final disposal. Gases and lighter dust particles pass through a flue gas purification system for dust removal and gas treatment.

The major subsystems used for minimizing environmental impacts of incineration are

- Combustion control
- Flue gas purification
- Material separation prior to incineration

In complete combustion, food waste and other organic matter form carbon dioxide (CO_2), water, and heat. The presence of carbon monoxide (CO) in the flue gas is considered an indication of incomplete combustion, as is the presence of other unwanted hydrocarbons, such as methane (CH_4). Hydrogen and many hydrocarbons are much easier to oxidize than CO; thus, CO is considered a good overall indicator of burning efficiency. The temperature of flue gas is another control indicator.

CONTROL

Efficient combustion in an MSW incinerator is achieved by controlling furnace temperature, providing sufficient air turbulence, and regulating O_2 content. These factors have been found to reduce the CO level in the flue gas. Although there is evidence that furans and dioxins tend to lose their chemical integrity at temperatures above 700°C, they have been detected at 1000°C. Full-scale plant studies show that furan and dioxin concentrations as well as other hydrocarbons are correlated to CO concentrations. Thus, it is claimed that conditions that maximize combustion efficiency will limit furan and dioxin emission. Studies of these plants found that a temperature of about 800°C was most effective in reducing CO concentration.

Flue gas purification systems, which are comprised of flue gas temperature control, limestone injection, and electrostatic precipitators or textile fabric filters, can remove dust and reduce the concentrations of gas and metal emissions. Metals, with the exception of mercury, bind to dust particles. Electrostatic precipitators remove 95% of the dust. Between 80 and 90% of mercury removal is achieved by thermal control of the flue gas, limestone injection, and efficient dust control operation. Lime injection also reduces the acidity of the flue gases.

Separation of potentially dangerous materials, including products containing heavy metals, occurs prior to incineration. For example, to reduce mercury emissions, household batteries would be removed prior to incineration.

AIR EMISSION STUDIES

Scientific evidence indicates that emissions from MSW incineration are a minor source of pollution and pose a minimum human health risk. The combination of optimized combustion and advanced flue gas cleaning has shown that emissions of less than 10^{-7} mg/m^3 TCDD can be achieved. There is evidence to suggest that PAHs and other halogenated organic compounds are minimized under proper operating conditions. Furthermore, the major combustion sources of dioxins and furans are steel and copper plants, automobiles, and pulp and paper plants. Concentration levels are not significantly elevated around MSW plants. Background dioxin and furan contamination account for 99% of human exposure to dioxin and furans around MSW plants.

SOCIETY'S REACTION

In spite of government regulatory efforts and campaigns by owners and operators, incinerators are considered bad neighbors. Not surprisingly then, individuals and citizen groups are more and more taking legal action to ban incineration in their community. Claims and counterclaims add to the public concern. The permit application process is beset with legal controversy. Environmentalists contend that the provisions for use of the best available technology are insufficient in controlling dioxins, furans, other organics, and heavy metals, and that the provisions for monitoring are too lax. They also argue that incinerator ash is often toxic and should be subject to regulations regarding hazardous waste disposal.

Even in the absence of scientific evidence as to dangers to the public, extra precautions are built into our systems and precautionary operating procedures are used. There is a fear of the unknown; thus, society prefers to err on the side of safety. However, society should be aware of the monetary costs involved.

Summary

Systems analysis is a goal-oriented methodology that relies on mathematical models to improve knowledge. As illustrated in this chapter, these models can simulate reality, reduce complex relationships into a more understandable form, and be used to establish decision rules. Systems analysis, however, is more than an exercise in mathematical formulation and analysis. A systems analyst must thoroughly understand the system under analysis, appreciate the impact that an action may have on society, anticipate controversy, and finally have patience and tolerance to deal with an emotionally charged public.

QUESTIONS FOR DISCUSSION

5.2.1 Define the following terms:

ADI	risk analysis
LD	TD_{50}
MCL	VSD
NOEL	

5.2.2 (a) Now that you have completed reading Example 5.4, answer Problem 5.1.2 (b) and (c) again. Have your perceptions about the risk associated with incineration been reinforced or changed?

(b) Compare the fears you listed in Problem 5.1.2(b) and (c) for incineration with the evidence presented in Example 5.4.

5.2.3 Suppose the following bill has been introduced into your state legislature.

Solid Waste Bill

An ACT clarifying the definition of public benefit relative to permitting solid waste facilities.

This bill enhances the standard for substantial public benefit which a solid waste facility must demonstrate to the division of waste management of the State department of environmental protection when applying for a permit to operate a facility. The bill places the burden on the applicant to demonstrate a public benefit and requires a public hearing relative to the pending facility application.

Would you support or reject this bill and why? Cite facts from this chapter to support your stance.

5.2.4 Suppose a politically active environmentalist presented you with the Solid Waste Bill given in Problem 5.2.3 and the following letter. You are requested to sign the petition and send a personal letter to your legislator. The sample letter is given to you to make the task as easy as possible.

Sample Letter

Dear Legislator:
I write to urge your support of the Solid Waste Bill that helps clarify the definition of the public benefit in the siting of solid waste facilities in our State. If we are

sincere in our efforts to reduce, reuse, and recycle we should need fewer of the large landfills and incinerators that are starting to dot our countryside here in our State. The fact that we are an importer (exporter) of trash is truly disturbing to me. This bill will force us to acknowledge the hierarchy in place at the state level for preferred treatment methods of solid waste and save us many local battles against unwanted neighbors. Please support the Solid Waste Bill and let me know how you have decided to vote.

(a) To your best knowledge, is your state an importer or exporter of solid waste?

(b) Would you sign the sample letter and send it to your legislator?

(c) The sample letter indicates that MSW management issues are within the jurisdiction of the state. However, federal regulations and policy apply to public health and environmental protection. In your opinion, should national or statewide regulation policies prevail? Explain your point of view.

QUESTIONS FOR ANALYSIS

Instructions: The solutions to questions marked with an asterisk (*) are most easily obtained with the aid of computer graphics.

5.2.5 (a) Determine the risk probability for a child continuously exposed to a background concentration level of 2,3,7,8 TCDD. Assume a safety factor of 1.

(b) Determine the risk probability for adults where $\rho = 10^{-7}$ mg per m^3 TCDD, the emission level observed for a well-operated MSW plant.

(c) Take an advocacy or adversary position on the construction of an incinerator in a populated section of your state. Support your position by using the answers to (a) and (b), together with other technical and scientific evidence data on 2,3,7,8 TCDD. The main thrust of your argument will be to choose one of the following classifications for describing the risk from 2,3,7,8 TCDD and then explaining why 2,3,7,8 TCDD fits into that classification.

- De minimus risk. The amounts and concentrations of chemicals released into the environment do not pose significant risks. Chemicals in this classification do not warrant control.
- Relative risk. Chemicals posing significant risks should be given priority for control.
- Uncertain risk. Preference should be given to known risks over uncertain ones of the same magnitude.

5.2.6 Consider a chemical with a unit risk of $\beta = 2 \times 10^{-2}$ (mg/day-kg body weight). For simplicity, β is assumed to be the same for exposure by drinking or inhaling this theoretical chemical. Assume $sf = 1$.

(a) Determine the MCL for water ingestion of a standard adult and child intake assuming $\theta(d) = 10^{-6}$.

(b) Determine the MCL for air intake of a standard adult and child intake assuming $\theta(d) = 10^{-6}$.

5.2.7* Consider the following dose-response functions with parameter estimates for DDT with dose d in mg per kg of animal body weight.

Linear model: $\theta(d) = 0.0109d$

One-hit model: $\theta(d) = 1 - e^{-0.00375d}$

Multistage model: $\theta(d) = 1 - e^{-0.00163d - 0.0000102d^2}$

Weibull model: $\theta(d) = 1 - e^{-0.00523d^{1.28}}$

(a) Plot $\theta(d) - d$ and label LD_{50} for each model where $\theta(d)$ ranges from 0 to 1 for each model

(b) Plot $\theta(d) - d$ where $\theta(d)$ ranges from 0 to 10^{-4} for each model.

(c) Complete the following table.

Model	LD_{50}	VSD at $\theta(d) = 10^{-4}$	VSD at $\theta(d) = 10^{-6}$
Linear			
One-hit			
Multistage			
Weibull			

5.3 RISK-BENEFIT-COST ANALYSIS

Risks are social costs and by our definition are the expected damage costs borne by human beings, property, and the environment caused by a failure of a man-made system or by some catastrophic events. When we use the definition of annual risk for year t, $R(t) = h\theta(t)$, and the principles of benefit-cost analysis, it is a straightforward procedure to introduce annual risk, a monetary measure, into the decision-making process of alternative selection. Our focus in this section is on risk-benefit-cost analyses of man-made systems, with particular emphasis on exposure risk.

Risk-benefit-cost analysis is conducted using the same principles as previously presented for capital investment analysis. Analyzed as a net present worth problem, the measure of effectiveness is

$$npw = \sum_{t=0}^{n} \frac{nw(t)}{(1+i)^t}$$

where

$$nw(t) = w(t) - R(t) - k(t) \text{ for year } t$$

All terms in this expression have been defined. The benefit term, $w(t)$, is the annual profit for a private investment, or user benefit for public investment. The annual risk, $R(t) = h\theta(t)$, is a social cost, $k(t)$ are real (out-of-pocket) costs representing annual variable and amortized costs, respectively.

Probability theory, for example, will be used to estimate the failure probability $\theta(t)$ of systems with a built-in redundancy. If the primary system fails, then the backup system will be put into service. The net effect of a backup is that failure probability $\theta(t)$ and risk $R(t)$ are reduced. Minimizing the risk in this way is not without cost. The additional costs will be reflected in $k(t)$ and the capital cost $k(0)$. The following example illustrates the principle.

EXAMPLE 5.5 *Designing for a Catastrophic Failure*

Consider the design of a storage reservoir. One proposal is designed for one failure event in 50 years, and the other one for one failure event in 100 years.

	Annual Risk Probability $\theta(t)$	Capital Cost $k(0)$, $	Annual Net Benefit $nw(t)$, $	Social Cost h, $	Design Life n, Years
A	1 in 50	$10M	$1M	$10M	20
B	1 in 100	$13M	$1M	$10M	20

The key question to ask is whether these are feasible alternatives; if so, is B worth the extra cost of $3M to recommend the selection of it over A? Note that net benefits excluding social cost, $nw(t)$, are the same for both alternatives.

Use the risk-benefit-cost method to analyze and recommend a solution. Assume a discount rate of $i = 5\%$ per year.

SOLUTION

The same procedures that are used for the benefit-worth method of selection apply here. The cash flow diagram for A and B are uniform series. The governing equations for A and B are

$$uspwf = \frac{(1+i)^n - 1}{i(1+i)^n} = \frac{(1.05)^{20} - 1}{0.05(1.05)^{20}} = 12.46$$

$$npw = 12.46nw(t) - k(0)$$

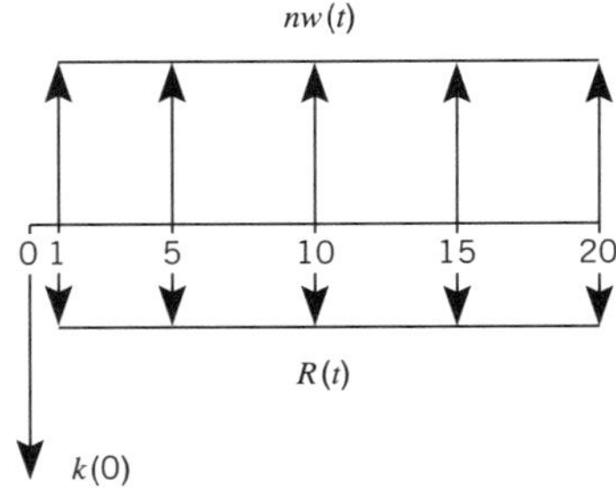

The events of the sample space $Z = \{S, S'\}$ are defined as

$$S = \{\text{flood}\}$$
$$S' = \{\text{no flood}\}$$

A.

$$R(t) = h\theta(t) = \frac{\$10\,\mathrm{M}}{50} = \$0.2\,\mathrm{M}$$

where

$$H = \begin{cases} h = \$10\mathrm{M} \text{ if } S \text{ and} \\ h = \$0\mathrm{M} \text{ if } S' \end{cases}$$

$$nw(t) = \$1\mathrm{M} - \$0.2\mathrm{M} = \$0.8\mathrm{M}$$

$$npw = 12.46(\$0.8\mathrm{M}) - \$10\mathrm{M} = -\$0.03\mathrm{M} < 0$$

B.

$$R(t) = h\theta(t) = \frac{\$13\,\mathrm{M}}{100} = \$0.13\,\mathrm{M}$$

where

$$H = \begin{cases} h = \$13\mathrm{M} \text{ if } S \\ h = \$0\mathrm{M} \text{ if } S' \end{cases}$$

$$nw(t) = \$1\mathrm{M} - \$0.13\mathrm{M} = \$0.87\mathrm{M}$$

$$npw = 12.46(\$0.87\mathrm{M}) - \$13\mathrm{M} = -\$2.16\mathrm{M} < 0.$$

Neither design is feasible; therefore, the do-nothing alternative is recommended.

DISCUSSION

If the discount rate i = 3% per year, then

A. $npw = \$1.19\mathrm{M} > 0$

B. $npw = -\$0.5\mathrm{M} < 0$

Design A is a feasible solution and is recommended if a discount rate of 3% per year can be justified.

Clearly, this example illustrates that the analytical procedure is straightforward. In a comparative sense, design A is a better choice than B. However, at $i = 5\%$ per year, it appears that the extra \$3M associated with B is not justifiable.

Risk-benefit-cost analyses are not without controversy. The question to build or do nothing is unclear. Here, by simply changing discount rate i from 5 to 3% per year, a feasible alternative is found. In the public forum, this change would be challenged by opponents of the project as well as other model parameter assignments, including $\theta(t)$.

QUESTIONS FOR ANALYSIS

5.3.1 Consider the following alternative:

	$k(0)$	$nw(t)$	n, years
A	\$2M	\$0.25M	20

Assume a discount rate of $i = 9\%$ per year.

(a) Determine the feasibility of A using the net present worth method.

(b) Suppose that there is an annual risk probability of 1 in 50 that flooding will result in a monetary loss equal to the capital investment $h = k(0)$. Determine the feasibility of A using a risk-benefit-cost analysis.

5.3.2 Two building sites are being considered. Site B is to be built in a flood plain. There is a 1 in 25 chance of flooding at site B. If flooding occurs, the damage costs are estimated to be $h = \$2M$.

	$k(0)$	$nw(t)$	t, years
A	\$1.2M	\$0.25M	6
B	\$0.6M	\$0.25M	6

Assume a discount rate of $i = 9\%$ per year.

Which site do you recommend? Perform a risk-benefit-cost analysis.

5.3.3* As equipment ages, it is more likely to fail. The estimated damage cost from equipment failure is h = \$1.0M. Assume that the annual failure probability $\theta(t)$ is a function of time t, or

t	1	2	3	4	5	6
$\theta(t)$	0.01	0.02	0.03	0.05	0.08	0.13

Consider the following alternatives:

	$k(0)$	$nw(t)$	n, years
A	\$1.2M	\$0.25M	6
B	\$0.8M	\$0.25M	3

For B, the equipment is replaced in year 3. Assume a discount rate of $i = 5\%$ per year. Recommend the better alternative using a risk-benefit-cost analysis.

5.3.4* A toxic chemical degrades over time. In other words, the annual dosage is a function of time,

$$d(t) = e^{-0.1t}$$

where

$$t = 0, 1, 2, \ldots, 20 \text{ years}$$

The annual risk probability is $\theta(d) = \frac{d}{1000}$ and the loss of life is equal to h = \$10M per person.

(a) Determine the annual risk R(t). Plot R(t)-t.

(b) Suppose that the chemical is stored in a tank that has annual probability of leaking of

$$q_T(t) = 1 - e^{-0.01t}$$

where

$$t = 0, 1, 2, \ldots, 20 \text{ years}$$

Plot $\theta_T(t)$–t to show that the probability of tank failure increases with time.

(c) Define the annual exposure risk to be $R_E(t) = R(t)\theta_T(t)$. Plot $R_E(t) - t$.

5.3.5 Risk assessment studies and risk-benefit-cost analyses are used for decision making. Obviously, chemicals that pose the greatest degree of risk should be subject to control. However, environmentalists and industrialists armed with the same information often interpret the information differently and draw different conclusions and decisions. These different points of view can lead to controversy.

Take a controversial issue that deals with a government policy, regulation, or some other issue dealing with a decision-making process involving a technological risk. Write a report that is comprised of the following elements:

- In the Introduction, state the issue and the controversy that surrounds it.
- Take a stance and support your point of view with facts and discussion. The discussion should contain properly referenced text, diagrams, tables, charts, and other material to support your stance.
- Conclusion.
- Bibliography containing referenced literature.

REFERENCES

Alborg, U. G., and K. Victorin (1987). "Impact on Health of Chlorinated Dioxins and Other Trace Organic Emissions," *Waste Management & Research*, **5,** 203–224.

Bruce, R. D. (1980). "Low Dose Extrapolation and Risk Assessment," *Chemical Times and Trends*, 19–23.

Clarke, M. J. (1988). *Minimizing Emissions from Resource Recovery*, Proceedings of International Workshop on Municipal Solid Waste Incineration. Montreal, Quebec.

Freeman H. M., ed. (1988). *Incinerating Hazardous Wastes*, Technomic Publishing Co., Lancaster, PA.

Glickman, T. S., and M. Gough, eds. (1990). *Readings in Risk*, Resources for the Future, Washington, DC.

Global Environment Monitoring System (1990). *Global Freshwater Quality, A First Assessment*, M. Meybeck, D. V. Chapman, and R. Helmer, eds. World Health Organization, Basil Blackwell, Oxford, England.

Goldsmith, E., and N. Hilyard, eds. (1990). *The Earth Report 2*, Mitchell Beazley International Limited, London.

Hasselriis, F. (1987). "Optimization of Combustion Conditions to Minimize Dioxin Emissions," *Waste Management & Research,* **5**, 311–326.

M. G. R. Cannell, and M. D. Hooper, eds. (1989). *The Greenhouse and Terrestrial Ecosystems of the UK.* National Environment Research Council, HMSO, London.

Lave L. B., and A. C. Upton, eds. (1987). *Toxic Chemicals, Health and the Environment*, Johns Hopkins University Press, Baltimore, MD.

Lind, N. C. ed. (1982). *Technological Risk.* University of Waterloo Press, Waterloo, Ontario.

Moolenar, R. J. (1992). "Overhauling Carcinogen Classification," *Issues in Science and Technology,* **8** (4), 70–75.

Mukerjee, D., and D. H. Cleverly (1987). "Risk from Exposure to Polychlorinated dibenzo-*p*-dioxins and Dibenzofurans Emitted from Municipal Incinerators," *Waste Management & Research*, **5**, 269–283.

Nieuwhof, G.W.E. (1985). "Risk: A Probabilistic Concept," *Reliability Engineering,* **10**, 183–185.

———Organization for Economic Co-operation and Development (1988). *Transport and the Environment.* Paris.

———(1989a). *Economic Instruments for Environmental Protection*, Paris.

———(1989b). *Environmental Policy Benefits: Monetary Evaluation*, Paris.

Park, C. N., and R. D. Snee (1983). "Quantitative Risk Assessment: State of the Art for Carcinogenesis," *Fundamental and Applied Toxicology*, **3**, 320–330.

Raiffa, H., and R. Schlaiffer (1961). *Applied Statistical Decision Theory*, MIS Press, Cambridge, MA.

Rau, J. G., and D. C. Wooten, eds. (1980). *Environmental Impact Analysis Handbook*, McGraw-Hill, New York.

Ricci, P. (1985). *Principles of Health Risk Assessment.* Prentice-Hall, Englewood Cliffs, NJ.

Rowe, W. D. (1983). *Evaluation Methods for Environmental Standards*, CRC Press, Boca Raton, FL.

Slovic, P. (1987). "Perception of Risk," *Science,* **236**, 280–285.

Tafler, S. (1982). "Cost-Benefit Analysis Proves a Tough Task," *High Technology*, **2**(4), 76–77.

Tihansky, D. P., and H. V. Kirby (1974). "A Cost-Risk-Benefit Analysis of Toxic Substances," *Journal of Environmental Systems*, **4**, 117–134.

Travis, C. C., and H. A. Hattemer-Frey (1989). "A Perspective on Dioxin Emissions from Municipal Solid Waste Incinerators," *Risk Analysis,* **9**(1), 91–97.

Travis, C. C., S. A. Richter, E.A.C. Crouch, R. Wilson, and E.D. Klema (1987). "Cancer Risk Management," *Environmental Science and Technology,* **21**(5), 415–420.

U.S. Environmental Protection Agency (1985a). *Chemical, Physical, and Biological Properties of Compounds Present at Hazardous Waste Sites*, Clement Associates, Arlington, VA.

———(1985b). *Principles of Risk Assessment, A Nontechnical Review.* Prepared by Environ Corp., Washington, DC.

———(1986). *Superfund Public Health Evaluation Manual*, EPA/540/1–86/060.

———(1992a). "New Approaches to Environmental Protection," *EPA Journal,* **18**(2).

———(1992b). "Setting Environmental Priorities: The Debate About Risk." *EPA Journal,* **17**(2).

U.S. General Accounting Office (1987). *Health Risk Analysis, Technical Adequacy in Three Selected Cases*, GAO/PEMD-87-14, Washington, DC.

Van Ryzin, J. (1980). "Quantitative Risk Assessment," *Journal of Occupational Medicine*, **22**, 321–326.

Victorin, K., M. Stahlberg, and U. G. Alborg (1988). "Emission of Mutagenic Substances from Waste Incineration Plants," *Waste Management & Research,* **6**, 149–161.

Wellburn, A. (1988). *Air Pollution and Acid Rain*. Longman Scientific & Technical, Essex, England.

Wilson, R., and E. A. C. Crouch (1987). "Risk Assessment and Comparisons: An Introduction," *Science,* **236,** 267–270.

World Commission on Environment and Development (1987). *Our Common Future*, Oxford University Press, New York.

Chapter 6

System Performance

Chance favors only those who know how to court her.

Charles-Jean-Henri Nicolle (1932)

In this chapter probability and reliability theory are used to consider the performance of systems with nonrepairable and repairable systems with and without backup. This and the remaining chapters have a twofold purpose: (1) to be able to analyze systems subject to the uncertainties of technological risk and uncertainty analyzed with risk-benefit-cost analysis; (2) to address design, control, operation, and management questions involving technological risk and uncertainty.

6.1 FUNDAMENTAL PRINCIPLES OF PROBABILITY

The annual risk probability is simply defined to be $P(S) = \theta(t)$ where t = time, generally given in years. For other systems involving backup systems, for example, it will be necessary to define different states of the system, such as $S_1, S_2, \ldots$, and then define the system state S in terms of these definitions. The Venn diagram is useful for interpreting the relationships between events.

Suppose a system can result in three possible events shown in the Venn diagram of Figure 1. The entire sample space is represented by $Z = \{S_1, S_2, S_3\}$. Note that state S_3 occurs if and only if neither S_1 nor S_2 occurs.

Since S_1 and S_2 do not overlap and have no outcome in common, they are said to be mutually exclusive events. That is, if S_1 occurs, it is impossible for S_2 to occur at the same time, and vice versa. For example, a die with sample space

$$Z = \{1, 2, 3, 4, 5, 6\}$$

is often used in games to determine the next move. The outcomes from a single roll of a die are mutually exclusive events. On a single roll, the probabilities of $S_1 = \{1\}$ and $S_2 = \{2\}$ are

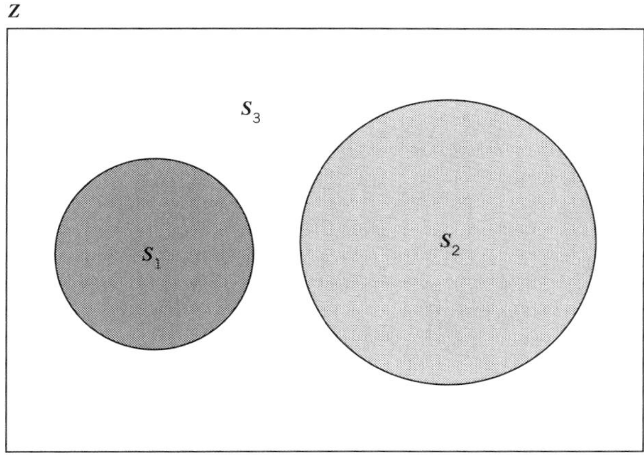

Figure 1 Venn diagram for two mutually exclusive events of S_1 and S_2.

the same, 1 chance in 6, or

$$P(S_1) = P(S_2) = 1/6$$

The *intersection* of the two mutually exclusive events, where $S_1 \cap S_2$ means that both events occur, forms the null set ϕ,

$$\phi = S_1 \cap S_2$$

Its probability is equal to zero, or

$$P(\phi) = P(S_1 \cap S_2) = 0$$

The probability of the union of two events, where $S_1 \cup S_2$ means that either event will occur, is the sum of the individual probabilities of S_1 and S_2,

$$P(S_1 \cup S_2) = P(S_1) + P(S_2)$$

Thus,

$$P(S_1 \cup S_2) = 1/3$$

Since $P(Z) = 1$, the following condition must hold:

$$P(Z) = P(S_1) + P(S_2) + P(S_3) = 1$$

or

$$P(S_3) = 1 - P(S_1) - P(S_2) = 2/3$$

where

$$S_3 = \{3, 4, 5, 6\}$$

The union of S_1 and S_2, $S_1 \cup S_2$, for two nonmutually exclusive events is shown in the Venn diagram of Figure 2. The probability of the union, where the outcomes can be S_1 exclusively, S_2 exclusively, or the intersection of S_1 and S_2, is

$$P(S_1 \cup S_2) = P(S_1) + P(S_2) - P(S_1 \cap S_2)$$

Adding the probabilities, $P(S_1) + P(S_2)$, results in double counting $P(S_1 \cap S_2)$. Thus $P(S_1 \cap S_2)$ is subtracted from the sum to obtain the correct formula for $P(S_1 \cup S_2)$. In order to be able to analyze $P(S_1 \cup S_2)$, it is necessary to know whether S_1 and S_2 are dependent or independent of each other.

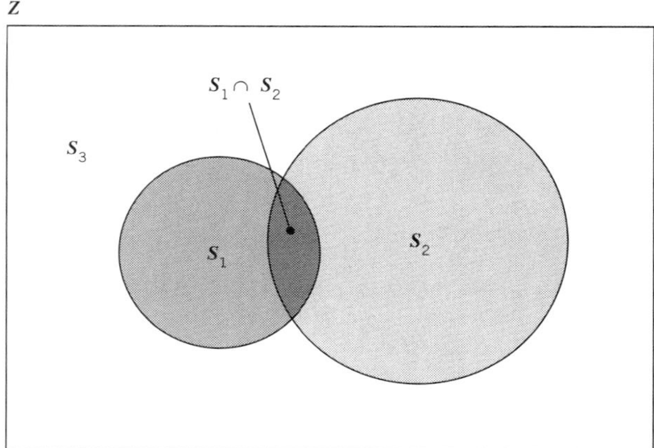

Figure 2 Venn diagram of two nonmutually exclusive events of S_1 and S_2.

Conditional, Independent, and Marginal Probability

Suppose that two experiments are run in succession. The outcome of the first experiment is represented by the event S_1 and the outcome of the second by S_2. The *conditional probability* of an event S_2 given that event S_1 has occurred, denoted as $S_2|S_1$, is defined to be the probability of the intersection of S_1 and S_2 divided by the probability of S_1,

$$P(S_2 | S_1) = \frac{P(S_1 \cap S_2)}{P(S_1)}$$

This definition requires that the marginal probability be nonzero, or $P(S_1) \neq 0$.

The *marginal probability* is defined as

$$P(S_1) = P(S_1 \cap S_2) + P(S_1 \cap S_2')$$

or as shown in the denominator of the conditional probability of $P(S_2|S_1)$, it can be expressed as

$$P(S_1) = P(S_1|S_2)P(S_2) + P(S_1|S_2')P(S_2')$$

In each equation, all states in $Z_2 = \{S_2, S_2'\}$ are evaluated.

If two events S_1 and S_2 are *independent*, then event S_2 will not depend on S_1 and its conditional probability is

$$P(S_2|S_1) = P(S_2)$$

From this definition and that of conditional probability it follows that the intersection of two independent events is

$$P(S_1 \cap S_2) = P(S_1)P(S_2)$$

If an assumption of independence can be justified, then it usually leads to modeling simplification. Consider the estimates of $P(S_1 \cap S_2)$ and $P(S_1 \cap S_{30})$ where the events are defined as

$$\begin{aligned} S_1 &= \{\text{rain on a first day}\} \\ S_2 &= \{\text{rain on the following day}\} \\ S_{30} &= \{\text{rain on the 30th day}\} \end{aligned}$$

Owing to the close proximity of events S_1 and S_2, it is reasonable to assume that observing rain on day 2 is somehow related to observing rain on day 1. For example, rains on successive days could be caused by a stationary weather front. Thus, the conditional probability $P(S_2|S_1)$ is assumed to exist. Suppose 100 observations of weather for two-day periods were investigated. One two-day period that occurred showed that it rained on the first day but not on the second day. A second two-day period showed rain lasting two days. Thus,

$$\begin{aligned} n_{S1} &= \text{number of days it rained on the first day} = 2 \\ n_{S1 \cap S2} &= \text{number of days it rained on the} \\ &\quad\ \text{first and following day} = 1 \end{aligned}$$

The probability estimates are

$$P(S_1) = \frac{n_{S1}}{n} = \frac{2}{100} = 0.02$$

and

$$P(S_2 \mid S_1) = \frac{n_{S1 \cap S2}}{n_{S1}} = \frac{1}{2} = 0.5$$

The probability forecast that rain will occur on two successive days is

$$P(S_1 \cap S_2) = P(S_2|S_1)P(S_1) = (0.5)(0.02) = 0.01$$

On the other hand, it seems most reasonable to assume that the events S_1 and S_{30} are independent events. The only rational explanation of rain on these days is chance. Thus,

$$P(S_{30}|S_1) = P(S_{30}) = P(S_1) = 0.02$$

is assumed. The probability of rain day 1 and day 30 is

$$P(S_1 \cap S_{30}) = P(S_1)P(S_{30}) = 0.02^2 = 0.0004$$

System Redundancy and Independent Events

Reliability $P(S')$ is defined as the probability that a system will perform its intended purpose. The complement of reliability is system failure,

$$P(S) = 1 - P(S')$$

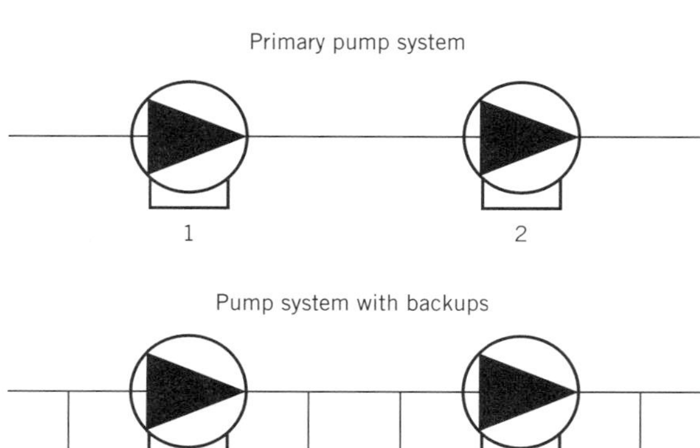

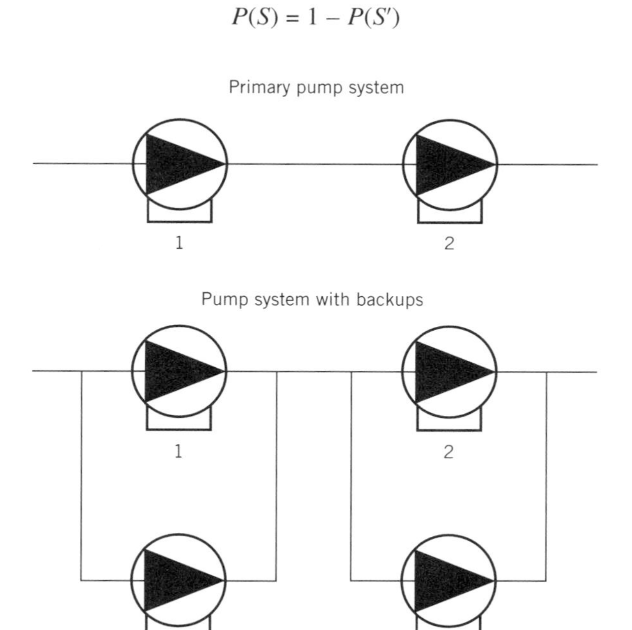

Figure 3 Series pumping systems with and without backups.

Figure 3 shows two pump systems, one with and one without backup pumps. Clearly, the reliability of the system with backup pumps is higher than that of the system without them.

However, the equipment cost, assuming no economy of scale for a system with backups, is twice that of a system without them. Let us compare the reliabilities of the two systems.

Assume that each pump has the same annual failure probability of 5 in 100 chances, or

$$P(S_1) = P(S_2) = P(S_3) = P(S_4) = \frac{5}{100} = 0.05$$

and their reliabilities are

$$P(S_1') = P(S_2') = P(S_3') = P(S_4') = 1 - 0.05 = 0.95$$

All pump failure events are assumed to be independent.

For the primary system without backup pumps, both pumps must operate for the entire year without failure. The system reliability is

$$P(S') = P(S_1' \cap S_2') = P(S_1')^2 = 0.95^2 = 0.9025 \approx 0.9 = 90\%$$

In order to simplify the analysis of the series system with backups, the system is decoupled into the pumping systems shown in Figure 4.

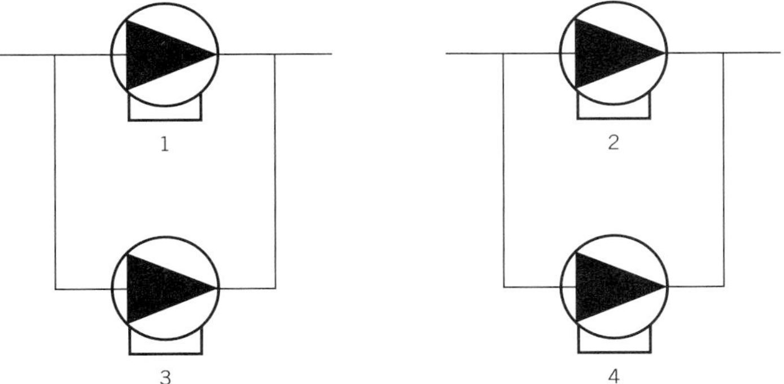

Figure 4 The decoupled series pump system with backups.

The reliability of the decoupled series pump system with backups is most easily determined by using the failure probabilities of each. They fail if both pumps in parallel fail.

$$P(S_1 \cap S_3) = P(S_2 \cap S_4) = P(S_1)P(S_3) = P(S_2)P(S_4) = 0.05^2 = 0.0025$$

Since the backup pump systems are in series, the reliability of the backup system is the product of the reliabilities of each of the backup pump systems.

$$P(S') = [1 - P(S_1 \cap S_3)][1 - P(S_2 \cap S_4)] = (1 - 0.0025)^2 = 0.995$$

In contrast, the system reliability without backup pumps is much less, only 90%. The selection of a pump system should not be based solely on the reliability probability. With a risk-benefit-cost analysis, it is possible to weigh the benefits of the system against the real costs of providing the system and the consequence of system performance.

EXAMPLE 6.1 *Human Error*

Heavy rains and flash flooding cause plant upset, resulting in the interruption of service and physical damage to the plant. Each episode costs h = \$1M and occurs about 1 in every 10 years.

The following \$0.5M pump system is proposed. If heavy rains and flash flooding are anticipated, the operator releases water from the storage tanks in advance of the anticipated storm. If this system is to avoid flood damage, it is imperative that the plant operator engage the pump in a timely fashion. Failure to do so can cause flooding and a loss of \$1M. The pump start-up reliability is 99%.

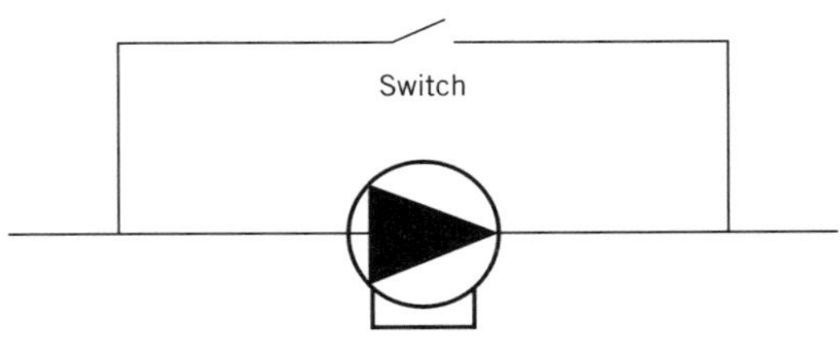

Owing to the uncertainty in weather prediction, suppose that the likelihood of plant operator error is estimated to be a 5 in 10 chance. This is a subjective assignment reflecting the difficulty the operator has in interpreting weather forecast information. This probability reflects the operator's reluctance to reduce the volume of water stored in the tanks because, if the operator is proven wrong, it reduces the volume of stored water needed for normal plant operation. The monetary losses from the lack of stored water are not considered in this analysis, only those associated with flooding.

Use a risk-benefit-cost analysis to determine the feasibility of installing this system. A discount rate of $i = 5\%$ per year is proposed. System design life is assumed to be $n = 20$ years.

SOLUTION

The benefit $nw(t)$ will be measured as an expected savings from flood avoidance,

$$nw(t) = R(t) - R'(t)$$

where

$R(t)$ = expected flood damage cost with no flood avoidance system

$R'(t)$ = expected flood damage cost with the flood avoidance system

with

$$H = \begin{cases} h = \$1\text{M if } S \text{ and} \\ h = \$0 \text{ if } S' \end{cases}$$

and

$$S = \{\text{system failure}\} \text{ and } S' = \{\text{no system failure}\}.$$

The cash flow diagram is

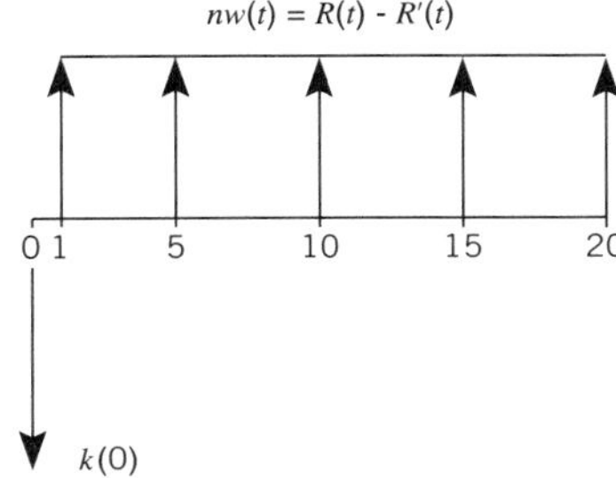

The governing equations for the risk-benefit-cost analysis are:

$$uspwf\,(i, n) = \frac{(1+i)^n - 1}{i(1+i)^n} = \frac{(1.05)^{20} - 1}{0.05(1.05)^{20}} = 12.46$$

$$npw = 12.46nw(t) - k(0)$$

The decision rule for recommendation is $npw \geq 0$, the feasibility test.

The conditions for S are defined in terms of the following events:

$$S_1 = \{\text{no or slow operator response causing plant flooding}\}$$
$$S_2 = \{\text{pump system failure causing plant flooding}\}$$
$$S_3 = \{\text{heavy rain and flash flooding}\}$$

The probabilities of these events are

$$P(S_1 | S_3) = 0.5$$
$$P(S_2 | S_3) = 1 - P(S'_2) = 1 - 0.99 = 0.01$$
$$P(S_3) = 0.1$$

where pump reliability is given as $P(S'_2) = 0.99$.

The failure state is the event that either the operator does not respond properly or the pump fails intersected with a flooding event, or

$$S = \{S_1 \cup S_2 \cap S_3\}$$

The probability of system failure is

$$P(S) = P(S_1|S_3 \cup S_2|\, S_3)P(S_3)$$

and the expected annual loss for the system is

$$R'(t) = \$1\text{M } P(S_1|S_3 \cup S_2|\, S_3)P(S_3)$$

Clearly, the determination of the conditional probability $P(S_1|S_2 \cup S_2|\, S_3)$ is the critical issue in determining $R'(t)$. Using fundamental definitions, it can be determined as

$$P(S_1|S_3 \cup S_2|\, S_3) = P(S_1|\, S_3) + P(S_2|\, S_3) - P(S_1|S_3 \cap S_2|\, S_3)$$

Assuming that human error and pump system failure are independent events, the conditional probability is calculated as

$$P(S_1|S_3 \cup S_2|\, S_3) = P(S_1|\, S_3) + P(S_2|S_3) - P(S_2|S_3)P(S_1|\, S_3)$$

The assumption of independence will be critically analyzed in Example 6.2.

Do not be misled by the notation for conditional events. The axioms and theorems of probability apply to conditional probabilities. For clarity, the conditional notation is dropped momentarily. The formula, using a reduced sample space, becomes

$$P(S_1 \cup S_2) = P(S_1) + P(S_2) - P(S_2)P(S_1)$$

where

$$P(S_1|\, S_3) = P(S_1) \quad \text{and} \quad P(S_2|\, S_3) = P(S_2)$$

It is understood that system failure occurs only when heavy rains and flash flooding occur. It might be helpful to refer to the Venn diagram of Figure 2 to get a visual interpretation of this result. The probability of system failure is

$$P(S_1 \cup S_2) = 0.5 + 0.01 - (0.5)(0.01) = 0.505$$

The risk as an expected damage cost with a flood damage avoidance system is

$$R'(t) = \$1\text{M } P(S_1 \cup S_2)\, P(S_3) = (\$1\text{M})(0.505)(0.1) = \$0.0505\text{M}$$

and the risk with no flood avoidance system is

$$R(t) = h\theta(t) = \$1\text{M}(0.1) = \$0.1\text{M per year}$$

where

$$\theta(t) = P(S_3) = 0.1$$

The expected annual saving is

$$nw(t) = \$0.1\text{M} - \$0.0505\text{M} = \$0.0495\text{M}$$

and the net present worth is

$$\begin{aligned} npw &= (12.46)(\$0.0495\text{M}) - \$0.5\text{M} \\ &= \$0.117\text{M} > 0 \qquad \text{(feasible)} \end{aligned}$$

Even with a relatively high probability of human error, $P(S_1) = 0.05$, the flood avoidance system is recommended.

EXAMPLE 6.2 *The Assumption of Independence for the Flood Avoidance System*

Given that weather conditions will trigger a plant flooding event S_3, the probability of pump system failure in Example 6.1, was given as

$$P(S_1 \cup S_2) = P(S_1) + P(S_2) - P(S_1 \cap S_2)$$

As discussed in the preceding example, the expression is written in a reduced form to emphasize the relationship between S_1 and S_2. It was also assumed that operator and pump system failure events were independent events: thus, $P(S_1 \cap S_2) = P(S_2)P(S_1)$. This assumption greatly simplified the analysis. In this problem, the assumption of independence of S_1 and S_2 is challenged. Consider the following scenario.

Plant flooding occurs when the capacity of the storage tanks is exceeded. Furthermore, assume that the pump system also becomes flooded. This can seriously jeopardize pump system start-up because the electrical switching system can become short-circuited. If the operator anticipates a flooding event, the pump will be engaged in the dry condition. If not, the plant, including the pump, will become flooded.

The working state of the pump system is dependent on the action of the operator; therefore, the probability of system failure $P(S)$ given above can be written as

$$P(S_1 \cup S_2) = P(S_1) + P(S_2) - P(S_2|S_1)P(S_1)$$

where

$$P(S_1 \cap S_2) = P(S_2|S_1)P(S_1)$$

Written in this form, the probability of system failure is dependent on the action of the operator.

Determine whether the assumption of independence can be justified for the solution given in Example 6.1.

SOLUTION

In order to emphasize the importance of the operator, consider marginal probability of S_2, or

$$P(S_2) = P(S_2|S_1)P(S_1) + P(S_2|S_1')P(S_1')$$

Here, the probabilities of $P(S_2|S_1)$ and $P(S_2|S_1')$ are interpreted to be the probabilities of engaging the pump in the wet (plant flooding) or dry (no plant flooding) condition, respectively.

When $P(S_2)$ is substituted into $P(S_1 \cup S_2)$, the relationship becomes

$$P(S_1 \cup S_2) = P(S_1) + P(S_2|S_1')P(S_1')$$

Since $P(S_1') = 1 - P(S_1)$,

$$P(S_1 \cup S_2) = P(S_1) + P(S_2|S_1')[1 - P(S_1)]$$

or

$$P(S_1 \cup S_2) = P(S_1) + P(S_2|S_1') - P(S_2|S_1')P(S_1)$$

The probability $P(S_2|S_1')$ is the probability of a pump start-up failure, given the pump is in a dry or an unflooded condition. Interestingly, if we assume that the pump reliability is based on tests run in a dry state, then the pump failure is

$$P(S_2|S_1') = P(S_2)$$

and the expression

$$P(S_1 \cup S_2) = P(S_1) + P(S_2) - P(S_2)P(S_1)$$

from Example 6.1 is justified.

QUESTION FOR DISCUSSION

6.1.1 Define the following terms:

conditional event	marginal probability
conditional probability	mutually exclusive event
independent event	

Use Venn diagrams to define these terms.

QUESTIONS FOR ANALYSIS

6.1.2 Define the events

$$S_1 = \{\text{heavy rain}\}$$
$$S_2 = \{\text{high winds}\}$$

The following probabilities are estimates:

$$P(S_1) = 0.02$$
$$P(S_2) = 0.05$$
$$P(S_1 \cap S_2) = 0.018$$

(a) Draw a Venn diagram.

(b) Determine the probability of a high wind given a heavy rain.

(c) Based on the answer from (b), does it appear that S_1 and S_2 are independent events? Why?

(d) Determine the expected monetary loss R if either high winds or heavy rains, or both, cause \$2M in damages.

6.1.3 A pumping system, consisting of three pumps, requires two of three pumps to operate for successful operation of the system.

(a) Draw a Venn diagram where S_1, S_2, and S_3 represent the failure state of each pump.

(b) Let $S = \{\text{pump system failure}\}$. Define S in terms of S_1, S_2, and S_3.

(c) Assuming S_1, S_2, and S_3 are independent events and $P(S_1) = P(S_2) = P(S_3) = p$, derive a probability relationship for system failure, $P(S)$.

(d) If $p = 0.05$, determine $P(S)$.

6.1.4 The reliability of an MSW incinerator is heavily dependent on furnace temperature control. If the gas temperature exceeds a threshold value of 1000°C, molten slag fusion builds up on the refractory material. To control the temperature, the operator must supply the proper amount of air to the furnace. Failure to do so could result in unwanted deposits of molten slag in the system and operational disturbance. Molten slag also reacts chemically with refractory material. Ultimately, the accumulation of slag will cause poor performance, which is considered a system failure. System failure is assumed to be dependent on the following:

$$S_1 = \{\text{incorrect temperature gauge readings}\}$$
$$S_2 = \{\text{improper control of temperature by operator}\}$$

Other pertinent information includes:

- The probability that the temperature gauge readings is incorrect is a 1 in a 100 chance.
- The operator will control the temperature properly with a frequency of 99 in 100 when a correct temperature gauge reading is given.
- The operator will improperly control the temperature with certainty when an incorrect temperature gauge reading is given.

(a) Draw a Venn diagram.

(b) Determine the probability that the operator improperly controls temperature $P(S_2)$.

6.1.5 Consider Problem 6.1.4 again; this time there are two temperature gauges. Let

$$S_1 = \{\text{incorrect temperature reading from gauge 1}\}$$
$$S_2 = \{\text{incorrect temperature reading from gauge 2}\}$$
$$S_3 = \{\text{operator improperly controls temperature}\}$$

with

$$P(S_1) = P(S_2) = 0.01$$
$$P(S_3|S_1 \cap S_2) = 0.99$$
$$P(S_3|S_1 \cap S_2') = P(S_3|S_1' \cap S_2) = 0.5$$
$$P(S_3|S_1' \cap S_2') = 0.01$$

Assume S_1 and S_2 to be independent events.

(a) Determine the probability that the operator will obtain two incorrect temperature readings and will improperly control temperature.

(b) Determine the probability that the operator improperly controls temperature $P(S_3)$.

6.1.6 A pump system consists of an electrical switch, pump, and valve.

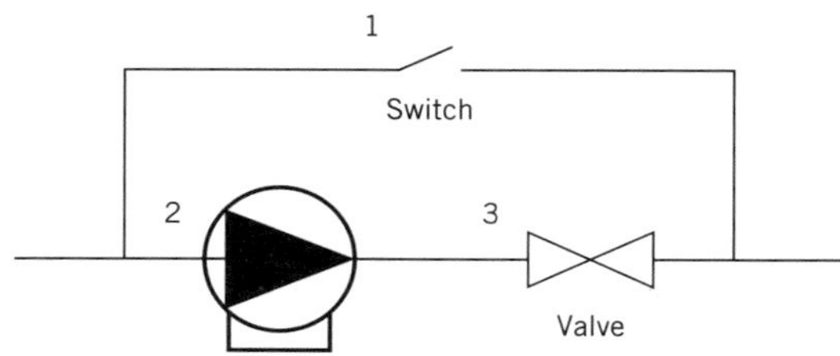

Let

$$S_1 = \{\text{electric switch fails}\}$$
$$S_2 = \{\text{pump fails to start}\}$$
$$S_3 = \{\text{valve fails to open}\}$$

(a) Determine the probability of system failure $P(S)$ with $P(S_1) = P(S_2) = P(S_3) = p = 0.01$ and S_1, S_2, and S_3 being independent events.

(b) Determine the probability of system failure $P(S)$ with $P(S_1) = p = 0.01$. In this formulation, assume that the system will depend on the performance of the switch. Assume

$$P(S_2 \cup S_3|S_1) = 1.0$$
$$P(S_2 \cup S_3|S_1') = 0.095$$

6.1.7 In the solution to Example 6.1, the monetary loss associated with an operator incorrectly removing water from the storage tanks was mentioned but not evaluated. Assume the actual monetary loss from an insufficient water supply is \$100K. In order to evaluate this loss, we need to investigate the weather forecast in more detail.

Use the data and information given in Example 6.1. It is necessary to introduce

$$S_4 = \{\text{flooding event is incorrectly forecast}\}$$
$$P(S_4) = 0.2$$

and replace $P(S_1)$ with $P(S_1|S_4') = P(S_1|S_4) = 0.5$.

(a) Show the probability of $P(S_1) = 0.5$, the probability of a flood given in Example 6.1.

(b) Determine the probability that the operator will pump out the storage tanks and the plant will have an insufficient water supply for normal plant operation.

(c) Estimate the annual monetary loss $R(t)$ due to insufficient water for normal plant operation.

6.2 SYSTEM RELIABILITY

Reliability theory is the study of product performance over time. With aging and use, a product deteriorates, and at some point, it will have to be repaired or replaced. This section discusses nonrepairable systems or systems that must be replaced after failure. Since an objective is to employ capital investment with the risk-benefit-cost analysis, part of the discussion focuses on determining $\theta(t)$, the risk probability function. Particular attention is given to the use of the uniform and exponential distributions for describing system reliability.

Basic Definitions

Since the exact time when the product will fail is unknown, the time to failure is treated as a random variable T. Define a sample space for a system as

$$Z = \{S, S'\}$$

where

$$S = \{\text{failure state}\} = \{0 \le T \le t\}$$

and

$$S' = \{\text{working state}\} = \{T > t\}$$

for given time t of interest. The system start-up time is $t = 0$.

The probability that a system survives until time t is

$$P(S') = P(T > t) = r(t)$$

where $r(t)$ is called the *reliability function*. Since $P(S) + P(S') = 1$, the probability of failure is

$$P(S) = 1 - r(t)$$

In order to better appreciate the concept, let us estimate $r(t)$ as a frequency. Suppose at time $t = 0$, $n(0)$ equal the number of identical systems in service at time 0 and $n(t)$ are the number of working units or survivors at time t. The reliability of the system is estimated to be

$$\hat{r}(t) = \frac{n(t)}{n(0)}$$

Cumulative Density Function

Since all values of $t \ge 0$ are of interest, T is treated as a continuous time variable. The laws of probability, given in Appendix A, apply to both discrete and continuous random variables.

However, there are important differences in calculating these probabilities. The probability that the random T lies between a and b is expressed as the integral

$$P(a \le T \le b) = \int_a^b f(x)\,dx$$

where $f(x)$ is called the probability density function of T.

The cumulative density function is defined as

$$F(t) = P(0 \le T \le t) = \int_0^t f(x)dx$$

For our purposes, the limit $t = 0$ is the system start time, and t is the time to system failure. Thus, $F(t)$ is a probability of system failure or *system failure probability*,

$$F(t) = P(0 \le T \le t)$$

Since $P(0 \le T \le t) + P(T > t) = F(t) + r(t) = 1$, the system reliability function can be expressed as

$$r(t) = 1 - F(t)$$

The Annual Failure Function

The annual failure probability $\theta(t)$ from the annual risk function, $R(t) = h\theta(t)$, is also an integral and equal to $P(S)$. Assuming that T is measured in years, the probability that a system will fail in year t has the state space $S = \{t - 1 \le T \le t\}$ and a probability,

$$\theta(t) = P(S) = \int_{t-1}^{t} f(x)dx$$

In terms of the failure function, it is

$$\theta(t) = F(t) - F(t-1)$$

In terms of the reliability function, it is

$$\theta(t) = r(t-1) - r(t)$$

The values of $\theta(t)$ are assumed to be constant values throughout year t.

The uniform and exponential distributions of $f(t)$ are commonly used to describe the characteristics of T, the time to failure.

The Uniform Distribution

The uniform distribution of T has a density function,

$$f(t) = \frac{1}{b-a} \quad \text{for } a \le t \le b$$

and

$$f(t) = 0 \quad \text{elsewhere}$$

It has a mean and variance of

$$\mu = \frac{a+b}{2}$$

$$\sigma^2 = \frac{(b-a)^2}{12}$$

Its failure probability function is

$$F(t) = \frac{1}{b-a}\int_a^t dx = \frac{t-a}{b-a} \quad \text{for } a \le t \le b$$

and its reliability function is

$$r(t) = \frac{b-t}{b-a}$$

EXAMPLE 6.3 ***Time to Fail***

It is estimated that a piece of equipment will fail with equally likely chance between three and six years.

(a) Determine $f(t)$ and plot $f(t)$–t.

(b) Derive the $F(t)$ and $r(t)$ from basic principles of probability and plot $F(t)$–t and $r(t)$–t.

(c) Determine $\theta(t)$ and plot the $\theta(t)$–t.

SOLUTION

(a) Let T be a random variable defined as

$$T = \{\text{time to failure}\}$$

with density function

$$f(t) = \frac{1}{6-3} = \frac{1}{3} \quad \text{for} \quad 3 \le t \le 6$$

The $f(t)$–t plot shows that there is an equal chance that the equipment will fail anywhere between three and six years.

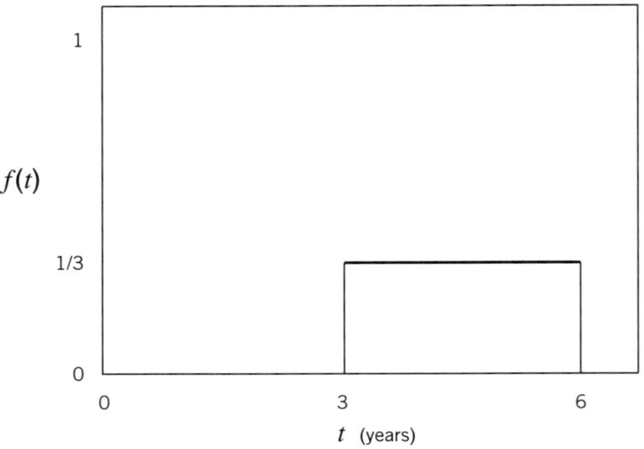

(b) The cumulative probability density function, failure probability, is

$$F(t) = P(T \le t) = \int_3^t \frac{1}{3}dx = \frac{t-3}{3} \qquad \text{for} \quad 3 \le t \le 6$$

and its reliability function is

$$r(t) = 1 - F(t) = \frac{6-t}{3} \qquad \text{for } 3 \le t \le 6$$

The $F(t)$–t and $r(t)$–t relations are:

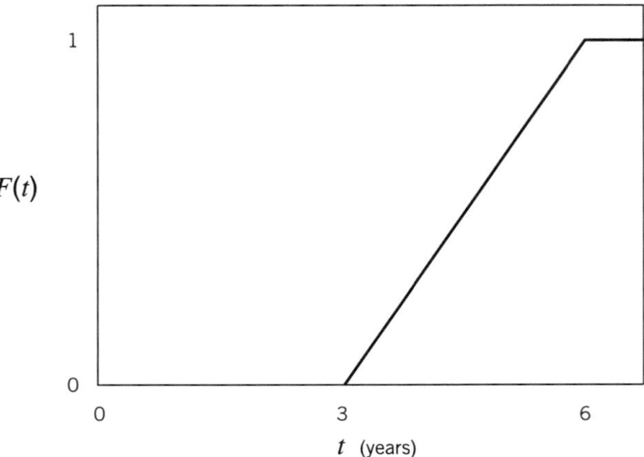

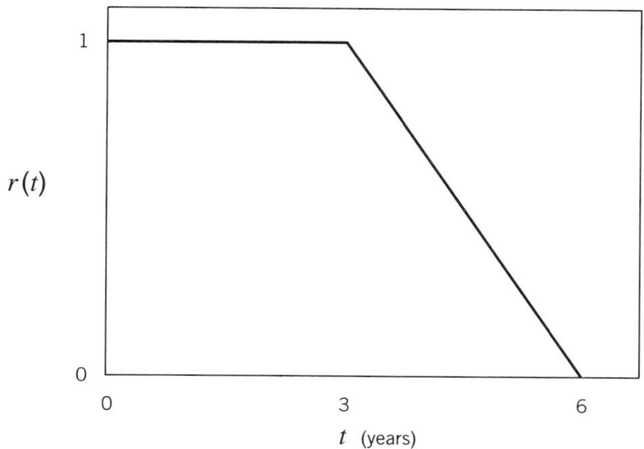

(c) The annual risk probability function is

$$\theta(t) = r(t-1) - r(t) = \frac{6-t-1}{3} - \frac{6-t}{3} = \frac{1}{3}$$

where

$$t = 4, 5, 6 \text{ years}$$

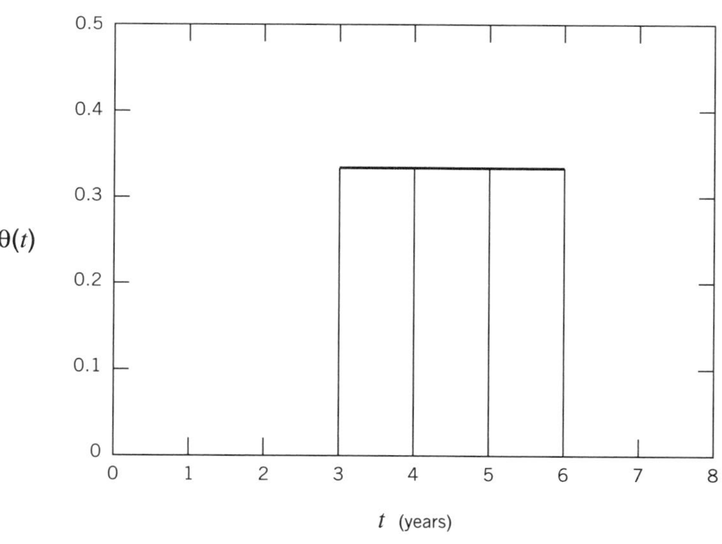

The Exponential Distribution

The exponential distribution for T has the density function,

$$f(t) = \rho e^{-\rho t} \quad \text{for } t \geq 0$$

where

ρ = failure rate in number of failures per unit time

Its cumulative density function is

$$F(t) = P(0 \le T \le t) = 1 - e^{-\rho t}$$

and its reliability function is

$$r(t) = 1 - F(T) = e^{-\rho t}$$

Its mean and variance are

$$\mu = \frac{1}{\rho}$$

$$\sigma^2 = \frac{1}{\rho^2}$$

The mean time to failure μ is equal to the reciprocal of the failure rate ρ.

EXAMPLE 6.4 *Design Life*

An incinerator using a roller grate system is currently in the design stage. No units of this design have been in service for their intended design lives of 20 years. However, experience with a unit of similar design revealed that three roller grates out of 35 units had to be replaced in 1.2 years.

(a) Determine $f(t)$ and plot $f(t)$–t, $F(t)$–t and $r(t)$–t.

(b) Determine $\theta(t)$ and plot $\theta(t)$–t.

SOLUTION

(a) The first task is to estimate the failure rate parameter ρ. Since $t = 1.2$ is given, it seems reasonable to estimate the probability of survival, where $T > 1.2$ years, as the frequency or ratio of the number of survivors at $t = 1.2$ years to the total number of units placed in service at time $t = 0$, or

$$\hat{r}(t) = \frac{n(t)}{n(0)} = r(t)$$

where

$n(t)$ = number of survivors at time t
$\hat{r}(t)$ = reliability estimate at time t
$n(0)$ = number of grates in service at time 0

$$\hat{r}(1.2) = \frac{n(1.2)}{n(0)} = \frac{35-3}{35} = \frac{32}{35}$$

If a constant rate of failure ρ is assumed, T has an exponential distribution with

$$\hat{r}(1.2) = e^{-1.2\rho} = \frac{32}{35}$$

After taking the natural logarithm of both sides and simplifying, the solution $\hat{\rho} = 0.075$ failure per year is calculated. The symbol ^ indicates that the parameter is an estimate.

The average failure rate is

$$\hat{\mu} = \frac{1}{\hat{\rho}} = \frac{1}{0.075} = 13.4 \text{ years}$$

The probability that a grate will fail in less than t years is

$$P(T \leq t) = F(t) = 1 - e^{-0.075t}$$

The frequency or density function is

$$f(t) = \frac{dF(t)}{dt} = 0.075\,e^{-0.075t} \quad \text{for t} \geq 0.$$

The density and cumulative frequency distributions for T, $f(t)$–t and $F(t)$–t are given in the following figures.

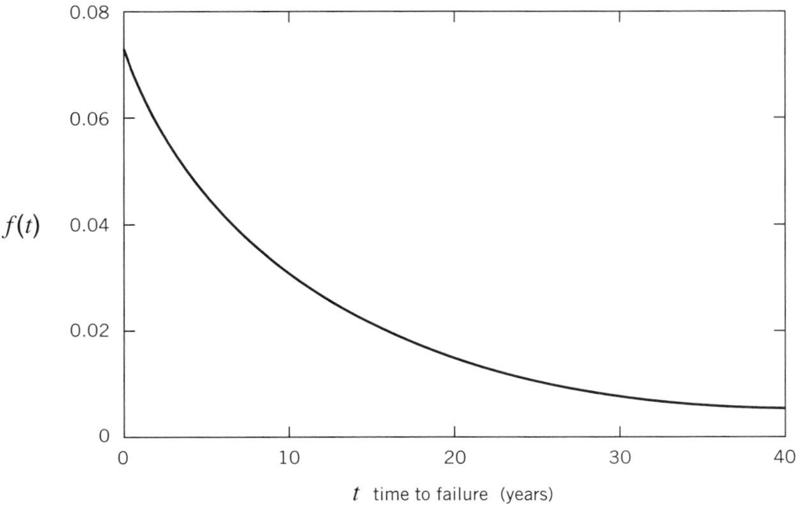

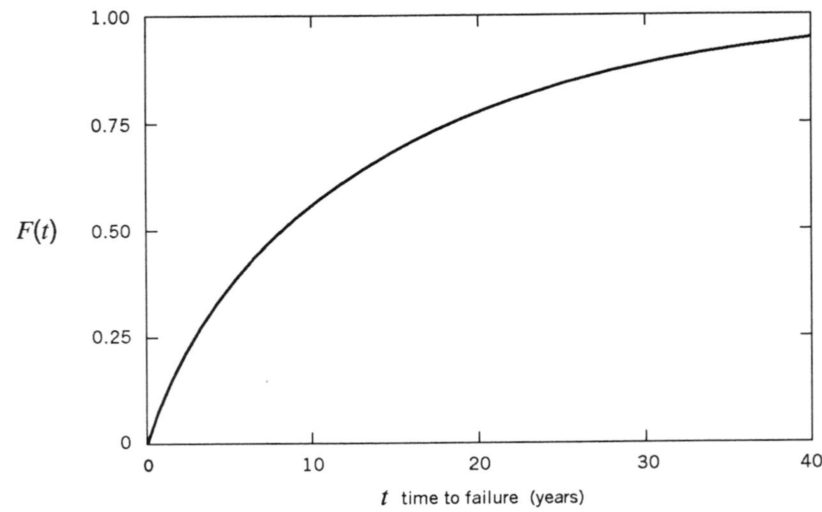

(b) The risk probability function is

$$\theta(t) = r(t-1) - r(t) = e^{-\rho(t-1)} - e^{-\rho t}$$

where

$$t = 1, 2, \ldots \text{ years}$$

The $\theta(t) - t$ plot shows that annual risk probability $\theta(t)$ decreases with time t.

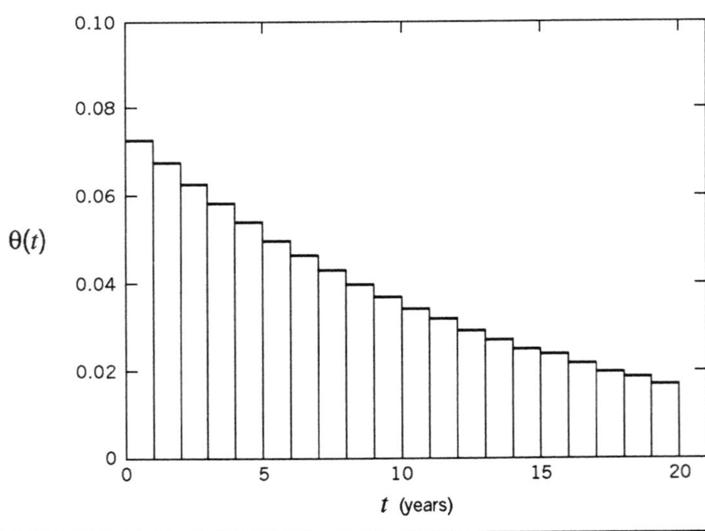

The Weibull Distribution

The *Weibull distribution* has application in reliability and risk assessment analyses.

$$F(t) = 1 - e^{-\rho t^{\gamma}}, \quad \text{for } t \geq 0$$

and

$$f(t) = \rho\gamma t^{(\gamma-1)}e^{-\rho t^{\gamma}}$$

A primary reason for using this distribution is its mathematical simplicity and adaptability. The model is similar to the exponential distribution, but it has an extra parameter γ, the shape factor. Figure 5 shows the effect that changing the shape factor γ has on the density function. When $\gamma = 1$, the distribution is an exponential distribution.

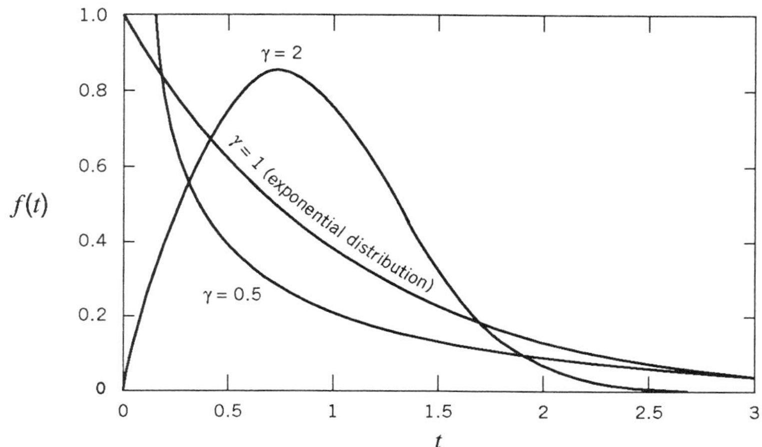

Figure 5 Weibull probability distributions where $\rho = 1$ and various γ values.

The Hazard Function

The *hazard function* $h(t)$ is a conditional probability density function. It is used to calculate the probability that a system will fail in the next time interval, given the system has remained in service for time t. Analytically, it is calculated to be the ratio,

$$h(t) = \frac{f(t)}{r(t)}$$

where

$$0 \leq t < \infty$$

Since the units of $h(t)$ are measured to be the number of failures per unit time, the hazard function is also a measure of the rate of failure of the system.

Assuming that the time to failure T is defined to be an exponential distribution, then $f(t) = \rho e^{-\rho t}$ and $r(t) = e^{-\rho t}$ are known. The hazard rate at time t is equal to

$$h(t) = \frac{\rho e^{-\rho t}}{e^{-\rho t}} = \rho$$

The resulting expression is independent of time t and is equal to a constant rate of failure parameter ρ. Thus, a system described by an exponential distribution of failure time T will fail at the same rate no matter how long a time it has been in service.

The probability that the system will fail in the time period from t to $t + \Delta t$ given the system has remained in service for time t is

$$P(t < T \leq t + \Delta t | T \geq t) = h(t)\Delta t$$

Assuming T has an exponential probability distribution, it is simply

$$P(t < T \leq t + \Delta t | T \geq t) = \rho\Delta t$$

Using the total probability theorem, we find that the probability that the system is in a working state is

$$P(T > t + \Delta t | T \geq t) = 1 - \rho\Delta t$$

These conditional probabilities for the exponential distribution have important application in analyzing repairable systems. This topic will be discussed further in Section 6.4 where system availability is investigated.

QUESTION FOR DISCUSSION

6.2.1 Define the following terms:

annual risk probability

cumulative density function

hazard function

probability density function

reliability

QUESTIONS FOR ANALYSIS

Instructions: The solutions to questions marked with an asterisk (*) and most easily obtained with the aid of computer graphics.

6.2.2 After five years, 50% of the systems are operating. Estimate the probability that a unit will be operating after 10 years. Let T be an exponentially distributed random variable.

6.2.3* The probability density function of T, the time to failure, increases with time t as a linear function,

$$f(t) = a + bt, \; t \geq 0$$

At time $t = 1$ year, 5% of the systems placed in service have failed. After 10 years, all systems have failed.

(a) Determine the model parameters a and b.

(b) Plot the $f(t)$–t, $F(t)$–t and $\theta(t)$–t.

6.2.4* Let T be a random variable defining the time to failure where T is distributed as a Weibull function. The failure function is

$$F(t) = 1 - e^{-\rho t^{\gamma}}$$

(a) Determine its density function $f(t)$.

(b) Determine its hazard function $h(t)$.

(c) Plot $f(t)$–t plots for $\rho = 1.0$ and $\gamma = 0.25$, 1, and 4.

(d) Plot $h(t)$–t plots for $\rho = 1.0$ and $\gamma = 0.25$, 1, and 4.

(e) Plot $F(t)$–t plots for $\rho = 1.0$ and $\gamma = 0.25$, 1, and 4.

6.2.5* Let T be a uniformly distributed random variable of time to failure in years, where $T = \{0 \leq t \leq 10 \text{ years}\}$. Assume the cost of failure declines as a linear function of T, $h = 1 - 0.1t$, where h has a unit of \$1M.

(a) Determine the expected loss, $R = E(H)$.

(b) Determine $R(t) = h(t)\theta(t)$ and plot the $R(t)$–t function. Let $h(t)$ be equal to the midpoint of the time interval. For example, for year $t = 2$, use $h(1.5)$.

6.2.6* Repeat Problem 6.2.5, where T is an exponential distribution and where the mean time to failure $\mu = 5$ years.

6.2.7* Bacterial waste is placed in secure landfill. Over time the concentration of bacteria, in units of mg per gm, decays as the exponential function of time t,

$$x = 100e^{-0.2t}$$

Let $T = \{\text{time to landfill failure in year 1}\}$ and T is an exponential distribution with mean time to failure $\mu = 15$ years.

(a) Determine the expected bacteria concentration $E(X)$ where the decay time is assumed to have the distribution T.

(b) Plot $R(t) - t$ for $0 \leq t \leq 30$ years, in steps of 1 year, where $R(t) = x(t)\theta(t)$. Here, the risk $R(t)$ is expressed as a bacteria concentration in lieu of a monetary value. Let $x(t)$ be equal to the midpoint of the time interval of t, or $x(t) = 100e^{-0.2(t-0.5)}$ with $\theta(t) = F(t) - F(t-1)$ where $t = 1, 2, \ldots, 30$.

6.3 DERIVED DISTRIBUTIONS

A model of the general form, $y = g(\mathbf{x})$, is called a deterministic model. The net worth model, $nw = px - k$, is an example of a deterministic model for a simple variable of x. If one or more model parameters are treated as random variables, we call the model a probability model and write it as

$$Y = g(\mathbf{X})$$

where $\mathbf{X}$ is a vector of random variables or $\mathbf{X} = [X_1, X_2, \ldots, X_n]'$. Since the response variable Y is a function of random variables, it too must be a random variable.

The problem is to derive the probability distribution of unknown random variable Y where the distributions of all random variables of $\mathbf{X}$ are assumed to be known. We will consider the cases where $Y = g(\mathbf{X})$ are single-value and multiple-value functions of X's. The profit models, $NW = pX - k$ and $NW = pX - K$, are examples of functions of the single and double random variable functions. The unit selling price, p, is a constant, and the number of items sold, X, is a random variable. In the first model, the fixed cost, k, is a constant, but in the second model, the fixed cost, K, is a random variable.

Functions of a Single Random Variable

Consider the case,

$$Y = g(X)$$

where $f_Y(y)$ is the unknown density function of Y and $f_x(x)$ is a known density function of X. $F_Y(y)$ and $F_X(x)$ represent the cumulative density functions of Y and X, respectively.

The cumulative density function of Y is most easily determined by a method of substitution. The procedure is first to find the cumulative density function of X or $F_X(x)$, and then to determine

$$X = g^{-1}(Y)$$

and substitute it into the cumulative distribution of $F_X(x)$

$$F_Y(y) = F_X[g^{-1}(y)]$$

The density function of Y is found by taking its derivative of $F_Y(y)$with respect to y,

$$f_Y(y) = \frac{d}{dy} F_X[g^{-1}(y)] = f_X[g^{-1}(y)] \frac{d}{dy} g^{-1}(y)$$

Although the overall procedure is straightforward, one must carefully establish the proper limits on y for Y. We evaluate the sample space of X and Y to find these limits. This is most easily shown by an example.

EXAMPLE 6.5 ***The Profitability of a Volunteer Recycling Plan***

A volunteer program shows considerable variation in the amount of paper that is recycled each day. The random variable X represents the supply of recycled paper received in tons per day where X has the distribution,

$$f_X(x) = 0.2e^{-0.2x} \quad \text{for } x \geq 0$$

The daily profit is

$$NW = pX - k$$

where

$$NW = \{\text{daily net worth or profit, \$ per day}\}$$
$$p = \text{selling price of recycled paper} = \$50 \text{ per ton}$$
$$k = \text{fixed daily operating cost for recycling paper} = \$200 \text{ per day}$$

Determine the probability that the recycling program will be profitable on any given day, $P(NW > 0)$ and that $P(X > x') = P(NW > 0)$ when x' = breakeven supply of paper.

SOLUTION

To answer the questions, we will find $F_{NW}(nw)$. First, treat NW and X as deterministic variables and determine the breakeven supply, x', for the net profit function,

$$nw = 50x' - 200 = 0$$

The following graph shows a profitable operation when $x \geq x' = 4$ tons of paper a day.

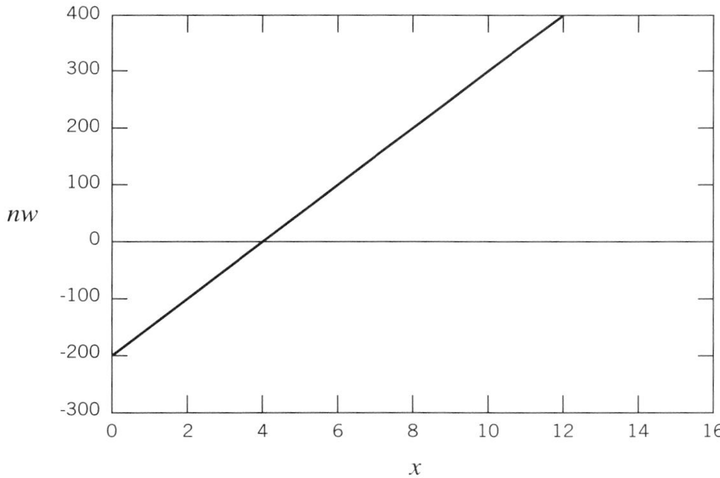

The probability of receiving more than x' each day is

$$P(X > x') = 1 - F_X(x') = e^{-(0.2)(4)}$$

$$= 0.45, \text{ or a 45 in 100 chance,}$$

where

$$F_X(x) = 1 - e^{-0.2x} \quad \text{for } x \geq 0$$

Establish the relationship for $x = g^{-1}(nw)$,

$$x = \frac{nw + 200}{50} = 0.02nw + 4$$

The sample space of NW and X is simply the $nw - x$ line. The sample space of X is $x \geq 0$; therefore, the sample space for NW is $nw \geq -200$. By substitution, the cumulative density distribution of NW is determined,

$$F_{NW}(nw) = F_X\left[g^{-1}(nw)\right] = F_X[0.02\,nw + 4] = 1 - e^{-0.2(0.02nw + 4)} = 1 - e^{-0.004nw + 0.8}$$

where

$$nw \geq -\$200 \text{ per day}$$

The probability of a daily profit is

$$P(NW > 0) = 1 - F_{NW}(0) = 0.45$$

Thus, $P(X > x') = P(NW > 0)$.

For completeness, the density function of NW is calculated.

$$f_{NW}(nw) = \frac{d}{dnw} F_X[g^{-1}(nw)] = \frac{d}{dnw}\left(1 - e^{-0.004\,nw - 0.8}\right)$$
$$= 0.004\,e^{-0.004\,nw - 0.8} \quad \text{for } nw \geq -\$200 \text{ per day}$$

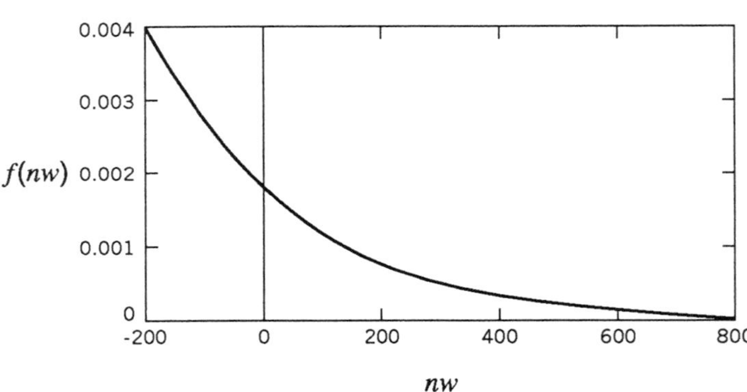

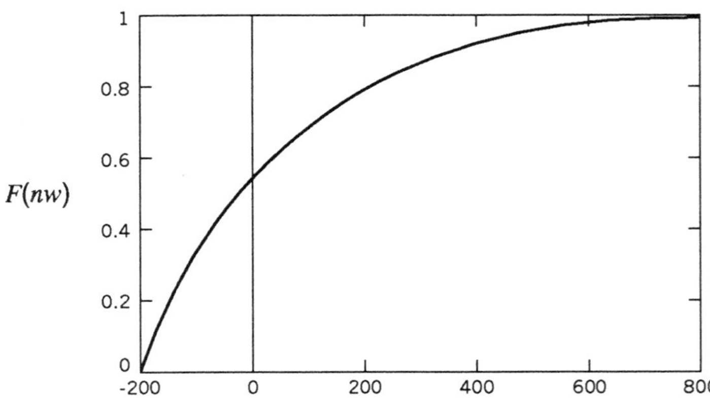

The density and cumulative density functions are shifted exponential distributions.

Functions of Two or More Random Variables

The same principles for deriving functions of a response for a function of a single variable is applied to functions of two or more random variables. Consider the two variable function,

$$Y = g(X_1, X_2)$$

where $f_Y(y)$ is the unknown distribution and both distributions of $f_{X_1}(x_1)$ and $f_{X_2}(x_2)$ are known.

First, the conditional density function $F_{Y|X_1}(y, x_1)$ will be calculated. For a given $X_1 = x_1$, Y thus becomes a function of a single random variable,

$$Y = g(x_1, X_2)$$

Following the steps outlined for a single variable function, we determine the conditional density function of Y given x_1, as

$$X_2 = g^{-1}(Y, x_1)$$

$$F_{Y|X_1}(y, x_1) = F_{X_2}[g^{-1}(y, x_1)]$$

$$f_{Y|X_1}(y, x_1) = \frac{d}{dy} F_{X_1}[g^{-1}(y, x_1)]$$

Second, the density function for the intersection of Y and X_1 is determined as the product,

$$f_{Y \cap X_1}(y, x_1) = f_{Y|X_1}(y, x_1) f_{X_1}(x_1)$$

The marginal function of Y is found by integrating over all values of x_1,

$$f_Y(y) = \int_{-\infty}^{\infty} f_{Y|X_1}(y, x_1) f_{X_1}(x_1)\, dx_1$$

The limits on Y are determined by considering the sample space of $Y = g(X_1, X_2)$. This is demonstrated by example. The approach is general and can be extended to functions of more than two random variables of X.

EXAMPLE 6.6 ***Time to Failure of a Backup System***

The system consists of primary and backup systems. System failure occurs when both the primary and backup systems fail. The total time to failure is the sum,

$$T = T_1 + T_2$$

where T_1 and T_2 represent the failure times of the primary and backup systems. The failure times T_1 and T_2 are assumed to be exponentially distributed with mean time to failure of $\mu_1 = 3$ years and $\mu_2 = 1$ year, respectively.

If the design life of the system is three years, determine the probability that the system will not fail over this time.

SOLUTION

The failure rates of T_1 and T_2 are

$$\rho_1 = \frac{1}{\mu_1} = \frac{1}{3} \quad \text{and} \quad \rho_2 = \frac{1}{\mu_2} = 1$$

Thus,

$$f_{T_1}(t_1) = \frac{1}{3}e^{-t_1/3}$$

$$F_{T_1}(t_1) = 1 - e^{-t_1/3}$$

$$f_{T_2}(t_2) = e^{-t_2}$$

$$F_{T_2}(t_2) = 1 - e^{-t_2}$$

First, let us establish the sample space for $T = T_1 + T_2$. When $t_2 = 0$, the sample space is the line $t = t_1$ and when $t_2 > 0$, it is the shaded region.

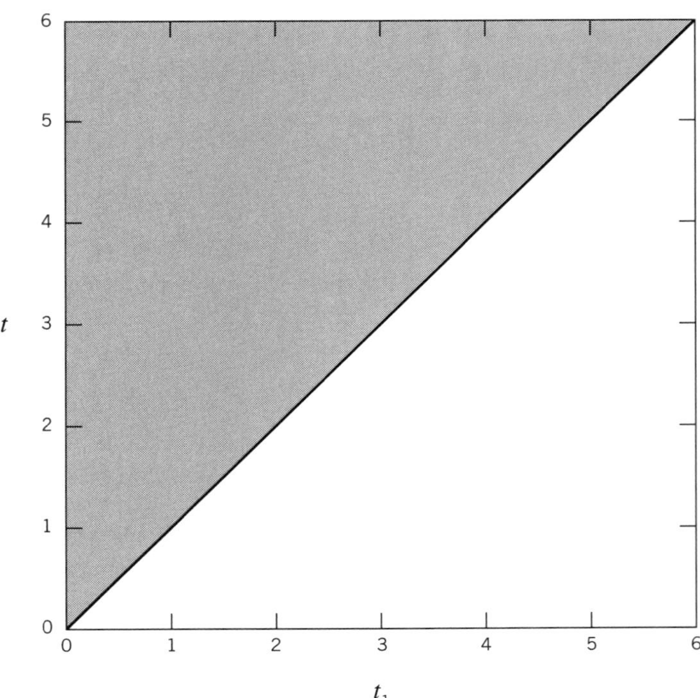

The conditional cumulative density function of T given $T_1 = t_1$ or $T_2 = T - t_1$ is

$$F_{T|T_1}(t, t_1) = F_{T_2}[g^{-1}(t_2)]$$
$$= F_{T_2}(t - t_1) = 1 - e^{-(t - t_1)}$$

and its density function is

$$f_{T|T_1}(t, t_1) = \frac{d}{dt} F_{T|T_1}(t, t_1)$$
$$= e^{-(t - t_1)}$$

for $t - t_1 \geq 0$.

The cumulative density function for the intersection of T and T_1 is

$$f_{T \cap T_1}(t, t_1) = f_{T|T_1}(t, t_1) f_{T_1}(t_1)$$
$$= e^{-(t - t_1)} \frac{1}{3} e^{-t_1/3} = \frac{1}{3} e^{-t}\, e^{2t_1/3}$$

for $t - t_1 \geq 0$.

The density function for T is the marginal distribution

$$f_T(t) = \int f_{T \cap T_1}(t, t_1)\, dt_1$$

$$= \frac{1}{3} e^{-t} \int_0^t e^{2t_1/3} dt_1 = \frac{1}{2}(e^{-t/3} - e^{-t})$$

for $t \geq 0$.

The cumulative density functions of T is

$$F_T(t) = \int_0^t f_T(x)dx$$

$$= \frac{1}{2} \int_0^t (e^{-x} - e^{-x})dt = 1 - 1.5\mathrm{e}^{-t/3} + 0.5\mathrm{e}^{-t}$$

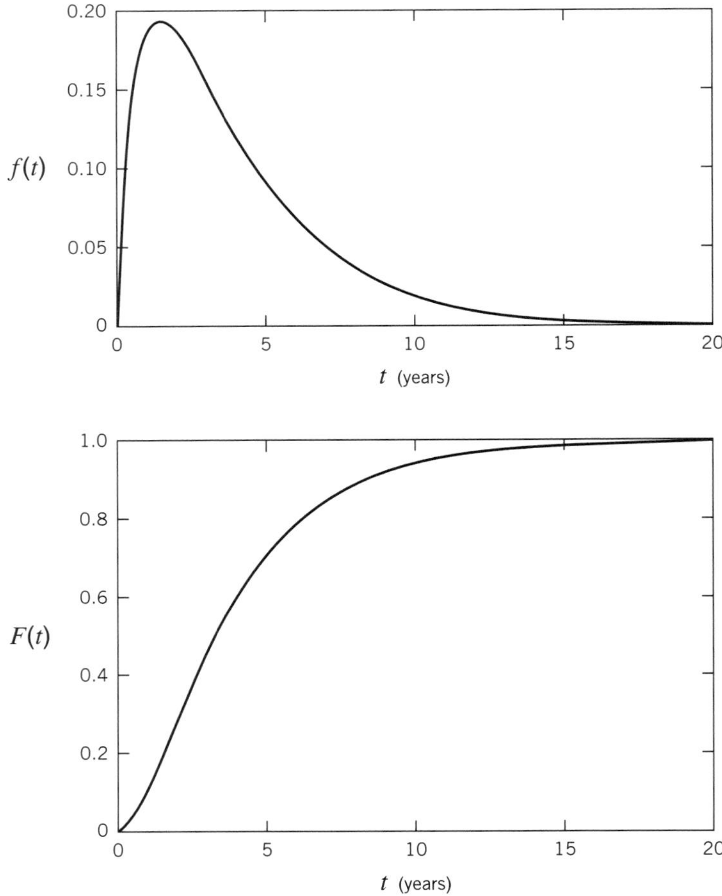

The probability that the system will not fail over its design life of three years is

$$P(T > 3) = 1 - F_T(3) = 0.53$$

In contrast, the probability that the primary system will survive three years is

$$P(T_1 > 3) = 1 - F_{T_1}(3) = 0.37$$

DISCUSSION

The cumulative density model of $T = T_1 + T_2$ will have application in Example 6.7. The model, written more generally, is

$$f_T(t) = \frac{\rho_1 \rho_2}{\rho_1 - \rho_2}\left(e^{-\rho_2 t} - e^{-\rho_1 t}\right) \quad \text{for } t \geq 0 \text{ and } \rho_1 \neq \rho_2$$

EXAMPLE 6.7 ***Evaluating a Secure Landfill Site***

A land area with property value of $k(0)$ = \$300,000 is being considered for a solid waste disposal site. In an effort to protect the public from drinking groundwater polluted with toxic chemicals, a secure landfill is required. The system consists of the following components:

- Building a landfill containment structure with a liner
- Providing a monitoring well system to detect contamination when the landfill containment structure fails

The next diagram shows a failed containment structure where the contamination has broken through the liner system, entered into the saturated soil layer, and formed a plume in the groundwater. The time for a contaminated plume to reach the monitoring wells is equal to the time at which the containment structure fails plus the time required for the plume to travel to the monitoring wells, or

$$T = T_1 + T_2$$

Plan View

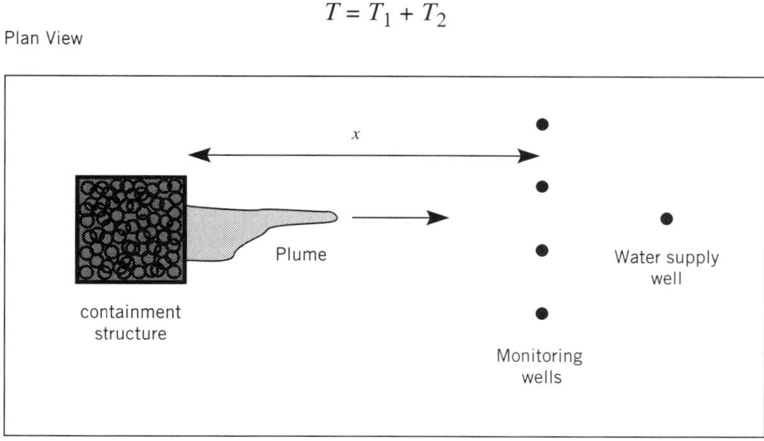

Cross-section view

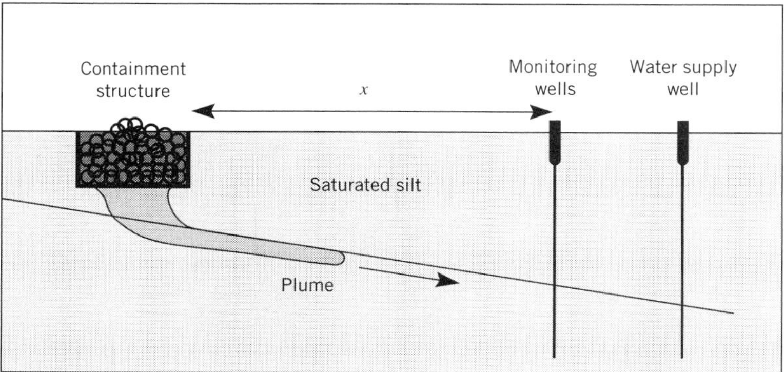

Source: Massman and Freeze (1987)

where T_1 and T_2 are assumed to be exponential distributions with mean times to failures of $\mu_1 = 15$ years, the average life of a protective liner, and $\mu_2 = 238$ years, the migration time for the plume to travel $x = 200$ meters in a silty soil, respectively.

The toxic chemicals are assumed to be slowly degradable, but for purposes of analysis, its toxicity is assumed to remain constant with time. The containment structure is assumed to be a nonrepairable system; therefore, it is only a matter of time before a toxic plume will form and will reach the monitoring wells. The ultimate success of the system in protecting the public will depend entirely on the monitoring well detection system. It is expected to detect a toxic plume with an 8 in 10 chance, or $P(S_3') = 0.8$.

In order to encourage careful operation and maintenance of the facility, a fine of \$5M will be imposed if groundwater contamination is discovered in the water supply well. The site restoration cost is estimated to be \$10M, bringing the total cost of a failure to h = \$15M per failure.

The site has enough land area to dispose of x = 10,000 tons of MSW per year and to be operational for a period of 25 years. The tipping fee and operation and maintenance costs are

$$\begin{aligned} p &= \text{tipping fee} = \$50 \text{ per ton of MSW} \\ c &= \text{unit disposal cost} = \$40 \text{ per ton of MSW} \\ c_1 &= \text{unit cost for liner} = \$12 \text{ per square meters} \\ a &= \text{liner area} = 1000 \text{ square meters per year} \end{aligned}$$

(a) Use a net present method to determine if the site is economically feasible for design life of $n = 25$ years. Ignore the environmental risks. All prices are assumed to keep pace with inflation and the discount rate i is assigned to be equal to the real interest rate of 4% per year.

(b) Use a risk-benefit-cost analysis to determine the feasibility of the site for (a).

SOLUTION

(a) The annual net revenue is

$$nw = (p - c)x - c_1 a = (50 - 40)10{,}000 - 12(1000) = \$88{,}000 \text{ per year}$$

The net present worth is

$$uspwf = \frac{(1+i)^n - 1}{i(1+i)^n} = \frac{(1.04)^{25} - 1}{0.04(1.04)^{25}} = 15.6$$

$$npw = nw\ uspwf(i, n) - k(0) = 88{,}000(15.6) - 300{,}000 = \$1.07\text{M} > 0$$

Since $npw > 0$, the site is recommended for a landfill.

(b) Since the environmental risk $R(t)$ is a function of time t, the net present worth is evaluated as

$$npw = \sum_{t=1}^{n} \frac{nw - R(t)}{(1+i)^t} - k(0)$$

where

$$nw = \$88{,}000$$
$$k(0) = \$300{,}000$$
$$R(t) = h\theta(t) = \$15\text{M}\theta(t) \text{ per year}$$

In order to calculate $\theta(t)$, the failure even S is defined in terms of the following failure states:

S_1 = {containment structure failure}
S_2 = {groundwater is contaminated with toxic material}

and

S_3 = {failure to detect toxic plume with the monitoring well detection system}

The failure state S is the intersection of the individual failure events

$$S = \{S_1 \cap S_2 \cap S_3\}$$

The probability of system failure $P(S)$, assuming that $S_1 \cap S_2$ and S_3 are independent events, is

$$P(S) = P(S_1 \cap S_2)P(S_3) = 0.2P(S_1 \cap S_2)$$

where

$$P(S_3) = 1 - P(S') = 1 - 0.8 = 0.2$$

The probability of $P(S_1 \cap S_2)$ depend on the failure times T_1 and T_2. The total time for the toxic plume to reach the monitoring wells is the sum

$$T = T_1 + T_2$$

where T_1 and T_2 are exponential distributions.

The distribution of T, given in Example 6.6, is

$$f_T(t) = \frac{\rho_1 \rho_2}{\rho_1 - \rho_2}(e^{-\rho_2 t} - e^{\rho_1 t})$$

$$= \frac{1}{223}(e^{-t/238} - e^{-t/15}) \quad \text{for } t \geq 0$$

and

$$F_T(t) = 1 + \frac{15e^{-t/15}}{223} - \frac{238e^{-t/238}}{223} \qquad \text{for } t \geq 0$$

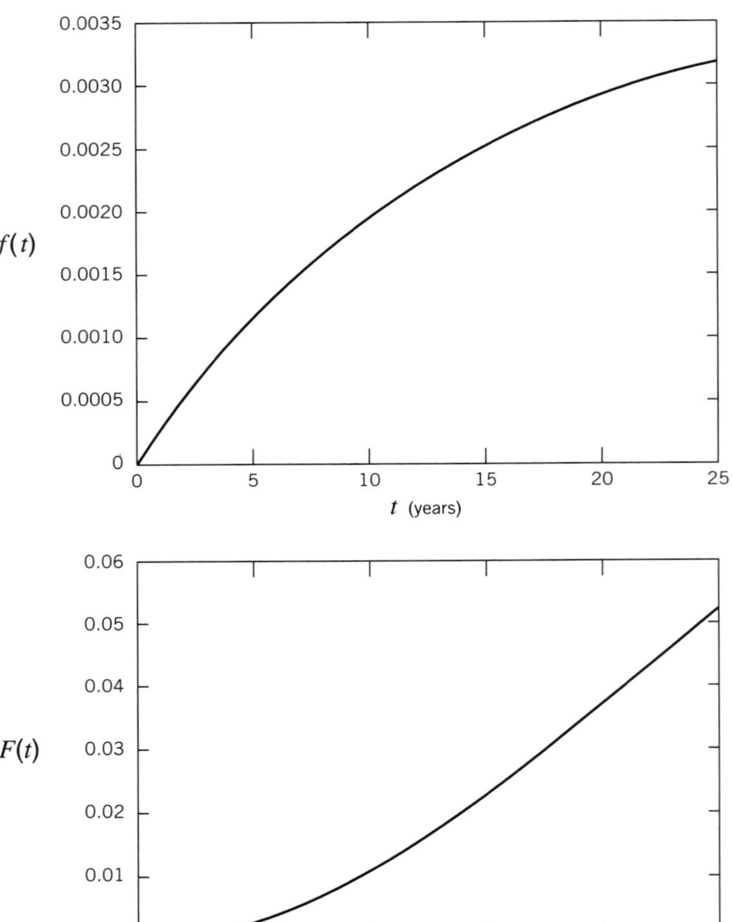

Summarizing these results gives

$$P(S) = 0.2P(S_1 \cap S_2) = 0.2P(T \leq t) = 0.2F_T(t)$$

The annual risk probability is

$$\theta(t) = 0.2[F_T(t) - F_T(t-1)]$$

The probability of system failure $\theta(t)$ increases with time.

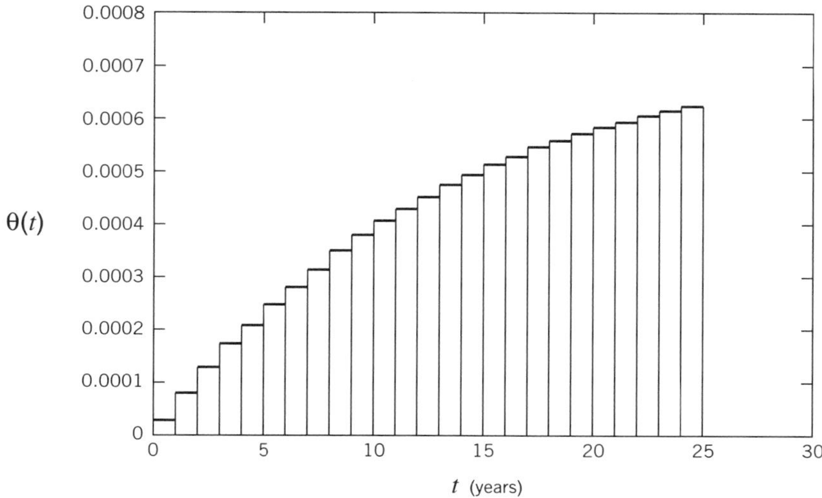

The net present worth is calculated to be

$$npw = \sum_{t=1}^{25} \frac{88,000 - 15\,\mathrm{M}\theta(t)}{1.04^t} - 300,000 = \$0.992\,\mathrm{M} > 0$$

The solution is an economically feasible solution; therefore, the site is recommended for construction.

DISCUSSION

This analysis considers only the economic feasibility of the project over its design life of 25 years. It can be argued that this site should not be recommended because long-term effects of groundwater pollution have not been evaluated. A feedback analysis that includes the degradability rates of the toxic chemicals being disposed of in the landfill seems warranted.

Secondly, it can be argued that the landfill design is poor and not recommended because the groundwater will be eventually contaminated. The timing of this event, whether or not it occurs within the design life of the project, is irrelevent. The landfill design shown in Example 1.2, a repairable system, may prove to be better because leakage caused by liner failure is detected and stopped before the groundwater is contaminated.

EXAMPLE 6.8 *Repair Costs*

The total cost to repair a system in \$1000 or \$1K units is

$$K = c_1T_1 + c_2T_2 = T_1 + 2T_2$$

where T_1 is the time in weeks to diagnose, order, and receive shipment of damaged parts, T_2 is the time in weeks to repair, test and return the system to service, c_1 is \$1K per week, and c_2 = \$2K per week to perform these respective services. The distributions of T_1 and T_2 are

$$f_{T_1}(t_1) = \frac{1}{2} \qquad 0 \le t_1 \le 2$$

and

$$f_{T_2}(t_2) = 1 \qquad 0 \le t_2 \le 1$$

(a) Determine the $f_K(k)$.

(b) Determine the probability that a system may be repaired for less than \$1K.

(c) Determine the mean and variance of K.

SOLUTION

(a) Establish the sample space for $K = T_1 + 2T_2$. The boundaries of the sample space in terms of k and t_1 are

$$\text{when } t_2 = 0, \quad k = t_1 \quad \text{and} \quad 0 \le t_1 \le 2 \quad \text{(A)}$$

$$\text{when } t_2 = 2, \quad k = t_1 + 2 \quad \text{and} \quad 0 \le t_1 \le 2 \quad \text{(B)}$$

The boundary lines A and B are shown in the figure. When t_2 is an intermediate value satisfying the inequality $0 < t_2 < 2$, the sample space is the shaded region. The sample space expressed as inequalities of t_1 and k are

$$0 \le t_1 \le 2 \text{ and } t_1 \le k \le t_1 + 2$$

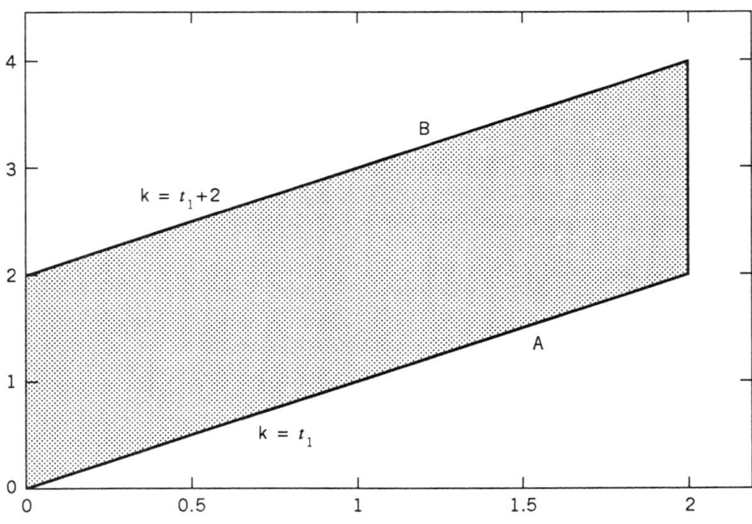

The conditional cumulative density function of K given $T_1 = t_1$ is established by solving for

$$T_2 = \frac{K - t_1}{2}$$

and

$$F_{T_2}(t_2) = \int_0^{t_2} dx = t_2$$

Thus, its density function is obtained by taking the derivative of the cumulative density with respect to t

$$F_{K|T_1}(t, t_1) = F_{T_2}\left(\frac{k - t_1}{2}\right) = \frac{k - t_1}{2}$$

$$f_{K|T_1}(t, t_1) = \frac{d}{dk}\left(\frac{k - t_1}{2}\right) = \frac{1}{2}$$

The density function of K is determined as the intersection of K and T_1

$$f_{K \cap T_1}(t, t_1) = f_{K|T_1}(t, t_1) f_{T_1}(t_1) = \left(\frac{1}{2}\right)\left(\frac{1}{2}\right) = \frac{1}{4}$$

for $0 \le t_1 \le 2$, and $t_1 \le k \le t_1 + 2$

The marginal density of K is determined by integrating over values of t_1. Since the sample space is bounded by a piece-wise linear boundary, the range of t_1 will depend on the value of k.

$$f_K(t) = \begin{cases} 0 \le k \le 2 \rightarrow \dfrac{1}{4}\displaystyle\int_0^k dt_1 = \dfrac{k}{4} \\ 2 \le k \le 4 \rightarrow \dfrac{1}{4}\displaystyle\int_{k-2}^2 dt_1 = \dfrac{4 - k}{4} \end{cases}$$

The density distribution is a triangular distribution.

(b) The probability that the system will be repaired for \$1K or less

$$P(K \le 1) = \int_0^1 \frac{k}{4} dk = \frac{1}{8} = 0.125$$

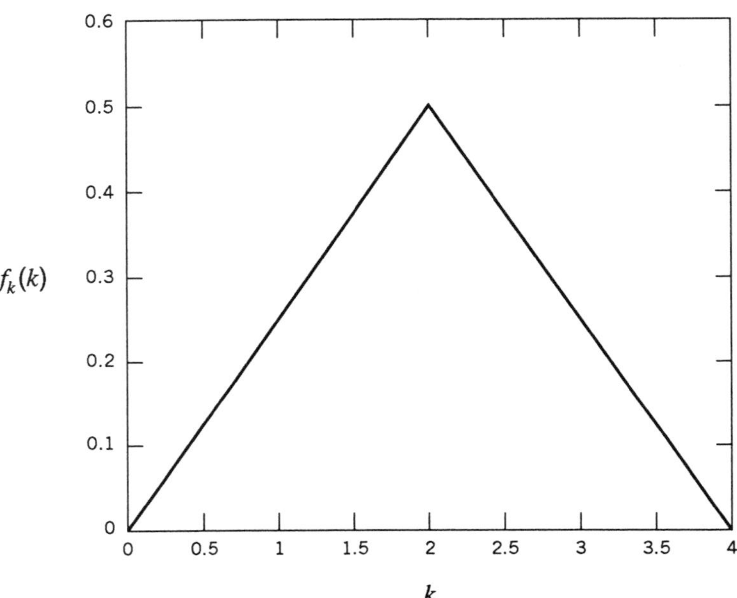

(c) The easiest approach to calculate the mean and variance of K is to use the properties of expectation.

$$\mu_K = E(K) = E(T_1 + 2T_2) = \mu_{T_1} + 2\mu_{T_2} = 1 + (2)(0.5) = \$1.5\text{K}$$

and

$$\sigma_K^2 = Var(K) = \sigma_{T_1}^2 + 2^2 \sigma_{T_2}^2 = \frac{1}{3} + 4(\frac{1}{12}) = 0.82\, K^2$$

or

$$\sigma_K = \$0.82\text{K}$$

where

$$\mu_{T_1} = \frac{a+b}{2} = \frac{0+2}{2} = 1 \text{ week}$$

$$\mu_{T_2} = \frac{0+1}{2} = 0.5 \text{ week}$$

$$\sigma^2_{T_1} = \frac{(b-a)^2}{12} = \frac{(2-0)^2}{12} = \frac{1}{3}, \quad \text{or} \quad \sigma_{T_1} = 0.58 \text{ week}$$

$$\sigma^2_{T_1} = \frac{(1-0)}{12} = \frac{1}{12}, \quad \text{or} \quad \sigma_{T_2} = 0.29 \text{ week}$$

QUESTIONS FOR ANALYSIS

6.3.1* A storage tank has a radius of 20 ft and a height of 40 ft. The distribution of wind speed V has exponential distribution with mean speed $\mu = 30$ mph. The wind force w, in kips, on a structure is a function of v, measured in tt/sec, where

$$w = \frac{1}{2} c_d \, a \rho v^2$$

$c_d = 2.0$ = drag coefficient for a square cylinder (unitless)
a = the projected area of the tank in ft^2
ρ = air density = 0.002 lb sec^2/ft^4

(a) Determine the cumulative density function for wind speed $F_V(v)$ where v is measured in mph.

(b) Derive the relationship between wind speed w, in kips, and v, in mph.

(c) Determine the cumulative density function for wind force $F_W(w)$.

(d) Plot the functions of $F_V(v)$ and $w(v)$ for the range $0 \le v \le 200$ mph.

(e) Plot $F_W(w)$ for the range $0 \le w \le 250$ kips.

6.3.2* (a) The time to failure for a system with a backup is

$$T = X + Y$$

where X and Y have exponential distributions with equal failure rates. Here, the primary and backup systems are of the same design. Derive the density function, $f_T(t)$.

(b) Derive the cumulative density function, $F_T(t)$. (Hint: Use integration by parts.)

(c) Compare the probabilities of failure for systems with and without a backup with design life of $n = 25$ years, $P(T < 25)$. Let the mean time to failure be $\mu = 15$ years.

6.3.3* In Example 6.7, the site under consideration has a soil type of silt with mean migration time of $\mu_2 = 238$ years. If a liner with mean failure time $\mu_1 = 15$ years is used, the probability that a plume will be released and reach the monitoring wells in 25 years or less is about 1%.

Suppose a site with sand is considered. Compute the probability that a plume will be released and will reach the monitoring wells in 25 years or less. Let $\mu_1 = 15$ and $\mu_2 = 32$ years.

6.3.4 The total time to repair and place a system back into service is

$$T = T_1 + T_2$$

where

T_1 = {time to diagnose damaged parts, order, and receive shipment of new parts}

T_2 ={time to repair, test, and replace the system back into service}

The distributions of T_1 and T_2 are

$$f_{T_1}(t) = \frac{1}{2}, \quad 2 \le t \le 4 \text{ weeks}$$

$$f_{T_2}(t) = \frac{1}{3}, \quad 0 \le t \le 3 \text{ weeks}$$

(a) Derive $f_T(t)$.

(b) Determine the mean and variance of T.

(c) Determine that the probability of repair will be less than 4 weeks.

6.3.5* Compare the following systems by plotting their annual risk probabilities $\theta(t)$ as a function time where $0 \le t \le 15$ years for

A. T = {time to failure for a primary system with no backup}

B. T = {time to failure for a system with backup}

The average times to failure for the primary backup systems are equal, $\mu = 5$ years. (Hint: See problem 6.3.2(a) and (b)).

6.4 SYSTEM AVAILABILITY

The term *availability* describes repairable systems; *reliability*, on the other hand, generally refers to systems that are not repairable. Availability $a(t)$, like reliability $r(t)$, is the probability that a system is in a working state at time t. While reliability and availability have this common feature, they differ in an analytical and a physical sense. For example, if a repairable system is working at time t, then it is available. In the period of time t, the system could have failed and been repaired one or more times. In contrast, a nonrepairable system must remain in a working state for the entire time period t to be available at time t.

Basic Assumptions

Define the sample space as $Z = \{S, S'\}$, where

$$S = \{\text{unavailable or in a system failure state at time } t\} = \{Y = 0\}$$

and

$$S' = \{\text{available or in a system working state at time } t\} = \{Y \ge 1\}$$

The number of units in the system operating at time t will be treated as a random variable Y. The availability function $a(t)$ is the probability that one or more units are working at time t, or

$$P(S') = P(Y \ge 1) = a(t)$$

Since $P(S) + P(S') = 1$,

$$P(S) = P(Y = 0) = 1 - a(t)$$

For a single-unit system, $Y = 1$, when the system is available.

Like reliability, the availability of a system can be estimated as a frequency, or

$$\hat{a}(t) = \frac{n(t)}{n(0)}$$

where $n(0)$ equals the number of identical systems in service at time 0 and $n(t)$ is the number of working units at time t. In the time period from 0 to t, some units could have failed, been repaired, and returned to service. In fact, multiple failures and repairs of some units are possible.

The discussion is limited to systems with time-to-failure and time-to-repair distributions with exponential functions of T,

$$f(t) = \rho e^{-\rho t} \quad \text{for } t > 0$$
$$g(t) = \eta e^{-\eta t} \quad \text{for } t > 0$$

where

ρ = failure rate in number of failures per unit time
η = repair rate in number of repairs per unit time

After a system is repaired, it is assumed to be as good as new. Thus, it will have a failure rate of ρ once again. The times to failure and repair are assumed to be independent random variables. The mean times to failure and repair are $\mu = 1/\rho$ and $\lambda = 1/\eta$, respectively.

In order to derive the availability function $a(t)$, we will analyze the transition of a system from a working state to a failure state and vice versa. We will formulate a system model and employ differential calculus to solve it.

Define

$P_Y(t) = P[Y(t) = y]$ = the probability that y units are in a working state at time t.

If we limit the discussion to a single-unit system, then

$$Y = \{y = 0 \text{ if } S \text{ and } y = 1 \text{ if } S'\}$$

The transition states at time t are shown in Figure 6. The failure ρ and repair η rates are assumed to be the same regardless of the number of times the system fails and is repaired.

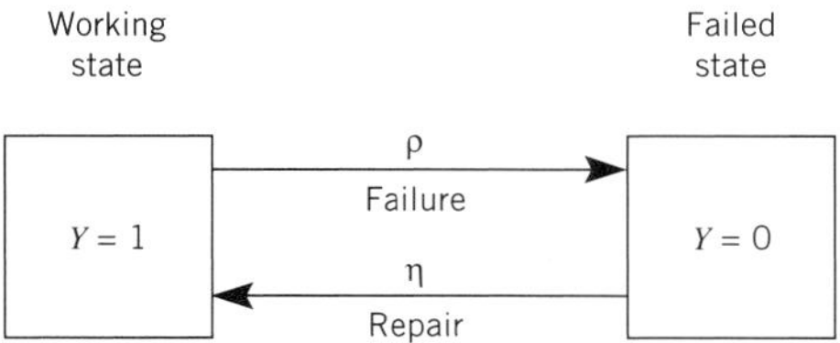

Figure 6 Transition states for a single repairable unit system.

Renewal Processes and the Poisson Process

Reliability and availability functions are derived by focusing on the transition from one state to another over a short time interval from t to $t + \Delta t$. Consider the probability of failure of a unit that fails at rate ρ. During this interval, we assume that

1. $P[Y(t) - Y(t + \Delta t) = 1] = \rho\Delta t$
2. $P[Y(t) - Y(t + \Delta t) > 1] = 0$
3. $P[Y(t) - Y(t + s) = n] = P[Y(t') - Y(t' + s) = n]$ where t and t' are two different start times and s is the length of a time interval.
4. $\text{Cov}[Y(t) - Y(t + s), Y(t') - Y(t' + s')] = 0$ where t and t' are two different start times and s and s' are the lengths of their time intervals, respectively. Additionally, the interval from t to $t + s$ does not overlap the interval from t' to $t' + s'$.

Assumptions 1 and 2 allow only one unit to fail in the time period between t and $t + \Delta t$. Assumption 3 states that the probability of n failures in any two equal-length time intervals is the same regardless of when the two periods begin. Assumption 4 states that any two nonoverlapping time intervals are independent. The failure events that happen in one time interval, t to $t + s$, are unaffected by the events in the time interval from t' to $t' + s'$. These same assumptions can be applied to repair of a unit where the rate of repair is η.

These assumptions are used to study various systems involving changes of state over time. In reliability and availability analysis, system unit failure and repair are studied. In transportation, the arrival and departure of vehicles are studied, generally for the purpose of analyzing and designing systems involving waiting lines. These systems, as well as other systems, share the common attribute of analyzing changes of state over time. Systems with this property are called *renewal processes*. Regardless of the application, the analysis of renewal processes uses these four assumptions, called the assumptions of the Poisson process, and employ the same strategy for model development. In this textbook, the application of these assumptions is limited to reliability and availability analyses.

The properties of the exponential distribution play an important role in the study of renewal processes. The probability that a single unit fails between times t and $t + \Delta t$ is stated as assumption 1 above. It is the probability that the unit is working at time t and will fail in the next time interval of Δt. This probability is calculated using the hazard function of $h(t) = \rho$ for the exponential distribution. It is

$$P[Y(t) - Y(t + \Delta t) = 1] = h(t)\Delta t = \rho\Delta t$$

Similarly, the probability of repair of a failed unit is $P[Y(t) - Y(t + \Delta t) = 1] = h(t)\Delta t = \eta\Delta t$, where the rate of repair is η.

We will use this result for unit failure to define the transition probabilities for the time interval between t and $t + \Delta t$. Assuming the unit is working at time t, the transition probabilities for the failure and working states at time $t + \Delta t$, written as condition probabilities, are

$$\text{Failure:} \quad P(t < T \leq t + \Delta t | T \geq t) = \rho\Delta t$$

and

$$\text{Working:} \quad P(T > t + \Delta t | T \geq t) = 1 - \rho\Delta t.$$

We can also apply this assumption to the transition probabilities for repair. Assuming the unit at time t is in a failed state, the transition probabilities for repaired and in-repair states at time $t + \Delta t$ are:

Repaired: $P(t < T \leq t + \Delta t \mid T \geq t) = \eta \Delta t$

In-repair: $P(T > t + \Delta t \mid T \geq t) = 1 - \eta \Delta t$

The transition probabilities for the failure, working, repaired, and in-repair states are assumed to be valid for a short Δt time interval only. As a consequence, during time interval Δt, no more than one failure or repair can occur. Two or more failures or repairs in time Δt are assumed impossible as given by assumption 2. Second, the failure and repair rates ρ and η are assumed to be the same or constant for any time t.

A Nonrepairable System

In order to demonstrate basic principles, consider a nonrepairable system first. We will show that $a(t) = r(t) = e^{-\rho t}$. Once basic concepts are presented, then more complicated cases of $a(t)$ are derived. For a nonrepairable system, it is impossible to have a transition from $Y = 0$ to $Y = 1$. In other words, in Figure 6 assume $\eta = 0$ and the repair of the system will not enter the derivation.

Now, imagine that at some arbitrary time t the system is working. In the Δt time interval from t to $t + \Delta t$, the unit may remain working or fail. The probabilities for the working and failure states at time t are defined to be

Working: $P_1(t) \rightarrow P_1(t + \Delta t)$

Failure: $P_1(t) \rightarrow P_0(t + \Delta t)$

Our discussion will focus on the probability that the system will remain working.

The probability of a working system at time $t + \Delta t$, $P_1(t + \Delta t)$, is the probability of the intersection of the system is working at time t and remains working for the time interval between t and $t + \Delta t$. Taking advantage of the transition probability for the working state, we obtain

$$P_1(t + \Delta t) = P_1(t)P(T > t + \Delta t \mid T \geq t) = P_1(t)(1 - \rho \Delta t)$$

Rearranging and taking the limit $\Delta t \rightarrow 0$, we derive a first-order differential equation,

$$\lim_{\Delta t \rightarrow 0} \frac{P_1(t + \Delta t) - P_1(t)}{\Delta t} = \frac{d}{dt} P_1(t) = -\rho P_1(t)$$

The solution for this equation with an initial condition of $P_1(0) = 1$ is

$$P_1(t) = e^{-\rho t} \qquad \text{for } t \geq 0$$

For a nonrepairable system, the availability and reliability functions are the same,

$$a(t) = r(t) = P_1(t)$$

A Repairable System

In deriving the availability function for a repairable system, we must consider the probabilities that the unit is in the working and failed states at times t and at $t + \Delta t$. The probability of being in a working state at time $t + \Delta t$, $P_1(t + \Delta t)$, is the sum of the following probabilities:

1. The system is working at time t and remains in a working state in the time period between t and $t + \Delta t$.
2. The system is not working at time t, and it is repaired in the time period between t and $t + \Delta t$.

Expressed in mathematical terms, it is

$$P_1(t + \Delta t) = P_1(t)(1 - \rho\Delta t) + P_0(t)(\eta\Delta t)$$

The probability that the system is in a failed state at time $t + \Delta t$, $P_0(t + \Delta t)$ is the sum of the probabilities:

1. The system is working at time t and fails in the time period between t and $t + \Delta t$
2. The system is not working at time t and is not repaired in the time period between t and $t + \Delta t$.

$$P_0(t + \Delta t) = P_1(t)(\rho\Delta t) + P_0(t)(1 - \eta\Delta t)$$

Rearranging these two equations and taking the limit $\Delta t \to 0$ for each one, we obtain the following system of linear equations.

$$\frac{d}{dt}P_1(t) = -\rho P_1(t) + \eta P_0(t)$$

$$\frac{d}{dt}P_0(t) = \rho P_1(t) - \eta P_0(t)$$

The initial conditions are $P_1(0) = 1$ and $P_0(0) = 0$. The solution is

$$P_1(t) = \frac{\eta}{\rho + \eta} + \frac{\rho}{\rho + \eta}e^{-(\rho+\eta)t}$$

$$P_0(t) = \frac{\rho}{\rho + \eta} - \frac{\rho}{\rho + \eta}e^{-(\rho+\eta)t}$$

The availability of the system is simply $a(t) = P_1(t)$.

Note that the total probability theorem is satisfied,

$$P_1(t) + P_0(t) = 1, \text{ or}$$

$$a(t) = 1 - P_0(t)$$

As $t \to \infty$, the probabilities reach a limiting state, or

$$P_1(\infty) = \frac{\eta}{\rho + \eta} = a(\infty)$$

$$P_0(\infty) = \frac{\rho}{\rho + \eta} = 1 - a(\infty)$$

EXAMPLE 6.9 ***Availability of Nonrepairable and Repairable Systems***

Consider repairable and nonrepairable systems with the same mean time-to-failure $\mu = 1$ year. The mean time-to-repair for the repairable system is $\lambda = 1$ month.

(a) In order to appreciate the advantage of a repairable system to a nonrepairable unit, plot $a(t) - t$ for each system.

(b) Perform a sensitivity analysis to determine the effect of time-to-repair on $a(t)$. Let λ = 0.5, 6, 12, and 24 months.

SOLUTION

(a) The availability function for the nonrepairable system is

$$a(t) = r(t) = e^{-\rho t}$$

with

$$\rho = \frac{1}{\mu} = 1 \text{ failure per year}$$

The availability function for the repairable system, using a time scale of years, is

$$a(t) = \frac{12}{1+12} + \frac{1}{1+12} e^{-(1+12)t} = 0.923 + 0.0769\, e^{-13t}$$

with

$$\rho = 1 \text{ and } \eta = \frac{1}{\lambda} = \frac{1}{1/12} = 12 \text{ repairs per year}$$

The $r(t)$–t and $a(t)$–t plots show that the repairable system gives significantly better performance.

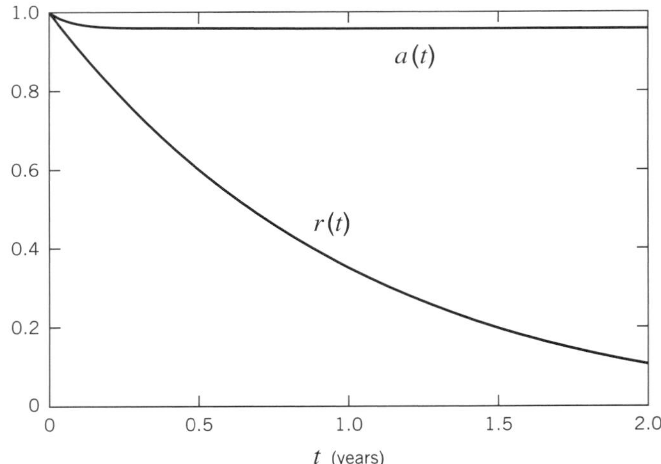

(b) The $a(t)$–t plot for the repairable system shows that the limit state probability $a(\infty)$ is a constant and a good summary statistic to characterize the annual risk probability, $\theta(t)$, for any year $t \geq 1$ year.

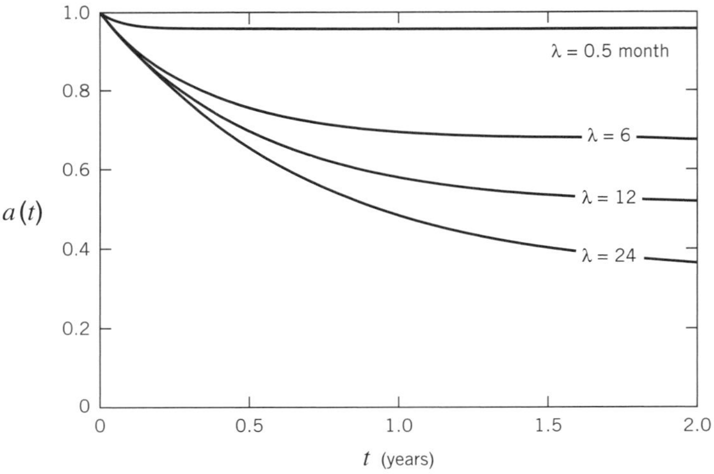

Availability of a Two-Unit Repairable System with a Single Repairman

Even when a system consists of a repairable component, its availability may not be sufficient to secure satisfactory service. The availability of a system, like reliability, can be improved by providing a backup system. The transition state diagram for a system consisting of two identical components and a single repairman is shown in Figure 7.

For the system with two units, define the total number of available units as

$$Y = \{y = 0 \text{ if } S \text{ and } y = 1, 2 \text{ if } S'\}$$

The transition probabilities for $Y = 2$, 1, and 0 will be determined in turn.

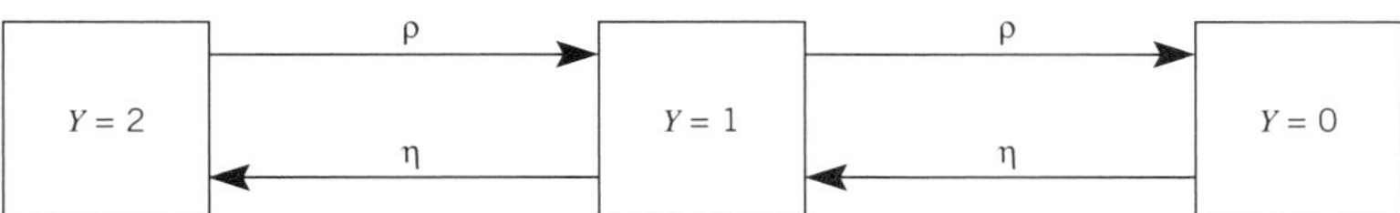

Figure 7 The transition state diagram for a repairable system with backup.

The probability of being in a working state with two working units at time $t + \Delta t$, $P_2(t + \Delta t)$ is the sum of the probabilities.

1. The system has two working units available at time t, and they remain in a working state in the time period between t and $t + \Delta t$.
2. The system has one available working unit at time t and is repaired in the time period between t and $t + \Delta t$.

$$P_2(t + \Delta t) = P_2(t)(1 - \rho\Delta t)^2 + P_1(t)(\eta\Delta t)$$

The expression, $(1 - \rho\Delta t)^2$, is the probability of the intersection that both units remain working in time interval Δt. Expanding the term, $(1 - \rho\Delta t)^2 = 1 - 2\rho\Delta t + \rho^2\Delta t^2$, and eliminating the Δt^2 term, gives

$$P_2(t + \Delta t) = P_2(t)(1 - 2\rho\Delta t) + P_1(t)(\eta\Delta t)$$

The probability of being in a working state with one working unit at time $t + \Delta t$, $P_1(t + \Delta t)$, is the sum of the probabilities

1. The system has two working units available at time t, and one fails in the time period between t and $t + \Delta t$.
2. The system has one available working unit at time t, the failed unit is not repaired, and the working unit does not fail in the time period between t and $t + \Delta t$.
3. The system has one available working unit at time t, the failed unit is repaired, and the working unit fails in the time period between t and $t + \Delta t$.
4. The system has no available working units at time t, and a failed unit is repaired in the time period between t and $t + \Delta t$.

$$P_1(t + \Delta t) = 2P_2(t)(\rho\Delta t)(1 - \rho\Delta t) + P_1(t)(1 - \rho\Delta t)(1 - \eta\Delta t) + P_1(t)(\rho\Delta t)(\eta\Delta t) + P_0(t)(\eta\Delta t)$$

The first term on the right of the equal sign contains a 2 because there are two units that can fail. The last three terms reflect the fact that there is only one repairman and a maximum of one unit can be repaired in the time interval Δt. Expanding and simplifying gives

$$P_1(t + \Delta t) = 2P_2(t)(\rho\Delta t) + P_1(t)(1 - \rho\Delta t - \eta\Delta t) + P_0(t)(\eta\Delta t)$$

The probability that the system is not working at time $t + \Delta t$, $P_0(t + \Delta t)$, is the probability that

1. One unit is working at time t and it fails between t and $t + \Delta t$.
2. The failed unit is not working at time t and is not repaired in the time period between t and $t + \Delta t$.

$$P_0(t + \Delta t) = P_1(t)(\rho\Delta t) + P_0(t)(1 - \eta\Delta t)$$

Rearranging these equations and taking the limit $\Delta t \to 0$ for each one, we obtain the following system of linear equations:

$$\frac{d}{dt}P_2(t) = -2\rho P_2(t) + \eta P_1(t)$$

$$\frac{d}{dt}P_1(t) = -2\rho P_2(t) - (\rho + \eta)P_1(t) + \eta P_0(t)$$

$$\frac{d}{dt}P_0(t) = +\rho P_1(t) - \eta P_0(t)$$

The initial condition is $P_2(0) = 1$ and $P_1(0) = P_0(0) = 0$.

The solution of this set of homogeneous, linear differential equations is considered beyond the scope of this book. Laplace transform and eigen-value, eigen-vector methods, are the most satisfactory methods for solving initial value problems of this kind. For risk-benefit-cost analysis applications, the limit state solution for this system is most useful.

$$P_2(\infty) = \frac{\eta^2}{2\rho^2 + 2\eta\rho + \eta^2}$$

$$P_1(\infty) = \frac{2\rho\eta}{2\rho^2 + 2\eta\rho + \eta^2}$$

$$P_0(\infty) = \frac{2\rho^2}{2\rho^2 + 2\eta\rho + \eta^2}$$

The derivations presented here for reliability analysis are also applicable to other processes, such as are encountered in transportation systems where arrival and service times are treated as random variables.

EXAMPLE 6.10 ***Availability of a Repairable System with a Backup***

In Example 6.9, where the mean times to failure and repair and $\mu = 1$ year and $\lambda = 1$ month, the limit state availability for a repairable system is $a(\infty) = 0.923$.

Determine the availability of this system if a backup unit is procured.

SOLUTION

The availability of the system is when there are one or two components in a working state, or

$$a(\infty) = P_1(\infty) + P_2(\infty)$$

or

$$a(\infty) = \frac{\eta^2 + 2\rho\eta}{2\rho^2 + 2\eta\rho + \eta^2} = \frac{12^2 + 2(1)(12)}{2\left(1^2\right) + 2(1)(12) + 12^2} = 0.988$$

EXAMPLE 6.11 ***System Design***

Learning from experience is a key element in successful design. Ideally, the performance of operating systems is analyzed in depth, and for availability analysis, an operation history of several plants is needed. The causes of failure can be identified, and the successes and failures can be used in making probability estimates. Although experience with MSW incineration facilities and energy recovery systems is limited, it is an important source of information.

FINDINGS AND DISCUSSION

Since the late 1970s, emphasis has been placed on the development of simple energy recovery systems. While mass-burn facilities are still being built, solid waste is being processed to obtain higher quality boiler fuel with refuse derived fuel (RDF) systems. Processing includes shredding waste to a more uniform particle size. Some systems remove noncombustible materials such as glass, metals, and grit prior to shredding. These efforts provide greater fuel homogeneity, higher thermal efficiency (due to reduced air requirements), as well as reduced boiler component corrosion.

There is some operating experience with these new systems. One study on waste-to-energy facilities, prepared by the Connecticut Resources Recovery Authority (1985), reported that 18 of 29 plants built in the 1970s and early 1980s were operating in 1985. The reasons for the 11 plant shutdowns were varied, but they can be broadly classified as:

- *Demonstration projects (three plants).* They were not intended to be operated for extended time periods.
- *Mechanical failures (four plants).* Material handling, slagging, and refractory failures were cited. One plant shutdown was caused by an explosion.
- *Economic (four plants).* Viable energy markets and MSW supply commitments were not found.

This summary clearly shows that good economic planning and engineering design are needed.

A study of the annual reports from one plant showed that over a 3-year period its availability exceeded 90%. An availability of over 90% is considered to be a good record.

This record can be viewed in perspective by comparing it to the availability of coal-fired systems. Since electrical energy is supplied on demand with no opportunity to store it and the utility industry ranks as one of the most capital intensive in terms of dollars of investment per dollars of revenue, the industry is quite concerned about availability and the causes of plant outages. Availability for plants operating from 1965 to 1975 was found to be a function of plant capacity as shown in the following graph.

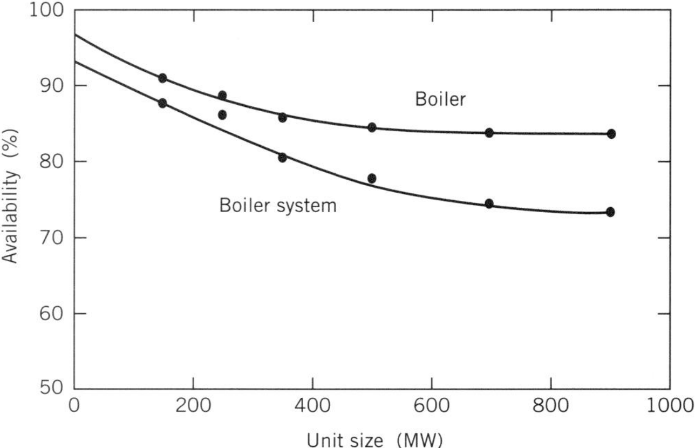

Source: Combustion Engineering.

Other plant components, including the steam turbine, the generator, and the condenser, are reported to have the same tendency toward reduced availability at larger unit sizes. The reasons cited for this trend are:

- In anticipation of a rapid escalation of electricity demand, larger units were designed based on extrapolation from smaller units without benefit of operating experience at larger unit size.
- In the period 1965–1975, units were ordered based on low bid and the hope that economy of scale benefits would be received. Little emphasis was given to design conservativism and reliability.
- The decline in coal quality required the use of more fuel. Extra demands were placed on coal pulverizers, furnace sootblowing and ash-removal equipment, and air emission systems.

This experience has led to the design of redundant systems throughout the electrical utility industry, including the design of MSW operations. It is standard practice to design and build two or more independent process trains, each with a capacity to meet the daily demand.

EXAMPLE 6.12 *Design of an MSW System*

Suppose an MSW plant consists of two parallel, waste stream lines, each consisting of a fuel processing unit and a boiler.

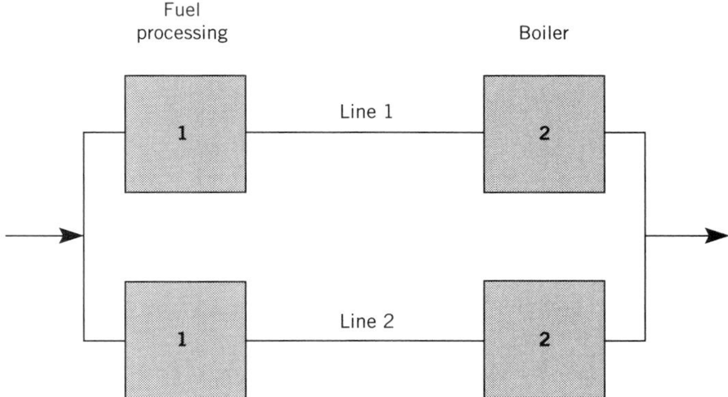

The following performance data are assumed:

Unit	μ, Mean Time to Failure	λ, Mean Time to Repair
Fuel processor	8 hours	1 hour
Boiler	26 weeks	2 weeks

(a) Determine the availability of a waste stream line.

(b) Determine the availability of the MSW incinerator system.

SOLUTION

(a) The limit state availabilities of fuel processor and boiler are:

$$\text{Unit } 1: \; a_1(\infty) = \frac{\eta_1}{\rho_1 + \eta_1} = \frac{1}{1/8 + 1} = 0.889$$

$$\text{Unit } 2: \; a_2(\infty) = \frac{\eta_2}{\rho_2 + \eta_2} = \frac{1/26}{1/26 + 1/2} = 0.928$$

Since an individual waste stream line contains no backup, the Y is represented as

$$Y = \{y = 0 \text{ if } S \text{ and } y = 1 \text{ if } S'\}$$

for units 1 and 2. Assuming the performance of units 1 and 2 are independent, we find that the availability of the waste stream lines 1 and 2 is the product

$$a_L(\infty) = a_1(\infty)a_2(\infty) = (0.889)(0.928) = 0.825$$

(b) The availability of the system $a(\infty)$ is the probability that one or two stream lines are available. The value of $a(\infty)$ is calculated as one minus the probability that both waste stream lines are unavailable.

$$a(\infty) = 1 - [1 - a_L(\infty)]^2 = 1 - (1 - 0.825)^2 = 0.969$$

DISCUSSION

According to the analysis, this parallel system provides high availability. However, there is little experience and lack of data for MSW facilities to check the accuracy of this estimate in greater detail. This is always a problem with new designs. Furthermore, only within the last 30 years has a concerted effort been made to compile operational and failure data for coal, oil, and nuclear power plants. Often it is necessary to construct probability models and make subjective assignments of model parameters to estimate the reliability of a system. The key point of this kind of analysis is to identify weak links in the system in order to make corrective action during the design and planning process.

Subjective probability assignments were made in this example. Fuel processing failure occurs most often because of clogging screw augurs and blocked feed bins. Therefore, it seemed reasonable to assume mean failure and repair times of 8 hours and 1 hour, respectively, for these types of failures. On the other hand, tube failure is a major cause of boiler shutdown. It is repaired by shutting down the boiler, removing the ruptured tubes, and welding new ones in their place. It seemed reasonable to assume mean failure and repair times of 26 weeks and 2 weeks, respectively, for this type of shutdown.

QUESTIONS FOR DISCUSSION

6.4.1 Define the following terms:

availability	renewal process
limit state	subjective probability
Poisson process	transition probabilities

6.4.2 In Examples 6.11 and 6.12, the problems dealt with system design, where there is little or no operating experience.

In your opinion, what is the most significant lesson learned from these discussions?

6.4.3 The exponential distribution is said to possess the "memoryless" property. In other words, the future performance of a system is independent of its past performance. Let $T = \{\text{time to failure}\}$ and ρ = failure rate of system.

(a) Calculate the reliability as a conditional probability,

$$P(T > t \mid T > t_0) \text{ where } t > t_0.$$

Show that

$$P(T > t \mid T > t_0) = e^{-\rho(t-t_0)} = e^{-\rho t'}$$

where $t' = t - t_0$ = time in service from time t_0 to time t.

The result shows that the future time in service t of a system is not affected by its past service time t_0.

(b) This problem and (c) are designed to show the importance of the memoryless property in model formulation. Consider two identical systems each having the same failure rate ρ.

A. No inspection

B. Annual inspection. If necessary, repairs are made and the system is returned to service.

It is assumed that a repaired system is as good as new; thus, the failure rate after repair is once again ρ.

Show that the reliabilities of the system operating with and without annual inspections are the same, or

$$r_A(t) = P(T_A > t) = P(T_B > t \cap T_B > t_0)$$

where

$$t = \text{time in service} = 2 \text{ years}$$
$$t_0 = \text{time of inspection} = 1 \text{ year.}$$

(c) Intuitively, the result from (b) can be argued to be unacceptable. It states that an annual inspection does no good in improving the reliability of the system. Discuss the assumption that $\rho_A = \rho_B = \rho$. Do you feel it is valid? Explain your reasoning.

QUESTIONS FOR ANALYSIS

6.4.4 (a) Determine the availability of a series system of two identical, nonrepairable pumps at the end of one year. The mean time to failure for each pump is $\mu = 5$ years.

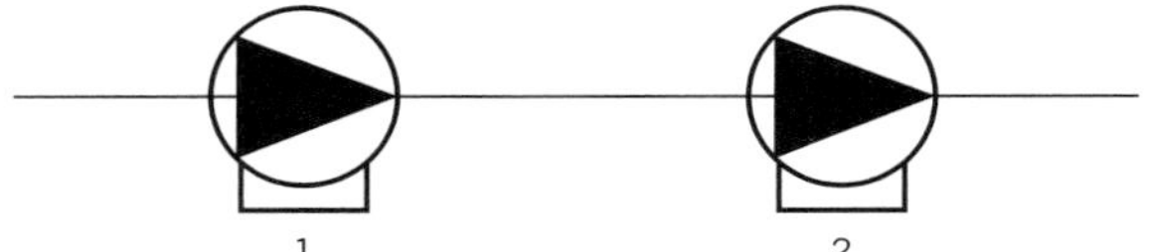

(b) Determine the availability of a series system of two identical, repairable pumps at the end of one year. The mean time to failure for each pump is $\mu = 5$ years and the mean time to repair is $\lambda = 1$ month. Assume a repairman is assigned to each pump.

(c) Two backup pumps with identical properties of the main pumps described in (a) have been procured. Determine the availability of this nonrepairable system with a backup pump assigned to main pumps 1 and 2 at the end of one year.

(d) Repeat (b) assuming the two repairable backup pumps are assigned to the main pumps 1 and 2. Assume limit state probabilities for estimating probabilities of the individual backup systems. Assume a single repairman is assigned to each primary and backup pump system.

6.4.5* Consider the following two single-unit repairable alternatives.

	$k(0)$ Capital Cost	μ, Average Time to Failure	η, Cost to Repair	λ, Average Time to Repair
A	\$8M	20 years	\$8M	1 year
B	\$10M	5 years	\$1M	0.1 year

The same annual benefits $nw(t) = nw =$ \$2M per year are received from each alternative. Assume the annual benefits are received when the system is available.

Assume the annual discount rate is $i = 5$ percent and that the project design life is $n = 10$ years. Use a risk-benefit-cost analysis and recommend an alternative. Use limit state availabilities for the analysis of A and B.

6.4.6* Consider alternative A from Problem 6.4.5. Since the cost to repair alternative A is equal to its capital cost of \$8M, assume that it is not replaced with a single-unit nonrepairable system if it fails.

	$k(0)$, Capital Cost	μ, Average Time to Failure
C	\$8M	20 years

Use a risk-benefit-cost analysis and compare the *npw* for a repairable system A of Problem 6.4.5 with the *npw* for a nonrepairable system C. Use limit state availabilities for the analysis of A.

6.4.7* Consider alternatives A and B from Problem 6.4.5. Alternative A is replaced with a two-unit with a single repairman system. Alternative B is the same.

	$k(0)$, Capital Cost	μ, Average Time to Failure	η, Cost to Repair	λ, Average Time to Repair
A	\$10M	2 years	\$1M	0.1 year
B	\$10M	5 years	\$1M	0.1 year

Use a risk-benefit-cost method to recommend a system. Use the limit state availability for the analysis of A.

6.4.8 Consider a nonrepairable system consisting of two units and use the Poisson distribution to estimate the probability of failure,

$$P_X(t) = \frac{(\rho t)^x e^{-\rho t}}{x!}$$

where

$$\begin{aligned} X &= \{\text{number of failures}\} = \{x = 0, 1, 2\} \\ \rho &= \text{failure rate} \\ t &= \text{time} \end{aligned}$$

(a) Use the Poisson distribution model to determine the equation for $P_0(t)$, $P_1(t)$, and $P_2(t)$.

(b) Derive the equations for $P_0(t)$, $P_1(t)$, and $P_2(t)$ from (a) as a Poisson process. The transition state diagram for X is

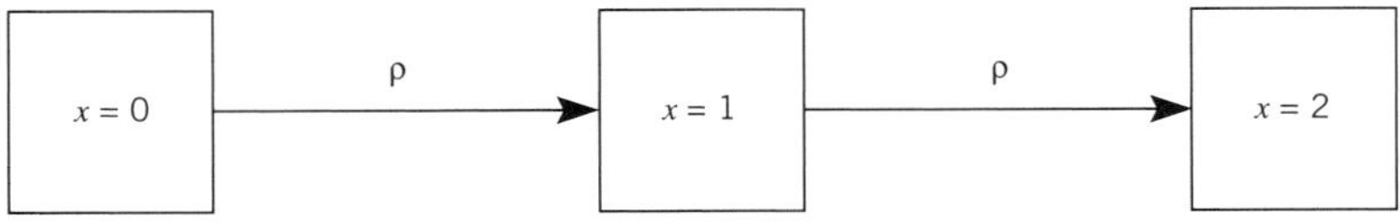

The solution of this set of equations can be found sequentially:

(1) Derive the governing differential equation and solve for $P_0(t)$.

(2) Derive the governing set of differential equations and solve for $P_1(t)$.

(3) Derive the governing set of differential equations and solve for $P_2(t)$.

By induction show that the governing equations for $P_0(t)$, $P_1(t)$, and $P_2(t)$ derived using the Poisson process are the same ones given by the Poisson distribution of (*a*).

REFERENCES

Ang, A., and W. H. Tang (1975). *Probability Concepts in Engineering Planning and Design*, John Wiley & Sons, New York.

Benjamin, J. R., and C. A. Cornell (1970). *Probability, Statistics and Decision for Civil Engineers*, McGraw-Hill, New York.

Billington, R., and R. N. Allan (1983). *Reliability Evaluation of Engineering Systems*, Plenum Press, Belfast, Northern Ireland.

Bogardi, I., W. E. Kelly, and A. Bardossy (1990). "Reliability Model for Soil Liner: Postconstruction," *Journal of Geotechnical Engineering*, **16**, 1502–1520.

Connecticut Resources Recovery Authority (1985). "Engineering Feasibility Report on the Connecticut Recovery Authority Mid-Connecticut System," Bechtel Civil and Mineral, Hartford, CT.

Grosh, D. L. (1989). *A Primer of Reliability Theory*, John Wiley & Sons, New York.

Henley, E. J., and H. Kumamoto (1981). *Reliability Engineering and Risk Assessment*, Prentice-Hall, Englewood Cliffs, NJ.

Klassen, K, B., and J. C. L. van Peppen (1989). *System Reliability Concepts and Applications*, Edward Arnold, Great Britain.

Massmann, J., and R. A. Freeze (1987). "Groundwater Contamination from Waste Management Sites: The Interaction Between Risk-based Engineering Design and Regulatory Policy 1. Methodology," *Water Resources Research*, **23**, 351–367.

McWhorter, D. B., and D. K. Sunada (1977). *Ground-Water Hydrology and Hydraulics*, Water Resources Publications, Fort Collins, CO.

Taguchi, G., E. A. Elsayed, and T. Hsiang (1989). *Quality Engineering in Production Systems*, McGraw Hill Book Company, New York.

U.S. Environmental Protection Agency (1990). *WHPA, A Modular Semi-Analytical Model for the Delineation of Wellhead Protection Areas*, prepared by T. N. Blandford and P. S. Huyakorn of HydroGeoLogic, Inc., Herndon, VA.

Chapter 7

Variability

Statistics are no substitute for judgment.

Henry Clay (1779–1855)

Sensitivity analysis and probability theory are proven methods to deal with questions of forecast uncertainty. Probability theory is particularly attractive because it permits us to express the predicted outcome with a level of success. For example, when dealing with the aging process, $P(T > 3 \text{ years}) = 90\%$ states that a system will survive for a period of three years with a 9 in 10 chance. The critical factors in successful forecasting are choosing a distribution and estimating its parameters in order to simulate a real world situation. These issues are discussed in this and the following chapter.

With sensitivity analysis, the questionable parameters of an optimization model are treated as variables and are varied over a specified range of values. The range of values is often subjectively determined. If the questionable parameters are treated as random variables with known probability distributions, then a more rational approach to specifying the range of each parameter is possible. Again, the distribution specified for these parameters is a critical consideration.

One of the main concerns of this chapter is to effectively incorporate the effects of variability into the decision-making process. To draw attention to the importance of variance, the notation, $\mu \pm \sigma$, is used.

In this chapter, statistics, the science of collecting, classifying, and interpreting information derived from experimental data, will be used to address and reduce forecast and model uncertainty. Prior to introducing topics in statistics in Sections 7.3 and 7.4, expectation of a derived function, $Y = g(\mathbf{X})$ where $g(\mathbf{X})$ is a sum of random variables $X_1, X_2, \ldots, X_m$, will be discussed in Section 7.1 and the properties of the normal distribution will be introduced in Section 7.2. In this chapter, we are concerned with the effects of both the mean μ and variance σ^2 of the random variable Y on decision making.

7.1 EXPECTATION OF DERIVED DISTRIBUTIONS

Let us obtain a basic understanding of the meaning of the terms *mean* μ and *variance* σ^2.

Basic Definitions

The first and second moments of the random variable X are

$$\mu = E(X) = \int_{-\infty}^{\infty} x f(x)\, dx$$

$$\sigma^2 = E[(X-\mu)^2] = \int_{-\infty}^{\infty} (x-\mu)^2 f(x) dx$$

where

$$\int_{-\infty}^{\infty} f(x)\, dx = 1$$

The first moment is centered at $E(X)$; therefore, it is a measure of central tendency and is called the mean μ of X or μ. Since the difference, $x-\mu$, is the distance between x and μ, the second moment is a measure of dispersion about μ; therefore, it is called the variance σ^2 of X or σ^2.

It is often convenient to calculate the variance as

$$\sigma^2 = \int_{-\infty}^{\infty} (x-\mu)^2 f(x) dx$$

Since $(x-\mu)^2 = x^2 - 2\mu x - \mu^2$, this function can be rewritten as

$$\sigma^2 = \int_{-\infty}^{\infty} (x^2 - 2\mu x - \mu^2)\ f(x) dx$$

It simplifies to

$$\sigma^2 = E(X^2) - \mu^2$$

It is often easier to compute $E(X^2)$ and solve for σ^2 with this identity.

Consider the random variable Y where it is a general function of a single random variable X,

$$Y = g(X)$$

and the distribution of X is $f(x)$. The expected value and variance of Y are

$$\mu_Y = E(Y) = E[g(X)] = \int_{-\infty}^{\infty} g(x) f(x)\, dx$$

and

$$\sigma_Y^2 = E[(g(X) - \mu_Y)^2] = E[g(X)^2] - \mu_Y^2 = \int_{-\infty}^{\infty} g(x)^2 f(x) dx - \mu_Y^2$$

Once again, these integrals show the importance of the $f(x)$ distribution in calculating the mean and variance of Y.

The same conclusions are drawn if the distribution function of X is discrete, where

$$f(x) = P(X = x)$$

and

$$\sum_{\text{all } x} P(X = x) = 1$$

We call a discrete distribution of X a *probability function* and a continuous distribution of X a *probability density function*.

Linear Functions and the Correlation Coefficient

When Y is a linear function of m random variables of X_i,

$$Y = a_0 + \sum_{i=1}^{m} a_i X_i$$

the expected value of Y calculations are relatively straightforward. For demonstration purposes, let Y be a linear function of two random variables,

$$Y = a_0 + a_1 X_1 + a_2 X_2$$

where μ_1, σ_1 and μ_2, σ_2 for X_1 and X_2 are known. The expected value of Y is simply

$$\mu_Y = E(Y) = E(a_0 + a_1 X_1 + a_2 X_2) = a_0 + a_1 \mu_1 + a_2 \mu_2$$

The variance of Y is

$$\sigma_Y^2 = E[(Y - \mu_Y)^2]$$

or

$$\sigma_Y^2 = E[(a_0 + a_1 X_1 + a_2 X_2 - a_0 - a_1 \mu_1 - a_2 \mu_2)^2]$$

It reduces to

$$\sigma_Y^2 = a_1^2 E[(X_1 - \mu_1)^2] + a_2^2 E[(X_2 - \mu_2)^2] + 2a_1 a_2 E[(X_1 - \mu_1)(X_2 - \mu_1)]$$

This function, written in terms of σ_1, σ_2, and the *correlation coefficient* ρ_{12} is

$$\sigma_Y^2 = a_1^2\sigma_1^2 + a_2^2\sigma_2^2 + 2a_1a_2\rho_{12}\sigma_1\sigma_2$$

The correlation coefficient ρ_{12} is a measure of the linear dependency between X_1 and X_2. It is defined to be

$$\rho_{12} = \frac{E[(X_1 - \mu_1)(X_2 - \mu_2)]}{\sigma_1\sigma_2}$$

It can be shown that $-1 \le \rho_{12} \le 1$.

When $\rho_{12} = +1$ or -1, there is a perfect, one-to-one linear relationship between X_1 and X_2. If the variables are uncorrelated, then $\rho_{12} = 0$ and there is no linear dependency between them. If two random variables are assumed to be independent or uncorrelated, then $\rho_{12} = 0$ and the variance relationship for Y reduces to

$$\sigma_Y^2 = a_1^2\sigma_1^2 + a_2^2\sigma_2^2$$

When $\rho_{ij} = 0$ for all i-j pairs X_i and X_j, the mean and variance of a linear function

$$Y = a_0 + \sum_{i=1}^{m} a_i X_i$$

simplify to

$$\mu_Y = a_0 + \sum_{i=1}^{m} a_i \mu_i$$

and

$$\sigma_Y^2 = \sum_{i=1}^{m} a_i^2 \sigma_i^2$$

The condition that $\rho_{ij} = 0$ must be justified for each i-j variable pair. This is most easily shown by example.

EXAMPLE 7.1 *The Effects of Variability on Decision Making*

The net worth is the difference between profits received from recovered energy of a waste-to-energy plant and unit shipping cost,

$$nw = p\left(1 - \frac{m}{100}\right) - sd$$

where

nw = the monetary value of processed MSW in $ per ton
s = unit shipping cost in $ per ton-mile
m = MSW moisture content in percent (%)
d = distance from the town to plant in miles
p = profit from recovered energy is $120 per ton of dry MSW

Since the shipping cost and moisture content are not known with certainty, they are treated as the independent random variables S and M, respectively,

S = {unit shipping cost in $ per ton-mile}
M = {MSW moisture content in percent (%)}

Now, the net worth measured is

$$NW = p\left(1 - \frac{M}{100}\right) - Sd = 120 - 1.20\,M - Sd$$

Suppose a survey of shipping costs and moisture contents has been conducted. The results are reported as follows:

Unit shipping cost: $\mu_S \pm \sigma_S$ = \$2.50 ± \$0.50 per ton-mile
Moisture content: $\mu_M \pm \sigma_M = 25 \pm 6\%$

Determine the feasibility of these variabilities on decision making. Determine the mean and variance of NW, or $\mu \pm \sigma$. The distance d is assumed to be known with certainty. Compare two routes:

Route 1: d = 15 miles

Route 2: d = 35 miles

Based on this comparison, establish a decision rule for recommending the selection of these routes based on maximizing net worth. Assume S and M are independent.

SOLUTION

Since the correlation between S and M is zero, the mean and variance of NW are

$$\mu = E(120 - 1.20M - Sd) = 120 - 1.2\mu_M - d\mu_S$$

$$\sigma^2 = \text{Var}(120 - 1.20M - Sd) = 1.2^2\sigma_M^2 + d^2\sigma_S^2$$

The standard deviation is $\sigma = \sqrt{\sigma^2}$. Substituting, we obtain:

$$\text{Route 1: } d = 15 \text{ miles} \rightarrow \mu \pm \sigma = \$52.50 \pm \$10.40$$

$$\text{Route 2: } d = 35 \text{ miles} \rightarrow \mu \pm \sigma = \$2.50 \pm \$18.92$$

The decision rule for feasibility from capital investment analysis is $nw \geq 0$. Using this rule and the assumption that $nw = \mu$, we find that the decision rule becomes $\mu \geq 0$. Both routes 1 and 2 are feasible alternatives.

Now, consider the variability of these estimates as measured by σ on this rule. The results show that the standard deviation for cost over route 1 is relatively small in comparison to its mean,

$$\text{Route 1: } \mu >> \sigma \rightarrow \$52.50 >> \$10.40$$

It seems reasonable to conclude that there is a good chance a profit will be secured for the range of shipping costs and moisture contents specified. For route 2, on the other hand, the standard deviation of NW is much greater than its mean.

$$\text{Route 2: } \mu << \sigma \rightarrow \$2.50 << \$18.92$$

There is a good chance that no or little profit will be made.

In conclusion, we must question the usefulness of the original decision rule for route 2. In this and following sections, we will be evaluating variability and then return to this problem in Section 7.4.

EXAMPLE 7.2 *The Effects of Variance*

Consider the net worth function discussed previously in Example 7.1, $NW = 120 - 1.20M - Sd$, where S and M are assumed to be independent.

(a) Justify the assumption of independence. In other words, the linear correlation between S and M is $\rho = 0$.

(b) Consider the effects of distance d on the variance of NW for routes 1 and 2.

SOLUTION

(a) The assumption of independence is justified on the grounds that there is no reason to believe that the moisture content of MSW has any relationship to the unit cost of shipping. The moisture content is dependent on the weather and the manner in which it is stored for pickup. The shipping fee, on the other hand, depends on the cost of fuel, equipment, and labor. Changes in any one of these variables will have no effect on the moisture content of MSW, and vice versa.

(b) Route 1: $d = 15$ miles

$$\begin{aligned}\sigma^2 &= 1.2^2\,\sigma_M^2 + d^2\sigma_S^2 \\ &= (1.2^2)(6^2) + (15^2)(0.5^2) = 51.84 + 56.25 \\ &= 108.09\end{aligned}$$

This calculation shows that the variance σ^2 is approximately equally divided between shipping and moisture content variances, $56.25 \approx 51.84$.

Route 2: $d = 35$ miles

$$\sigma^2 = 35^2(0.5^2) + 1.2^2\,(6^2) = 306.25 + 51.84 = 358.09$$

The large variance of σ^2 is caused by a relatively large variability in shipping costs as compared to the moisture content, $306.25 >> 51.84$. Shipping distance d is the important factor in determining a profit or loss for route 2.

EXAMPLE 7.3 ***The Expected Cost of Earthquake Damage***

Suppose the property damage cost increases with earthquake magnitude as

$$h = x^2$$

where h is measured in millions of dollars. Let

$$X = \{\text{earthquake magnitude, measured on the Richter scale}\}$$

with

$$\begin{aligned}P(X = 0) &= 0.99 \\ P(X = 1) &= 0.005 \\ P(X = 2) = P(X = 3) = P(X = 4) &= P(X = 5) = P(X = 6) = 0.001\end{aligned}$$

Determine $\mu_H \pm \sigma_H$.

SOLUTION

Since X is a discrete random variable, summation equations in lieu of integral equations are used in these calculations. The governing equations are

$$\mu_H = E(H) = E[g(X)] = \sum_x g(x)P(X = x)$$

$$\sigma_H^2 = E[(H - \mu_H)^2] = E(H^2) - \mu_H^2 = \sum_x x^2\, P(X = x) - \mu_H^2$$

$$\mu_H = (0^2)(0.99) + (1^2)(0.005) + (2^2)(0.001) + (3^2)(0.001) + (4^2)(0.001) + (5^2)(0.001) + (6^2)(0.001) = \$0.095M$$

$$E(H^2) = (0^2)^2(0.99) + (1^2)^2(0.005) + (2^2)^2(0.001) + (3^2)^2(0.001) + (4^2)^2(0.001) + (5^2)^2(0.001) + (0.6^2)^2(0.001) = 2.279$$

$$\sigma_H^2 = E(H^2) - \mu_H^2 = 2.279^2 - (0.095)^2 = 2.270$$

or

$$\sigma_H = \sqrt{\sigma_H^2} = \$1.507M$$

$$\mu_H \pm \sigma_H = \$0.095M \pm \$1.507M$$

QUESTION FOR DISCUSSION

7.1.1 Define the following terms:

correlation coefficient	variance

mean

QUESTIONS FOR ANALYSIS

7.1.2 Let

$$L = \{\text{HCl concentration entering an air scrubber}\}$$

where $\mu_L \pm \sigma_L = 200 \pm 100$ and HCl concentration is measured in ppmv dry gas CO_2,

Let

$$R = \{\text{HCl concentration removed from the scrubber}\}$$

where $\mu_R \pm \sigma_R = 200 \pm 20$ and $Y = \{\text{HCl concentration entering the atmosphere}\} = L - R$. L and R are assumed independent.

(a) Calculate $\mu_Y \pm \sigma_Y$.

(b) Calculate $\mu_H \pm \sigma_H$, where $H = \$100Y$, the fine for noncompliance.

7.1.3 Let

$$X = \{\text{magnitude of an earthquake}\}$$

where $x = 0, 1, \ldots, 8$ and earthquake magnitude is measured on a Richter scale.

The probability of no earthquake during a year is 0.92, and earthquakes with $x > 0$ are assumed to be a uniform discrete distribution of $x = 1, 2, \ldots, 8$.

The loss function h from earthquake damage is

$$h = \$0 \quad \text{for } x = 0, 1, 2, 3, 4$$

and

$$h = \$5e^{(x-4)} \text{ for } x = 5, 6, 7, 8$$

where h is measured in \$M.

(a) Calculate $\mu_X \pm \sigma_X$.

(b) Calculate $\mu_H \pm \sigma_H$.

7.1.4

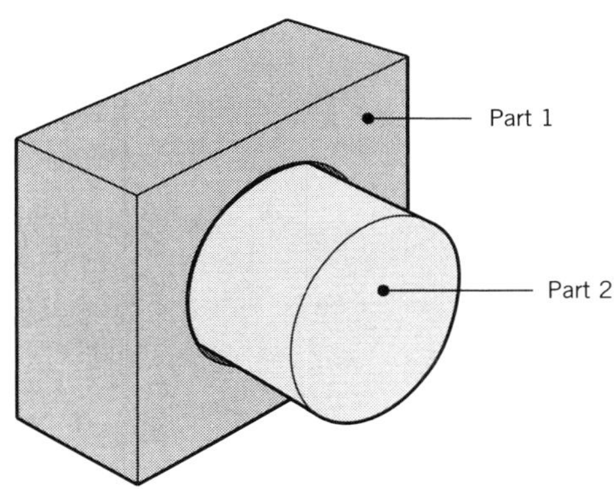

Let

$$D_1 = \{\text{diameter of part 1}\}$$

where $\mu_1 \pm \sigma_1 = 10 \pm 0.25$ inches and

$$D_2 = \{\text{diameter in inches of part 2}\}$$

where $\mu_2 \pm \sigma_2 = 9.75 \pm 0.25$.

Define the difference between these dimensions as a measure of lack of fit or manufacturing error, $e = d_1 - d_2$.

(a) Calculate $\mu_E \pm \sigma_E$.

(b) Calculate $\mu_H \pm \sigma_H$ where $h = \$100e$ and where h is assumed to be the monetary loss due to lack of fit.

7.1.5 Property damage costs are functions of rainfall X_1 and wind speed X_2 intensities. The total damage cost is $h = h_1 + h_2$, where

$$h_1 = \$1x_1^3$$

and

$$h_2 = \begin{cases} \$0 & \text{for } x_2 \le 40 \text{ mph} \\ \$0.2x_2^2 & \text{for } x_2 > 40 \text{ mph} \end{cases}$$

and h_1 and h_2 are measured in \$M. Let

$$X_1 = \{\text{rainfall intensity measured in inches}\}$$

with

$$P(X_1 = 0) = 0.98$$
$$P(X_1 = 1) = 0.01$$
$$P(X_1 = 2) = P(X_1 = 3) = 0.005$$

and

$$X_2 = \{\text{wind speed measured in mph}\}$$

with

$$P(X_2 \le 40) = 0.99$$
$$P(X_2 = 50) = 0.009$$
$$P(X_2 = 60) = 0.001$$

(a) Calculate the expected total damage cost $\mu_H \pm \sigma_H$ where $\rho_{12} = 0$.

(b) Repeat (a), assuming a reasonably strong correlation between rainfall intensity and wind speed where $\rho_{12} = 0.8$.

(c) Interpret the meaning of the correlation coefficient of $\rho_{12} = 0.8$ in terms of the damage cost H.

7.2 THE NORMAL DISTRIBUTION

The normal distribution is the most important distribution in statistics. The normal or Gaussian distribution, shown in Figure 1, is used to describe a random variable whose values are clustered around its mean μ. The width of $f(y)$ is determined by the standard deviation σ. The distribution is denoted as $N(\mu, \sigma^2)$.

Material strength, height and weight measurements, and dimensions of machined parts often have normal probability distributions. Machined parts, for example, are manufactured to have a specified dimension μ. Owing to process variability, even when a manufacturing process is stable, most measurements will not be exactly equal to μ. They tend to cluster around μ with a bell-shaped, symmetrical distribution, whose width is determined by σ. In the manufacturing of machine parts, it is desirable to minimize σ^2 or σ so that assembled pieces will fit together tightly.

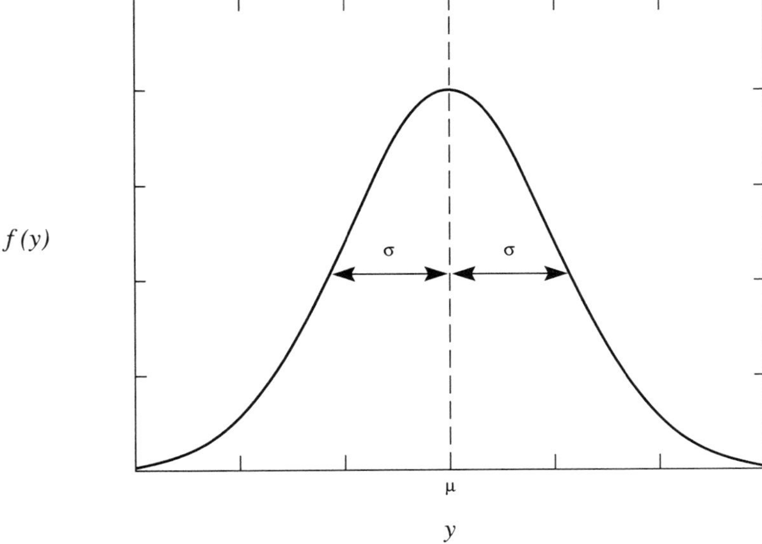

Figure 1 A normal (Gaussian) distribution about μ.

The Probability Density Function

The probability density function of a normally distributed random variable Y with mean μ and variance σ^2, $N(\mu, \sigma^2)$, is

$$f(y) = \frac{1}{\sqrt{2\pi\sigma^2}} \exp\left(-\frac{1}{2}\left(\frac{y-\mu}{\sigma}\right)^2 \right)$$

where

$$-\infty < y < \infty$$

The cumulative distribution function is

$$F(y) = P(Y \le y) = \int_{-\infty}^{y} \frac{1}{\sqrt{2\pi\sigma^2}} \exp\left(-\frac{1}{2}\left(\frac{x-\mu}{\sigma}\right)^2 \right) dx$$

The integral does not have a closed-form expression. One of the primary aims of this section is to describe how to calculate $F(y)$ by using a cumulative standard normal table, $\Phi(z)$.

The Standard Normal Distribution

The standard normal distribution is a normal distribution with zero mean and unit variance, $N(0, 1)$, as shown in Figure 2.

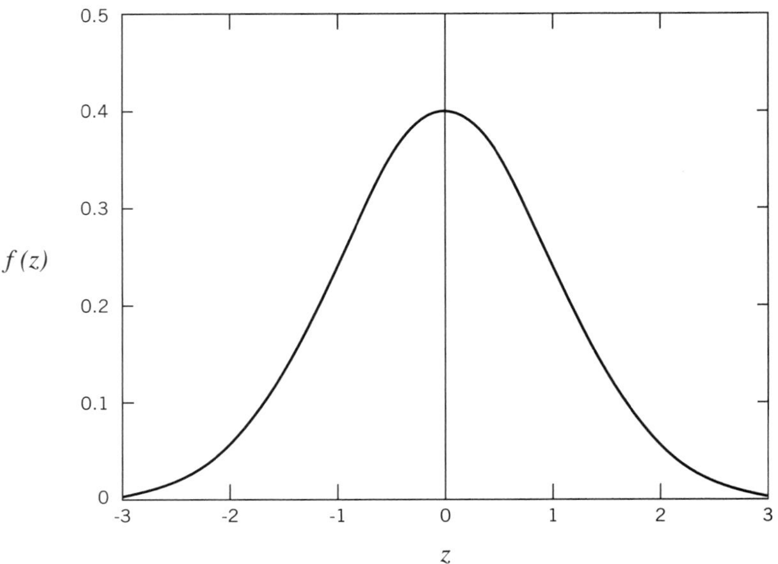

Figure 2 Standard normal $N(0,1)$ density fucntion.

$$f(z) = \frac{1}{\sqrt{2\pi}} \exp\left(-\frac{1}{2}z^2\right)$$

The area under the curve $f(x)$ from $-\infty$ to z is denoted as

$$\Phi(z) = \int_{-\infty}^{z} \frac{1}{\sqrt{2\pi}} \exp\left(-\frac{1}{2}x^2\right) dx$$

The function $\Phi(z)$ is shown in Figure 3, and a table of $\Phi(z)$ values is given in Table III of Appendix B. The values in the $\Phi(z)$ table are found by numerical integration.

The Calculation of $P(Y \leq y)$

$F(y) = P(Y \leq y)$ is calculated by transforming Y to Z where Z is defined to be

$$Z = \frac{Y - \mu}{\sigma}$$

The transformed variable, $\frac{Y-\mu}{\sigma}$, has an $N(0, 1)$ distribution. The equality

$$P(Y \leq y) = P\left(Z \leq \frac{y-\mu}{\sigma}\right) = F(y) = \Phi\left(\frac{y-\mu}{\sigma}\right) = \Phi(z)$$

is the key relationship for finding $P(Y \leq y)$. Basically, it states that the area under the $N(0, 1)$ curve from $-\infty$ to $z = \frac{Y-\mu}{\sigma}$ is the same area for the $N(\mu, \sigma^2)$ curve from $-\infty$ to y.

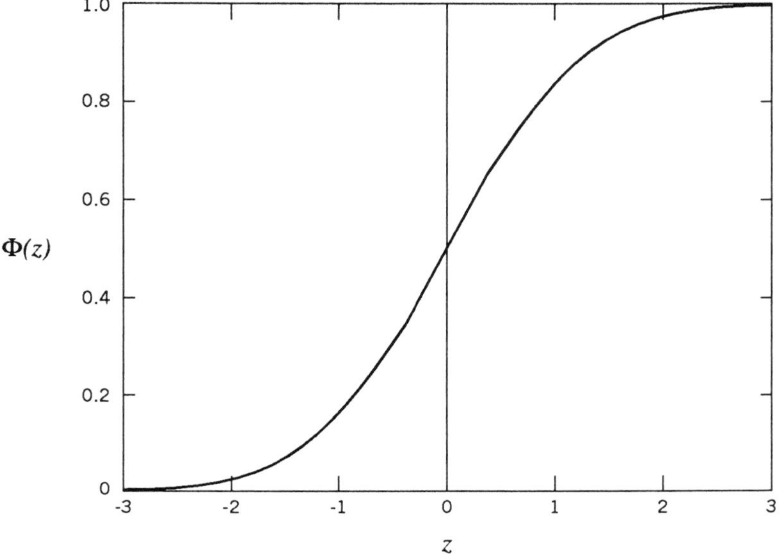

Figure 3 The cumulative density function for the standard normal distribution.

To find $F(y)$, use the following steps:

1. Transform Y to Z by calculating $z = \dfrac{y-\mu}{\sigma}$.
2. Find $\Phi(z)$ from the cumulative standard normal table for the calculated value of z.
3. Then equate $F(y) = \Phi(z)$.

The probability that $P(Y \le 30)$ for $N(20, 10^2)$ is

$$F(30) = P(Y \le 30) = \Phi\left(\frac{30-20}{10}\right) = \Phi(1) = 0.841$$

If $F(y) = P(Y \le y)$, μ and σ are known, y may be determined by

1. Finding $z = \Phi^{-1}[F(y)]$ where $\Phi(z) = F(y)$ for the given value of $\Phi(z)$ from the cumulative standard normal table.
2. Solving for y by calculating $y = \mu + z\,\sigma$.

The value of y satisfying the relationship, $P[Y > y] = 0.9$, for $N(20, 10^2)$ is desired. The cumulative probability is calculated using the total probability theorem:

$$F(y) = P(Y \le y) = 1 - P(Y > y) = 0.1$$

Since $\Phi(z) = F(y) = 0.1$,

$$z = \Phi^{-1}(0.1) = -1.28$$

and

$$y = \mu + z\,\sigma = 20 - 1.28(10) = 7.2$$

or

$$P(Y > 7.2) = 0.9$$

EXAMPLE 7.4 ***Regulatory Limits on Air Emissions***

An MSW incineration study was conducted to improve the economic plant efficiency and to reduce HCl (hydrogen chloride) emissions. A plant was retrofited with dry lime scrubbers in order to reduce HCl emissions to a compliance limit of 175 ppmv dry gas at 12% CO_2. HCl, a surrogate measure of overall plant efficiency, is used as an indicator as to how effective lime dosage is in removing acidic gases and mercury from the flue gas.

The study results show that an increase in lime dose reduces the average HCl concentration.

d Lime Dose (kg per hour)	$\mu \pm \sigma$ $Y = \{$HC1 Concentration$\}$ (ppmv dry gas CO_2)
75	335 ± 66
90	282 ± 44
110	230 ± 66
210	88 ± 22

Source: Bergström and Lundqvist (1983).

At higher dosages of d, σ is reduced, indicating a more stable plant operation in removing HCl.

Consider two strategies for determining an assigned lime dosage d^*. Assume Y, the flue HCl concentration, to have a normal probability distribution. Let the probability, $P(Y \le 175)$, be a measure of effectiveness in meeting the HCl compliance limit. Consider two plans:

Plan A: Establish a dosage of $d^* = 210$ kg per hour.

Plan B: Use linear interpolation to determine d^* where $\mu^* = E[Y] = 175$ppmv, the compliance limit. Assume that d^* decreases linearly between lime dosages of 110 and 210 kg per hour. The variance of HCl is not considered in the determination of d^* by linear interpolation.

(a) How effective are these plans in meeting a maximum of 175 ppmv?

(b) How effective is plan A in meeting a maximum of 50 ppmv?

SOLUTION

(a) Plan A: Given a dosage of $d^* = 210$ kg per hour, it is reasonable to assign the distribution of Y to be $f(y)$ to be $N(88, 22^2)$. The measure of effectiveness is

$$P(Y \le 175) = \Phi\left(z = \frac{175-88}{22}\right) = \Phi(3.95) = 0.999 \approx 100\%$$

Plan B: By using linear interpolation, the assigned lime dosage is

$$\frac{d^* - 110}{210-110} = \frac{175-230}{88-230} \quad \text{or} \quad d^* = 148 \text{ kg per hour}$$

Given $d^* = 148$ kg per hour, the distribution of Y is assumed equal to $N(175, \sigma^2)$, where the value of σ can only be crudely estimated and, in this special case, is not needed in the calculation of $P(Y \le 175)$.

$$P(Y \le 175) = \Phi\left(z = \frac{175-175}{\sigma}\right) = \Phi(0.00) = 0.50 = 50\%$$

At this lime dosage there is a 50–50 chance of compliance regardless of the variation σ^2 or σ for the process.

(b) The probability of meeting a 50 ppmv compliance standard at lime dosage of 210 kg per hour is

$$P(Y \le 50) = \Phi\left(z = \frac{50-88}{22}\right) = \Phi(-1.73) = 0.043 \approx 4\%$$

The retrofited plant would be successful in complying with this standard 4 percent of the time.

The key point in this illustration is to demonstrate the importance of the emission compliance limit as well as plant performance variability. Emission compliance standards are often given as a single value. For example, the HCl standard is specified as "a maximum of 50 ppmv corrected to 12% CO_2 or 80% removal efficiency." Other standards recognize plant variability by specifying an average time for calculating the mean emission level. According to the 1990 Clean Air Act, the 8-hour average for carbon monoxide CO of 9 ppmv should not be exceeded one time per year.

EXAMPLE 7.5 *Margin of Structural Design Safety*

Consider a structural weld that is essential to overall safety. If the weld fails, catastrophic failure will possibly cause loss of life, human injury, and financial loss. Structural failure tests have been performed on the weld and the mean failure strength has been found to be 20 ksi (20,000 pounds per square inch), with a standard deviation of 3 ksi. The load on the weld is known to vary with a mean value of 20 kips and a standard deviation of 10 kips. The cross-sectional area of the weld is 2 in^2.

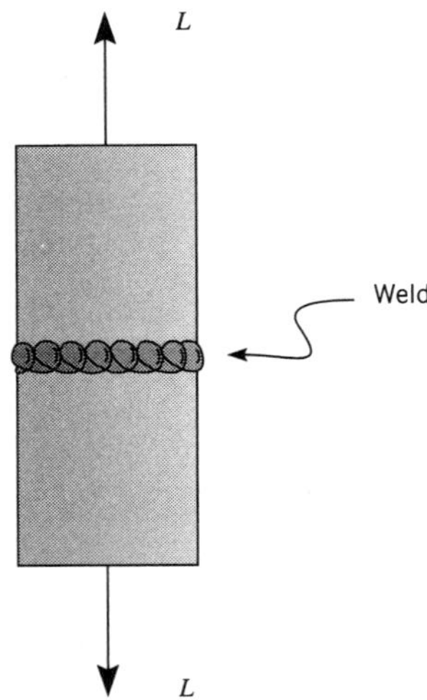

The weld strength and load are treated as normal random variables:

$$L = \{\text{weld loading}\}$$

where

$$L = N(\mu_L, \sigma_L^2) = N(20, 10^2)$$

and

$$S = \{\text{failure strength of the weld}\}$$

where

$$N(\mu_S, \sigma_S^2) = \mathrm{N}(20, 3^2)$$

with

$$a = \text{cross-sectional area of the weld} = 2 \text{ in}^2$$

The margin of safety, m, is defined to be the difference between the average failure strength and the average weld stress, or

$$m = \mu_S - \frac{\mu_L}{a} = 20 - \frac{20}{2} = 10\, ksi$$

where the weld stress is equal to the load divided by the cross-sectional area of the weld. There appears to be a substantial difference between the failure stress and weld stress. The nominal failure strength is twice the load stress.

This definition of margin of safety ignores the variability in weld failure strength and loading stress. Now, consider the margin of safety as a random variable,

$$M = S - \frac{L}{a}$$

Determine the probability of failure, $\theta = P(M \le 0)$.

SOLUTION

The expected margin of safety and its standard deviation are

$$\mu_M = \mu_S - \frac{\mu_L}{a} = 20 - \frac{20}{2} = 10 \text{ ksi}$$

$$\sigma_M^2 = \sigma_S^2 + \frac{\sigma_L^2}{a^2} = 3^2 + \frac{10^2}{2^2} = 9 + 25 = 34 \text{ ksi}^2$$

or

$$\sigma_M = 5.83 \text{ ksi}$$

The weld strength and loading are assumed to be independent events, $\rho = 0$. The variance equation for M shows the variability in the weld load to be a significant contribution to the overall variability in the margin of safety.

Since M is the difference of two normally distributed random variables, it makes sense that M is a normally distributed random variable also with $M = N(\mu_M, \sigma_M^2) = N(10, 5.83^2)$. Failure occurs when the loading stress exceeds the failure stress, or

$$\frac{L}{a} > S$$

or, equivalently, when the margin of safety is less than zero,

$$M = S - \frac{L}{a} < 0$$

The failure probability θ is defined to be

$$\theta = P(M < 0) = P\left(S - \frac{L}{a} < 0\right)$$

or

$$\theta = \Phi\left(\frac{0 - 10}{5.83}\right) = \Phi(-1.715) = 0.043 \approx 4\%$$

There is a 4 in 100 chance of failure; this risk probability is considered too great. Increasing the weld area will reduce θ. The increased cost of the weld could certainly not outweigh the cost of a catastrophic failure. The use of a structural safety backup system may prove more cost effective.

QUESTION FOR DISCUSSION

7.2.1 Define the following terms:
normal distribution standard normal distribution

QUESTIONS FOR ANALYSIS

Instructions: The solutions to questions marked with an asterisk (*) are most easily obtained with the aid of computer graphics.

7.2.2 What are the probabilities that a value of a random variable Y that is $N(\mu, \sigma^2)$ will lie within plus or minus one, two, and three standard deviations of the mean? In other words, what are the probabilities of (a) $P(\mu - \sigma < X \leq \mu + \sigma)$, (b) $P(\mu - 2\sigma < X \leq \mu + 2\sigma)$ and (c) $P(\mu - 3\sigma < X \leq \mu + 3\sigma)$?

7.2.3 (a) Four of five samples meet the maximum contamination limit of 25 ppmv for some toxic chemicals. The average concentration of these five samples is $\mu = 18$ ppmv. Estimate the variance of the process. Assume that the concentration of the chemical has a normal probability distribution.

(b) Based on the information from (a), what is the probability that a compliance limit of 21 ppmv will be met?

7.2.4 Let

$$L = \{\text{HCl concentration entering an air scrubber}\}$$

where $N(\mu_L, \sigma_L^2) = N(200, 100^2)$ and HCl concentration is measured in ppmv dry gas CO_2.

$$R = \{\text{HCl concentration removed from the scrubber}\}$$

where $N(\mu_R, \sigma_R^2) = N(200, 20^2)$ and

$$Y = \{\text{HCl concentration entering the atmosphere}\} = L - R$$

What is the probability that the air scrubber is in noncompliance and exceeds a limit of 175 ppmv dry gas CO_2? Assume L and R are independent.

7.2.5 Let

$$D_1 = \{\text{diameter of part 1}\}$$

where $N(\mu_1, \sigma_1^2) = N(10, 0.25^2)$, and

$$D_2 = \{\text{diameter in inches of part 2}\}$$

where $N(\mu_2, \sigma_2^2) = N(9.75, 0.25^2)$.

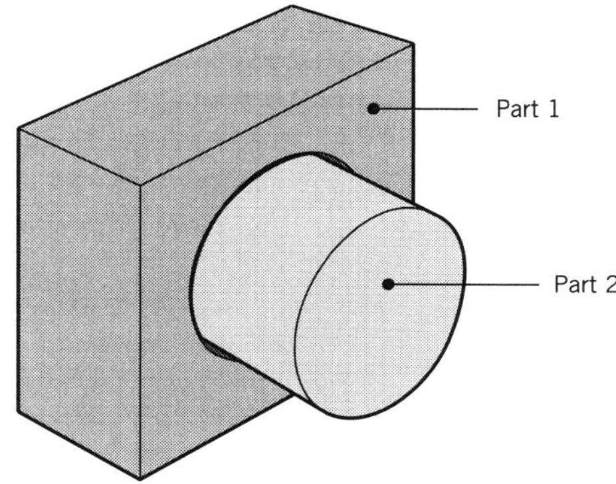

All dimensions are measured in inches.

(a) What is the probability of manufacturer error due to a lack of fit; in other words, part 2 will not fit inside part 1.

(b) What is the expected monetary loss? A lack of fit costs h = \$100 to repair.

(c) Repeat (a) and (b) where $D_1 = D_2 = N(9.75, 0.05^2)$.

7.2.6* Let

$$T = \{\text{time to system failure}\}$$

where T has an $N(10, 5^2)$ distribution and time is measured in years. The project life is n = 30 years.

(a) Plot reliability function, $r(t) - t$.

(b) Plot hazard function, $h(t) - t$.

(c) Plot the annual risk probability, $\theta(t)–t$ where $t = 0, 1, 2, \ldots$.

7.3 STATISTICAL SUMMARIES

By collecting a representative set of data about a process, we aim to identify an $f(y)$ distribution for Y that accurately models the process being observed. In this section, the sample mean, sample variance, and histogram are used for this purpose.

The Random Sample

The goal of any statistical experimentation is to obtain a better understanding of a process. Conducting an experiment requires careful planning. In order to maximize the amount of information derived from an experiment, three possible sources of data variation should be considered,

- Process
- Measurement
- Sampling

Process variability is a result of changes in the process state caused by some inherent unsteady or inconsistent condition or conditions of the system under study. Measurement and sampling variability is a function of unsteady or inconsistent condition or conditions of the measurement instrument being used or of the samples being collected.

Measurement and sampling error and variability are often minimized by carefully conducting experiments and collecting data. Improperly calibrated instruments and improperly recorded data are examples of measurement errors that can be avoided. Sampling error can be caused by collecting unrepresentative samples. Whenever possible, random samples should be used.

A *random sample* from a population is one in which each individual from the population is given an equal chance of being selected. Suppose the total population consists of 100 individuals and their opinion on a particular matter is sought. Since it is impractical to interview everyone in the population, a sample of five individuals is chosen. In order to avoid bias, each person in the group is assigned a number from 1 to 100 and five randomly chosen numbers are drawn. The individuals with assigned numbers matching the randomly drawn numbers are selected for interviews.

Clearly, care should be exercised so that each member of the population and each condition under investigation have an equal chance of being selected. In many applications, the goal of random sampling has to be sacrificed for practical and economic considerations. Random sampling does not always have to be employed to obtain meaningful results.

The Histogram

When the sample size is sufficiently large, the *histogram* or frequency diagram can be used to summarize and investigate the distributional pattern in the data. If the shape of a histogram closely resembles a theoretical probability distribution, then it can be put forward as being representative of the parent population.

For example, the energy production from an MSW plant, expressed as an average power output (MW) per shift, is observed. The sample size is n = 101. The first step in drawing a histogram is to choose a sampling interval and count the number of observations in each interval as shown in the following table.

TABLE 1 Frequency Table

k	Interval (MW)	y_k, interval midpoint	n_k, frequency
1	4.0–4.9	4.5	1
2	18.0–18.9	18.5	1
3	19.0–19.9	19.5	1
4	20.0–20.9	20.5	3
5	21.0–21.9	21.5	0
6	22.0–22.9	22.5	2
7	23.0–23.9	23.5	6
8	24.0–24.9	24.5	22
9	25.0–25.9	25.5	26
10	26.0–26.9	26.5	19
11	27.0–27.9	27.5	13
12	28.0–28.9	28.5	6
n = 13	29.0–29.9	29.5	1

Source: Bergström and Lundqvist (1983).

The choice of interval size is arbitrary. The number of intervals is usually 10 to 20. Here, the sample size is relatively large so an interval of 1 MW seemed appropriate. The histogram, shown in Figure 4, is a plot of the frequency versus the midpoints of the sampling interval.

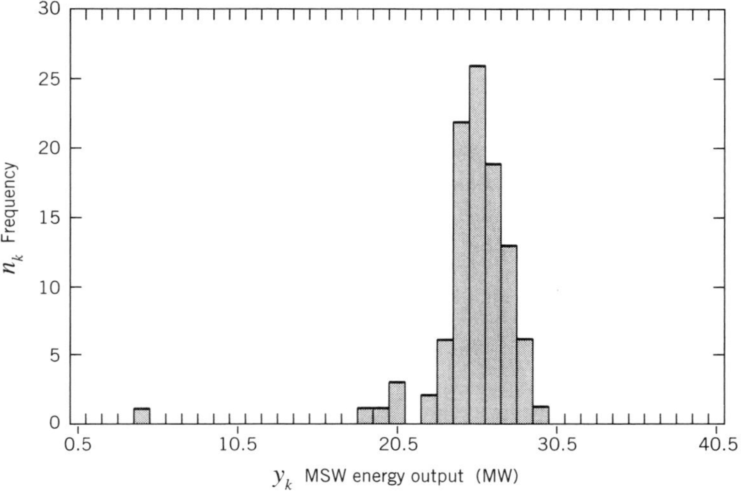

Figure 4 Histogram.

With the exception of the observation in the 4.0 to 4.9 interval, the shape of the histogram closely resembles that of a normal probability density function. Since it falls outside the region of central tendency, this point may be an *outlier* and may be removed from the data set. An outlier may have significant impact on sample mean and variance calculations.

Classifying a data point as an outlier must be done with care. First, find the reason why it is an outlier. If it is, say, due to measurement or recording error, remove it. If no explanation can be given, then the observation should remain in the data set.

Sample Mean and Variance

The sample mean $\bar{y}$ and variance s^2 are estimates of the theoretical mean $\mu = E(Y)$ and variance $\sigma^2 = E[(Y-\mu)^2]$. The sample mean $\bar{y}$, like μ, is a measure of central tendency, and the sample variance s^2 and sample standard deviation s, like σ^2 and σ, are measures of dispersion. Given a set of data $y_1, y_2, \ldots, y_n$ with n denoting sample size, its sample mean $\bar{y}$, and variance s^2, and standard deviation s are calculated as

$$\bar{y} = \frac{\sum_{j=1}^{n} y_j}{n}$$

$$s^2 = \frac{\sum_{j=1}^{n} (y_j - \bar{y})^2}{n-1}$$

EXAMPLE 7.6 ***Moisture Content***

The moisture content data, in percent, for the random variable M are: 22, 28, 25, 23, 27, 32, 18, 35 and 15.

What are its sample mean, variance, and standard deviation?

SOLUTION

$$\bar{m} = \frac{\sum_{j=1}^{9} m_j}{9} = \frac{22+28+25+23+27+32+18+35+15}{9} = 25\%$$

$$s_M^2 = \frac{\sum_{j=1}^{9} \left(m_j - 25\right)^2}{(9-1)} = [(22-25)^2 + (28-25)^2 + (23-25)^2 + (27-25)^2$$

$$+ (32-25)^2 + (18-25)^2 + (35-25)^2 + (15-25)^2] / 8 = 36$$

where $n = 9$ observations

$$s_M = \sqrt{s_M^2} = 6\%$$

$$\bar{m} \pm s_M = 25\% \pm 6\%$$

The values of $\bar{m}$ and s_M were used as the best estimate of μ_M and σ_M as illustrated in Examples 7.1 and 7.2. Consider, however, that in a single day hundreds of people in the study region discard MSW. Because of time constraints and other economic considerations, the size of the sample is limited to nine observations. Even though the sample size $n = 9$ is relatively small, these statistics may adequately represent μ_M and σ_M.

For some applications, it may be also important to estimate a probability function $f(y)$ of the parent distribution of the process under study. Since $n = 9$, a histogram is of limited use. The central limit theorem, discussed in Section 7.4, and the quantile-quantile plot, to be introduced in Chapter 8, are often employed to justify the choice of frequency distribution.

A Weighted Mean and Variance

When a population can be broken into m subgroups so that a measure of some quantity is available for each subgroup, it is often convenient to calculate the sample mean and variance as

$$\bar{y} = \sum_{k=1}^{m} \omega_k y_k$$

and

$$s^2 = \sum_{k=1}^{m} \omega_k \left(y_k - \bar{y}\right)^2$$

where

$$k = \text{classification index} = 1, 2, \ldots, m,$$

$$\omega_k = \text{relative frequency} = \frac{n_k}{n} \text{ with } n_k = \text{number of observations in interval } y_k$$

and

$$n = \text{total number of observations in the sample} = \sum_{k=1}^{m} n_k$$

The sample mean and sample standard for the MSW energy production plant of Table 1 using the weight formulas give

$$\bar{y} \pm s = 25.3 \pm 2.82 \quad \text{for } n = 101 \text{ observations}$$

If the $y_1 = 4.5MW$ is treated as an outlier and removed from the data set, the sample mean and standard deviation is

$$\bar{y} \pm s = 25.5 \pm 1.91 \quad \text{for } n = 100 \text{ observations}$$

The point affects the sample standard deviation estimate s most significantly.

EXAMPLE 7.7 ***Estimating the Heat, Moisture, and Ash Contents of a Single Sample***

The effect of MSW moisture content in Section 7.1 was shown to have an important bearing in estimating unit cost from recovering heat energy from combustion. With the aid of the weighted average, it is possible to straightforwardly calculate the moisture content, heat value, and ash content of an MSW sample. Suppose the sample contained the following fractions.

k	Component	ω_k
1	Paper	0.45
2	Vegetable and animal waste	0.22
3	Rag and textiles	0.05
4	Plastics, mixed	0.03
5	Metal	0.09
6	Glass	0.06
7	Dust and cinder	0.10

Estimate the moisture content, heat value, and ash content of the sample. In addition, estimate the heat value on a dry weight basis.

SOLUTION

The following table gives the gross heat values and the proximate moisture and ash contents of various components of MSW.

k	Component	m_k, Moisture Content (% as collected)	h_k, Heat Value (kcal/kg, as collected)	a_k, Ash Content (% as collected)
1	Paper	10	3778	5.4
2	Vegetable and animal waste	72	1317	4.5
3	Rag and textiles	10	3833	2.4
4	Plastics, mixed	2	7833	10
5	Metal	0	—	100
6	Glass	0	—	100
7	Dust and cinder	3.2	2039	70

These data were obtained from calorimetric bomb tests conducted on MSW. Since no heat value for glass was reported, it is assumed to be zero. See Table I in Appendix A for more detailed information.

The ash content a, as-collected heat value h, and moisture content m of an MSW mixture can be estimated as weighted averages. Since the calculations are easily determined, the governing equations and numerical results are given.

$$\bar{a} = \sum_k \omega_k a_k = 25\%$$

$$\bar{h} = \sum_k \omega_k h_k = 2620 \text{ kcal per kg}$$

$$\overline{m} = \sum_k \omega_k m_k = 21\%$$

The heat content on a dry weight basis is

$$\overline{h}_d = \frac{\overline{h}}{1-\overline{m}} = 33\,16 \text{ kcal per kg}$$

EXAMPLE 7.8 *Historical Data and Tipping Fee*

The study of balance sheets and other financial data from two existing MSW facilities gives the following data.

j, Plant	t, Construction Year	$k(0)$, Capital Cost in Construction year t	b, Plant Capacity (tons/day)	$k(t)$, 1987 Total Annual Operating Cost	u, Utilization (% of plant capacity)
1	1980	$3.3M	100	$1.5M	100%
2	1985	$4.1M	100	$2.0M	70%

Calculate the tipping fee using amortized capital debt and the plants are operating at $u = 100\%$ (full capacity). The design life of the project is 20 years with a discount rate of $i = 4\%$ per year. Assume that the current year ($t = 0$) is 1987, that all prices are to be calculated in 1987 dollars, and there are 250 working days per year.

SOLUTION

The tipping fee is calculated as the unit cost in $ per ton,

$$u = \frac{k(t) + k}{250\,b}$$

where

$k(t)$ = total annual operating cost for a plant running at capacity in year t

k = amortized capital cost k(0) in $ per year

b = plant capacity

$k = k(0)\,crf(i, n)$

where $k(0)$ = construction cost in 1987 and the capital recovery factor is

$$crf(i,n) = \frac{i(1+i)^n}{(1+i)^n - 1} = \frac{0.04(1.04)^{20}}{(1.04)^{20} - 1} = 0.0736$$

The capital costs are in 1987 dollars, using the CCI price index.

Plant 1: $k(0) = \frac{\text{CCI}(1987)}{\text{CCI}(1981)} k(1981) = \frac{4406}{3535} \$3.3\text{M} = \$4.1\text{M}$

Plant 2: $k(0) = \frac{\text{CCI}(1987)}{\text{CCI}(1985)} k(1985) = \frac{4406}{4195} \$4.1\text{M} = \$4.3\text{M}$

The CCI is used because these facilities were constructed primarily by unskilled laborers.

The annual amortized capital debt, $k = k(0)crf(i, n)$, is

Plant 1: $k = \$4.1\text{M}(0.0736) = \$0.302\text{M}$ per year

Plant 2: $k = \$4.3\text{M}(0.0736) = \$0.316\text{M}$ per year

The tipping fees, $u = \frac{k(t)+k}{250b}$, for the two plants are

Plant 1: $u = \frac{\$1.5\text{M} + 0.302\text{M}}{(250)(100)} = \$72.08$

Plant 2: $u = \frac{(\$2\text{M}/0.7) + 0.316\text{M}}{(250)(100)} = \$126.93$

where the $k(t)$ estimate for plant 2 is based on the assumption that annual operating cost is a linear function of plant utilization u.

The average tipping fee for the two plants is

$$\bar{u} = \frac{u_1 + u_2}{2} = \frac{\$72.08 + \$126.93}{2} = \$99.51 \text{ per ton}$$

$$s^2 = \frac{(u_1 - \bar{u})^2 + (u_2 - \bar{u})^2}{1} = (72.08 - 99.51)^2 + (126.93 - 99.51)^2 = 1504$$

$$s = \$38.78 \text{ per ton}$$

or

$$\bar{u} \pm s = \$99.51 \pm \$38.78 \text{ per ton}$$

DISCUSSION

Valuable information can be gained from using historical data. Clearly, more confidence would be placed in the average tipping fee if the sample size $n = 2$ was larger. These calculations show how it is possible to adjust the data for different years and plant utilization.

In spite of this care in evaluating the data, there are possible sources of measurement and sampling error. For example, the actual loan periods, interest rates, and tipping fees for the two plants based on 1980 and 1985 dollars may not coincide with estimates made with 1987 dollars. The key factor is that the data are transposed to the same base (1987) year through discounting factors. Thus, the two tipping fee estimates for plant 1 and 2 are comparable, and their average may be calculated.

Furthermore, it is tacitly assumed that the accounting methods for compiling the data are the same. Any major difference in accounting methods may bias the data.

QUESTION FOR DISCUSSION

7.3.1 Define the following terms:

measurement error	random sample
process variability	sample error

QUESTIONS FOR ANALYSIS

7.3.2 The number of pounds of newspapers discarded each week from 100 households is as follows:

13	6	15	7	13	10	8	14	14	5
9	17	10	5	9	12	12	16	5	10
13	10	8	11	16	20	14	16	10	13
7	10	7	16	12	14	12	8	16	16
15	17	14	7	4	16	9	7	11	19
16	11	12	15	10	19	18	13	6	7
15	18	11	13	19	14	15	11	11	13
11	17	10	13	13	13	4	16	16	16
12	17	13	20	18	12	15	6	15	9
18	11	8	7	12	6	15	11	13	12

Plot a histogram using intervals of two units in width starting with zero. Does the data set seem to have a normal distribution? If not, which distribution do you feel is more suitable?

7.3.3 The number of pounds of aluminum discarded each week from 100 households is as follows:

0.1	0.4	0.5	0.7	0.1	1.5	0.0	0.7	0.0	0.0
0.0	0.0	0.0	0.8	0.0	0.5	0.8	0.8	0.0	1.3
1.1	0.0	1.2	0.4	0.0	0.6	0.6	1.1	0.7	0.0
0.0	0.0	1.4	0.4	0.0	0.6	1.4	0.5	0.0	1.1
1.0	0.0	0.0	1.4	0.5	0.0	0.0	0.0	1.3	0.0
0.1	0.5	0.7	0.3	0.0	0.0	1.2	1.2	0.7	0.0
0.0	1.0	1.2	0.8	0.9	0.5	0.0	0.0	1.3	0.0
0.0	0.9	0.0	1.0	0.4	0.8	1.3	0.8	0.0	0.1
0.1	0.0	1.3	0.0	0.0	0.0	1.5	1.2	0.6	0.0
0.0	0.6	0.9	0.6	0.1	0.4	0.7	0.0	0.1	0.1

Plot a histogram using intervals of 0.1 unit in width starting with zero. Does the data set seem to have a normal distribution? If not, which distribution do you feel is more suitable?

7.3.4 Consider the following data compiled for mass-burn MSW facilities.

$k(0)$ Capital Cost	b, Capacity (tons per day)	t, Year	Products	Location
\$3.2M	50	1984	Steam	AL
\$35M	480	1986	Electricity	CA
\$24M	450	1986	Electricity	CT
\$38M	510	1987	Steam and electricity	FL
\$1.5M	50	1982	Steam	ID
\$10.8M	240	1981	Steam	MA
\$3.3M	100	1980	Steam	NH
\$253M	3000	1985	Electricity	NJ
\$42M	500	1987	Electricity	TX

Source: Neal and Schubel (1987).

(a) Use 1988 dollars and the BCI price index to estimate $k(1988)$.

(b) Calculate $\bar{k} \pm s_K$ for (a).

(c) Calculate $\bar{c} \pm s_C$ where c is defined as the unit capital cost, $c(1988) = k/b$ on a dollar per unit ton of capacity for each plant. Use the data set from (a).

(d) If a plant has a capacity of 480 tons per day, estimate its expected capital cost and standard deviation.

(e) List possible sources of error for estimating the capital cost of a 480-ton-per-day plant built in Boston from (d). In your opinion, which is the best estimate, $\bar{k}$ or $k(1988)$ for the 480-ton-per-day unit from (a).

7.3.5* The following table shows a dramatic difference in the composition of MSW between different regions of the world:

k	Component	Weight Fraction, ω_k Europe	USA	Middle East
1	Paper	0.27	0.46	0.25
2	Vegetable and animal waste	0.21	0.23	0.62
3	Rag and textiles	0.04	0.05	0.01
4	Plastics, mixed	0.03	0.03	0.06
5	Metal	0.09	0.09	0.03
6	Glass	0.10	0.06	0.01
7	Dust and cinder	0.27	0.10	0.01

Source: Buekens and Patrick (1974).

Compare the average and standard deviations, $\bar{y} \pm s$, for each region for the following properties:

(a) Ash content

(b) Moisture content

(c) Heat value, as collected

Use the appropriate data given in Example 7.7.

7.3.6* (a) Calculate the as-collected heat values for each region, assuming that all paper and plastics are removed for recycling. Use the appropriate data given in Example 7.7 and Problem 7.3.5. Compare them to the values with no plastics and paper removed.

(b) The average American is estimated to dispose of over 3.5 pounds of MSW per day and other industrialized countries, less than one-half this figure. For purposes of comparison, assume European and Middle Eastern countries dispose of 1.7 pounds of MSW per day. Compare the expected as-collected heat and standard deviation values, measured in kcal per person-day, for the regions. 1 lb = 0.4536 kg.

7.4 STATISTICAL INFERENCE

In this section, we address questions of uncertainty associated with model parameter estimation and forecasting. The size of the data set from which these estimations are made is an important consideration. Suppose, for instance, that the exponential distribution, $f(t) = \rho e^{-\rho t}$, is to be used as a time-to-failure model for a new machine design. If a small number of machines have been in service for a small fraction of the design life of the equipment, then a lack of operational experience and data could prove a serious obstacle in making a reliable estimate of ρ. With the use of sampling theory and the central limit theorem, problems like these will be addressed. Our ultimate aim is to establish $100(1-\alpha)\%$ confidence intervals for parameter estimates and forecasts.

The Sample Mean $\bar{Y}$

Let $Y_1, Y_2, \ldots, Y_n$ be a random sample of size n from a distribution with mean μ and variance σ^2. A random sample means that $Y_1, Y_2, \ldots, Y_n$ are drawn from the same distribution and that $Y_1, Y_2, \ldots, Y_n$ are independent random variables. The sample mean is the random variable

$$\overline{Y} = \frac{\sum_{j=1}^{n} Y_j}{n}$$

Its mean and variance are

$$E(\overline{Y}) = E\left(\frac{1}{n}\sum_{i=1}^{n} Y\right) = \frac{n\mu}{n} = \mu$$

and

$$\text{var}\,(\overline{Y}) = E[(\overline{Y}-\mu)^2\,] = E\left[\left\{\frac{1}{n}\sum_{i=1}^{n}(Y_i-\mu)^2\right\}\right] = \frac{n\sigma^2}{n^2} = \frac{\sigma^2}{n}$$

Two important conclusions follow from this result.

1. The estimate of the population mean is unbiased because

$$\text{E}(\overline{Y}) = \mu$$

The expected value of the sample mean is equal to mean μ of the parent distribution, which is the underlying probability or probability model from which samples are drawn.

2. The variability of $\overline{Y}$ around μ decreases as the sample size grows. In fact, as $n \to \infty$, $\text{var}(\overline{Y}) \to 0$.

The Central Limit Theorem

The sample mean $\overline{Y}$ can be expressed as a standardized random variable,

$$Z = \frac{\overline{Y}-\mu}{\sigma/\sqrt{n}}$$

The *central limit theorem* states that, regardless of the individual distribution of Y_j, the distribution of

$$Z = \frac{\overline{Y}-\mu}{\sigma/\sqrt{n}}$$

approaches an $N(0, 1)$ distribution as n becomes large.

The most remarkable property of the central limit theorem is that $\overline{Y}$ tends to be normally distributed for large n even when the parent distribution is not normal. If the parent distribution has a skewed distribution such as an exponential, for example, the distribution of the sample mean of the sample sums of this distribution will tend to be normal for a sufficiently large n. In Figure 5, the sum of two exponentially distributed random variables shows signs of this property. Of course, as the sample size n increases, the distribution becomes more bell-shaped.

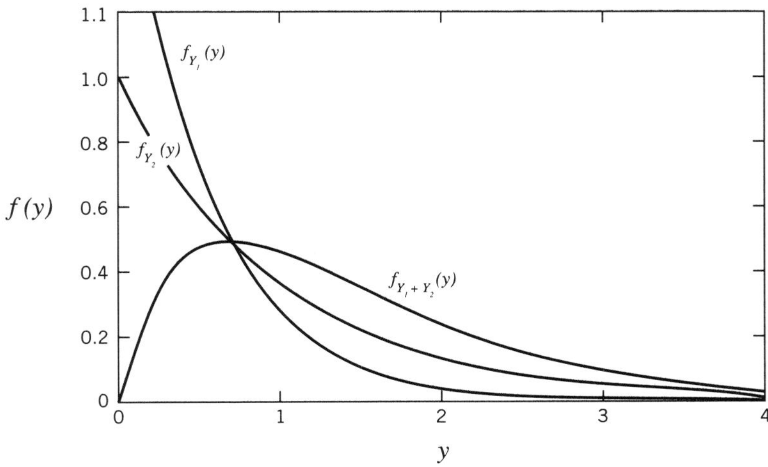

Figure 5 The distribution for the sum of two exponential distributions Y_1 and Y_2.

The 100(1 – α)% Confidence Intervals with σ Known

A 100(1 – α)% confidence interval is defined as an interval with random endpoints, L_Y and U_Y, with the property that the interval includes the unknown, theoretical mean μ. The constant α is called the "significance level" or the probability that the interval does not contain μ. For a confidence interval, 1 – α is called the "confidence level." A two-sided confidence interval is shown in Figure 6.

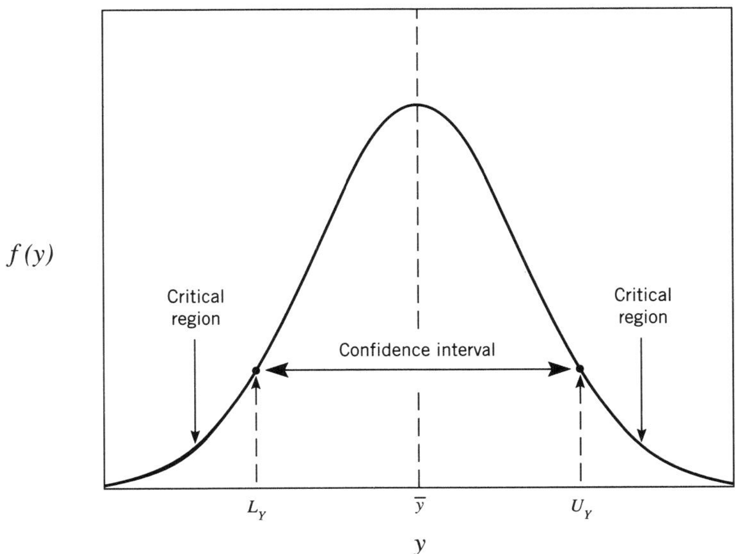

Figure 6 A two-sided 100(1 – α)% confidence interval.

We want to find the random variables, L_Y and U_Y, so that

$$P(L_Y \le \mu \le U_Y) = 1 - \alpha$$

Assume that $\overline{Y}$ has the distribution $N(\mu, \sigma^2/n)$. Taking advantage of the symmetry of the normal distribution and assuming that

$$\frac{\overline{Y} - \mu}{\sigma / \sqrt{n}} \sim N(0,1)$$

then

$$P\left(-z_{\alpha/2} \le \frac{\overline{Y} - \mu}{\sigma / \sqrt{n}} \le z_{\alpha/2} \right) = 1 - \alpha$$

It follows that

$$P(\overline{Y} - \frac{z_{\alpha/2}\sigma}{\sqrt{n}} \le \mu \le \overline{Y} + \frac{z_{\alpha/2}\sigma}{\sqrt{n}}) = 1 - \alpha$$

$$L_\mu = \overline{Y} - \frac{z_{\alpha/2}\sigma}{\sqrt{n}}$$

and

$$U_Y = \overline{Y} + \frac{z_{\alpha/2}\sigma}{\sqrt{n}}$$

where

$$z_{\alpha/2} = \Phi^{-1}(1 - \alpha/2)$$

A two-sided $100(1 - \alpha)\%$ confidence interval for μ with known σ is thus

$$[L_Y, U_Y] = \overline{y} \pm \frac{z_{\alpha/2}\sigma}{\sqrt{n}}$$

where n is large, typically $n > 30$ observations. This comment about σ being known and $n > 30$ will be clearer after we discuss the student's t distribution.

Using a similar approach, we find that the lower and upper bound one-sided $100(1 - \alpha)\%$ confidence intervals for μ with known σ are

$$P(\mu > L_Y) = P\left(-z_\alpha > \frac{\overline{Y} - \mu}{\sigma / \sqrt{n}} \right) = 1 - \alpha$$

$$L_Y = \overline{y} - \frac{z_\alpha \sigma}{\sqrt{n}}$$

and

$$P(\mu \le U_Y) = P\left(\frac{\overline{Y}-\mu}{\sigma/\sqrt{n}} \le z_\alpha\right) = 1-\alpha$$

$$U_Y = \overline{y} + \frac{z_\alpha \sigma}{\sqrt{n}}$$

The Student's *t* Distribution

In our discussion of the confidence interval, the variance σ^2 of the parent distribution model is assumed to be known. In practice, however, σ^2 is often not known but is estimated from sampling. The sample variance is calculated as

$$S^2 = \frac{\sum_{j=1}^{n} (Y_j - \overline{Y})^2}{n-1}$$

If we replace σ in $\dfrac{\overline{Y}-\mu}{\sigma/\sqrt{n}}$ with S, the resulting distribution has a student's t distribution. It is similar in shape to the normal distribution—bell-shaped—but unlike the normal distribution, the spread of the distribution is dependent on the sample size n. The random variable T, defined by

$$T = \frac{\overline{Y}-\mu}{S/\sqrt{n}}$$

where S is defined above, is said to have a student's t distribution with $n - 1$ degrees of freedom.

The degrees of freedom df is a parameter associated with the distribution. It is defined as the number of observations needed to define the data set of **y** of sample size n. For example, if $\overline{y}$ and the first $n - 1$ values of y are known, then this information is sufficient to calculate the remaining point, y_n; thus, $df = n - 1$.

As df or n increases, the T distribution becomes more like the standard normal. For $df > 30$, the two distributions are, for all practical purposes, the same. For small df, generally $df < 15$, the normal distribution underestimates the probability in the tails. The percentiles of the student's t distribution are given in Appendix B, Table IV.

A two-sided $100(1 - \alpha)\%$ confidence interval for μ with unknown σ is

$$[L_Y, U_Y] = \overline{y} \pm \frac{t_{(1-\alpha/2,\ df)} S}{\sqrt{n}}$$

where

$$df = n - 1$$

s = sample standard deviation

The lower and upper bound one-sided $100(1-\alpha)\%$ confidence intervals for μ with unknown σ are

$$L_Y = \bar{y} - \frac{t_{(1-\alpha,\, df)}s}{\sqrt{n}}$$

and

$$U_Y = \bar{y} + \frac{t_{(1-\alpha,\, df)}s}{\sqrt{n}}$$

EXAMPLE 7.9 ***Confidence Intervals for Moisture Content, Unit Shipping Cost, and Net Worth***

This is a follow-up question to Example 7.1 on establishing a decision rule for route selection. Determine the 95% two-sided confidence intervals for the mean estimates of

- Moisture content M: $\bar{m} \pm s_M = 25\% \pm 6\%$ with sample size $n = 9$ (See Example 7.6)
- Unit shipping cost S: $\bar{s} \pm s_S = \$2.50 \pm 0.50$ per ton-mile with sample size $n = 30$
- Net worth NW: $NW = 120 - 1.20M - Sd$ for route 1 with $d = 15$ miles and route 2 with $d = 35$ miles

SOLUTION

The 95% two-sided confidence interval for moisture content M is

$$[L_M, U_M] = 25 \pm \frac{2.306(6)}{\sqrt{9}} = 25 \pm 4.61$$

$$[L_M, U_M] = [20.4\%, 29.6\%]$$

where $t_{(1-\alpha/2,\, df)} = t_{(0.975,\, 8)} = 2.306$ where $df = n - 1 = 9 - 1 = 8$. Note that in this calculation and the following one, the standard deviations σ are treated as unknowns and estimated as the sample means s.

The 95% two-sided confidence interval for unit shipping cost S is

$$[L_S, U_S] = 2.50 \pm \frac{1.980(0.50)}{\sqrt{30}} = 2.50 \pm 0.18$$

$$[L_S, U_S] = [\$2.32, \$2.68 \text{ per ton-mile}]$$

where

$$t_{(1-\alpha/2,\, df)} = t_{(0.975,\, 29)} = 1.980 \text{ where } df = n - 1 = 30 - 1 = 29$$

The 95% two-sided confidence interval for net worth *NW* for routes 1 and 2 are

$$\text{Route 1:} \qquad [L_{NW}, U_{NW}] = [\$44.28, \$60.72 \text{ per ton}]$$

where

$$L_{NW} = 120 - 1.20L_M - 15L_S \text{ and } U_{NW} = 120 - 1.20U_M - 15L_S$$

$$\text{Route 2:} \qquad [L_{NW}, U_{NW}] = [-\$9.32, \$14.32 \text{ per ton}]$$

where

$$L_{NW} = 120 - 1.20L_M - 35L_S \text{ and } U_{NW} = 120 - 1.20U_M - 35U_S$$

There is a 95% chance that the net worth of routes 1 and 2 will lie in the ranges of \$44.28 and \$60.72 and –\$9.32 and \$14.32 per ton, respectively. Since the range of net worth values for route 1 at a 95% confidence level excludes \$0 per ton and no monetary losses, route 1 may be regarded as a feasible alternative. This claim cannot be made for route 2, however. Clearly, the variance plays an important role in decision making.

EXAMPLE 7.10 ***Design Life Revisited***

Example 6.4 analyzes a failure of a roller grate system. The time to failure T is assumed to be an exponential distribution with annual failure rate ρ. The model parameter is estimated to be $\hat{\rho} = 0.075$ failure per year. It is based on the assumption that the survival frequency and reliability are equal at time $t = 1.2$ years. The number of roller grates at time zero is $n(0) = 35$ units.

The failure times are: 0.5, 1, and 1.2 years. Use these data to estimate ρ and to construct a 95% confidence interval for the probability that a failure is observed within the 20-year design of the roller grate system. In Example 6.4, this probability is calculated to be $P(T \le 20) = 1 - e^{-20\rho} = 0.781$ where $\rho = \hat{\rho} = 0.075$.

SOLUTION

The governing equation for estimating ρ is determined by equating the survival frequency with the reliability function as

$$\frac{n(t)}{n(0)} = r(t) = e^{-\rho t}$$

or solving for ρ, the relationship is

$$\rho = \frac{-1}{t} \ln\left[\frac{n(t)}{n(0)}\right]$$

where

$$n(t) = \text{number of survivals at time } t$$

This same approach is used in Example 6.4, except that here all observed failure times of 0.5, 1, and 1.2 years are used in the estimation of ρ.

Since only a small portion of the roller grates have failed, 3 of 35 units, an estimate of the mean time to failure μ will be estimated with the three failure time observations, or

$$\bar{\mu} = \frac{\sum_{j=1}^{3} \mu_j}{3}$$

where

$$\mu_j = \frac{1}{\rho_j}$$

The ρ_j estimates are calculated as

$$\rho_j = \frac{-1}{t_j} \ln\left[\frac{n(t_j)}{n(0)}\right]$$

where

$$t_j = \text{time when the } j\text{th unit fails}$$

and

$$n(t_j) = \text{number of units operating at time}$$

or

$$n(t_j) = n(0) - j = 35 - j$$

t_j	$n(t_j)$	ρ_j	μ_j
0.5	34	0.058	17.3
1.0	33	0.059	17.0
1.2	32	0.075	13.4

Since the expected failure time is $E(T) = 1/\rho$, its mean and variance are used to estimate its lower and upper 95% confidence interval.

$$\bar{\mu} = \frac{\sum_{j=1}^{n} \mu_j}{n} = 15.9$$

and

$$s_{\mu}^2 = \frac{\sum_{j=1}^{n} (\mu_j - \bar{\mu})^2}{n-1} = 4.71, \text{ or } s_{\mu} = 2.17$$

with $n = 3$ observations.

The 95% confidence interval of the mean failure time μ is

$$[L_{\mu}, U_{\mu}] = \bar{\mu} \pm \frac{t_{(0.975,\ 2)} s_{\mu}}{\sqrt{n}} = 15.9 \pm \frac{(4.303)(2.17)}{\sqrt{3}}$$

$$[L_{\mu}, U_{\mu}] = [10.5, 21.3]$$

Since the failure rate is equal to $\rho = \frac{1}{\mu}$ and $\bar{\mu}$ is an estimate of the theoretical mean μ, the value of ρ is estimated to be $\hat{\rho} = \frac{1}{\bar{\mu}} = 0.063$. It follows that the lower and upper bounds of ρ are

$$[L_{\rho}, U_{\rho}] = \left[\frac{1}{U_{\mu}}, \frac{1}{L_{\mu}} \right]$$

$$[L_{\rho}, U_{\rho}] = [0.047, 0.095]$$

Using these estimates, we find that the probability of failure in a 20-year design life is

$$P(T \le 20) = 1 - e^{-0.063 \text{ x } 20} = 0.716$$

with

$$[P_L(T \le 20), P_U(T \le 20)] = [1 - e^{-(0.47)(20)}, 1 - e^{-(0.095)(20)}]$$

$$[P_L(T \le 20), P_U(T \le 20)] = [0.609, 0.850]$$

The probability estimate of $P(T \le 20) = 0.781$ for $\rho = 0.075$, given in Example 6.4, lies within the confidence interval. Since the true population parameter ρ is unknown, it cannot be said that one estimate is better than the other one.

The 100(1 – α)% Two-Sided Tolerance Intervals

Suppose a forecast model is derived to forecast the daily tonnage of MSW to be collected and processed. The model provides a point estimate of $\bar{y}$. This statistic is useful in assigning the number of workers that must be available each day. It is better to have a lower and upper forecast bound $[L_Y, U_Y]$ than the statistic $\bar{y}$ only. In a one-year period with 250 working days, it would be advantageous to claim that at least in 95% of these 250 days (about 238 days) there is a 9 in 10 chance (90% of the time) that the forecast will fall within these bounds. A tolerance interval is used for this purpose.

A tolerance interval is an interval that one can claim to contain at least a specified proportion 100p% of the forecast population with a $100(1-\alpha)\%$ confidence. In terms of the example, the minimum proportion $p = 0.95$ and the confidence probability $100(1-\alpha)\% = 95\%$.

The tolerance interval is constructed with the use of the following equation,

$$[L_Y, U_Y] = \bar{y} \pm g_{(1-\alpha,\, p,\, n)} s$$

where $g_{(1-\alpha,\, p,\, n)}$ = tolerance factors, given in Table V of Appendix B. Its application is most easily shown by example.

EXAMPLE 7.11 ***Confidence and Forecast Models***

Consider a revenue model for a waste-to-energy facility.

$$W = 120 - 1.20M$$

where the revenue decreases with an increase in MSW moisture content M.

(a) If the random variables M is assumed to have a normal distribution with $\bar{m} \pm s_M$ = 25% ± 6% with sample size $n = 9$ (see Example 7.9), determine a 95% two-sided tolerance interval that will contain at least 95% of the predictions of M.

(b) Use the tolerance interval of M to construct a tolerance interval for expected revenue W.

SOLUTION

(a) The 95% two-sided tolerance interval for M where $(1-\alpha) = 0.95$, $p = 0.95$, and $n = 9$ is calculated as

$$[L_M, U_M] = \bar{m} \pm g_{(0.95,\, 0.95,\, 9)} s_M$$

or

$$[L_M, U_M] = 25\% \pm (3.546)(6\%) = 25\% \pm 21.28\%$$

$$[L_M, U_M] = [3.72\%, 46.28\%]$$

(b) The tolerance for expected revenue $E(W)$ is

$$[L_W, U_W] = [120 - 1.20U_M, 120 - 1.20L_M]$$

$$[L_W, U_W] = [\$64.46, \$115.54 \text{ per ton}]$$

DISCUSSION

In Example 7.9, the 95% two-sided confidence interval for parameter μ_M is $[L_M, U_M]$ = [20.4%, 29.6%]. It is a much tighter interval than the 95% two-sided tolerance interval for M given in (a).

EXAMPLE 7.12 ***Process Variability and Sensitivity Analysis***

This problem illustrates how some of the concepts developed in this chapter can be put to practical use. In particular, introducing a tolerance interval into a sensitivity analysis gives more meaning to the result. A sensitivity analysis is to be performed where the effects of the selling price of recovered energy and moisture content are studied.

The tipping fee, calculated as the unit production cost minus the price of recovered energy, will be used as the measure of effectiveness,

$$u = c - p\,h\,e\,(1 - m)$$

where

c = unit production cost, including collection, operation, and maintenance = \$100 per ton of as-collected MSW

p = selling price of delivered energy (\$ per kW-h)

h = 3317 kcal/kg or 3500 kWh per ton of MSW dry weight

e = thermal efficiency of plant and delivery system = 0.5

m = moisture content containing 90% of the forecast population with a 90% two-sided tolerance interval, $[L_M, U_M]$ = [6.25%, 43.8%]

(a) Perform a sensitivity analysis to determine the tipping fee range when p and m are varied from $0 \le p \le 0.20$ and $0 \le m \le 50$.

(b) Determine the tipping fee range given the 90% two-sided tolerance interval for moisture content.

(c) Compare the ranges from parts (*a*) and (*b*) for p = \$0.07 per kW-h.

SOLUTION

(a) After substituting the given values of c, e, and h, the tipping fee equation reduces to

$$u = \$100 - 1750p(1 - m)$$

where p and m are treated as variables in a sensitivity analysis. The results of the analysis are shown in the following figure. Negative values of u indicate a profit.

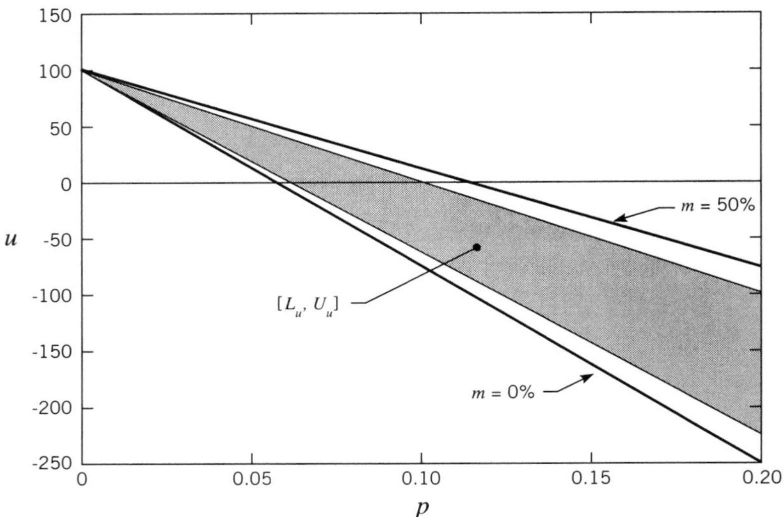

(b) Assuming the moisture content is a random variable M, the equation for tipping fee is

$$U = \$100 - 1750p(1 - M)$$

The boundaries of its tolerance interval or tipping fee range are calculated as

$$[L_U, U_U] = [100 - 1750p(1 - U_M), 100 - 1750p(1 - L_M)]$$

with $[L_M, U_M] = [0.0625, 0.438]$ and p is varied over a range of $0 \leq p \leq 0.20$. The tolerance region for U is shown as the shaded region.

(c) If the price of delivered energy is $p = \$0.07$ kW-h, the sensitivity analysis gives the following range of values:

$$-\$22.50 \leq u \leq \$38.75$$

In comparison, using the tolerance interval for moisture content, the 90% tolerance range for the tipping fee is tighter,

$$[L_U, U_U] = [-\$14.84, \$31.16]$$

or

$$-\$14.84 \leq u \leq \$31.16$$

DISCUSSION

The figure shows both p and m having a major effect on the tipping fee. In addition, it is worth mentioning that e also has an important bearing upon economic feasibility. It can range from as low as 30% and to as high as 75%. However, as compared to *c*, *p*, and *e*, the values of *h* and *m* are expected to demonstrate high variability.

The moisture contents of MSW in the USA and western European countries are reported to range from 25 to 40% by weight, and in Japan and Mediterranean countries, 50% by weight and more. The MSW mixture can change on a daily and seasonal basis. Paper and plastic have a relatively large heat values, thus removing them for the purposes of recycling, say, could have a significant bearing on the overall MSW management plan.

Sample Size Determination

The central limit theorem can be used to determine the sample size n needed to estimate the mean μ within $\pm d$ with degree of confidence equal to $100(1 - \alpha)\%$. Let the deviation d between the sample and theoretical means be equal to

$$d = \bar{Y} - \mu$$

The value *d* is set equal to the width of the confidence interval thus,

$$z_{\alpha/2} = \frac{\bar{Y} - \mu}{\sigma / \sqrt{n}} = \frac{d}{\sigma / \sqrt{n}}$$

and the function solved for *n*,

$$n = \left(\frac{\sigma z_{(1-\alpha/2)}}{d} \right)^2$$

In practice, since the variance σ^2 is not known, it is estimated as s^2.

For example, suppose that we wanted to be 99% certain in Example 7.6 that $\overline{m}$ is within ±2%, that is, $d = \pm 2$. Since the best information available on σ is $s_M = 6\%$, it is used in the calculation. For $z_{0.005} = -2.58$. The sample size is calculated to be

$$n = \left[\frac{(6)(-2.58)}{2} \right]^2 \approx 60 \text{ samples}$$

In Example 7.6, the sample size is $n = 9$. Clearly, there is a cost to gaining a higher degree of confidence in the estimate of the mean.

QUESTION FOR DISCUSSION

7.4.1 Define the following terms:

central limit theorem	student's *t* distribution
confidence interval	tolerance interval
sample mean and variance	

QUESTIONS FOR ANALYSIS

7.4.2 A new mass-burn MSW plant with electrical power generation is being considered for construction in a metropolitan area. The following data are considered representative of the plant being considered for construction.

i	b, Design Capacity (tons/day)	$k(0)$, Capital Cost ($M)
1	3000	253
2	2250	200
3	400	40
4	3000	290
5	400	30
6	2250	250
7	518	40
8	1650	113

Source: Neal and Schubel (1987).

(a) Calculate $\bar{u} \pm s_U$ where the unit cost $u = k(0)/b$ for each plant based on its plant capacity.

(b) Estimate the 90% upper confidence bound for unit construction cost.

7.4.3 Compare the 95% two-sided confidence intervals for the estimate of $\mu = \bar{y} = 100$ with known standard deviation $\sigma = 20$ and estimated standard deviation $s = 20$, which is based on a sample of $n = 10$ observations.

7.4.4* The number of pounds of newspapers discarded each week by 100 households is as follows:

13	6	15	7	13	10	8	14	14	5
9	17	10	5	9	12	12	16	5	10
13	10	8	11	16	20	14	16	10	13
7	10	7	16	12	14	12	8	16	16
15	17	14	7	4	16	9	7	11	19
16	11	12	15	10	19	18	13	6	7
15	18	11	13	19	14	15	11	11	13
11	17	10	13	13	13	4	16	16	16
12	17	13	20	18	12	15	6	15	9
18	11	8	7	12	6	15	11	13	12

This is the same data set given in Problem 7.3.2.

(a) Calculate the means and standard deviations for

1. $n = 10$ (first row)

2. $n = 20$ (first two rows)

3. $n = 30$ (first three rows)

4. $n = 100$

(b) Calculate the two-sided 95% confidence interval for each case in (a).

7.4.5* The number of pounds of aluminum discarded each week by 100 households is as follows:

0.1	0.4	0.5	0.7	0.1	1.5	0.0	0.7	0.0	0.0
0.0	0.0	0.0	0.8	0.0	0.5	0.8	0.8	0.0	1.3
1.1	0.0	1.2	0.4	0.0	0.6	0.6	1.1	0.7	0.0
0.0	0.0	1.4	0.4	0.0	0.6	1.4	0.5	0.0	1.1
1.0	0.0	0.0	1.4	0.5	0.0	0.0	0.0	1.3	0.0
0.1	0.5	0.7	0.3	0.0	0.0	1.2	1.2	0.7	0.0
0.0	1.0	1.2	0.8	0.9	0.5	0.0	0.0	1.3	0.0
0.0	0.9	0.0	1.0	0.4	0.8	1.3	0.8	0.0	0.1
0.1	0.0	1.3	0.0	0.0	0.0	1.5	1.2	0.6	0.0
0.0	0.6	0.9	0.6	0.1	0.4	0.7	0.0	0.1	0.1

This is the same data set given in Problem 7.3.3.

(a) Calculate the sample mean and standard deviation for

1. $n = 10$ (first row)

2. $n = 20$ (first two rows)

3. $n = 30$ (first three rows)

4. $n = 100$

(b) Calculate the two-sided 95% confidence interval for each case in (a).

(c) With the exception of $x = 0$, the histogram of these data, Problem 7.3.3, suggests that the distribution for these data may be uniform. Calculate the theoretical mean and standard deviation, $\mu \pm \sigma$, of a uniform distribution,

$$f(x) = \frac{1}{b-a} \quad \text{for } a = 0 \le x \le b = 1.5$$

with the sample mean and standard deviation, $\bar{x} \pm s$, for $n = 100$ observations. Does a uniform distribution of $f(x)$ seem to be a reasonable mathematical model for these data? Explain your answer.

7.4.6* Suppose after six years of operation 10 of 20 machines have failed The failure times in years are:

0.5, 0.6, 1.0, 1.1, 1.3, 3, 4.5, 5, 5.5, and 5.9

Assume

$$T = \{\text{time to failure}\}$$

where T is an exponential distribution.

(a) Estimate the failure rate ρ.

(b) Calculate the 95% confidence interval for the failure rate estimate.

(c) Estimate the 95% confidence interval for $P(T > 10)$.

7.4.7 Let

$$P = \{\text{selling price of recycled goods}\}$$

with $\overline{p} \pm s_P = \$50 \pm \$15$ per ton for $n = 5$ observations and

$$C = \{\text{cost of processing recycled goods}\}$$

with $\overline{c} \pm s_C = \$45 \pm \$5$ per ton for $n = 15$ observations and

$$x = 1000 \text{ tons of recycled goods per year}$$

(a) Estimate the expected annual net profit.

(b) Calculate the lower and upper 90% tolerance bounds that the expected annual net profit will be exceeded 90% of the time.

7.4.8 How many animal test samples are needed, if $\theta(d)$ = probability of death at dose d has a standard deviation of 0.01 and the deviation from $\theta(d)$ should be within ± 0.001? Assume a 95% confidence level.

REFERENCES

Bergvall, G., and J. Hult (1985). *Technology, Economics, and Environmental Effects of Solid Waste*, National Swedish Protection Board, No. 33, Solna, Sweden.

Bergström J., and J. Lundqvist (1983). *Operational Studies at the SYSAV Energy from Waste Plant Malmö*, National Swedish Protection Board, No. 11, Solna, Sweden.

Box, G. E. P., W. G. Hunter, and J. S. Hunter (1978). *Statistics for Experimenters*, Wiley Interscience, New York.

Buekens, M., and P. K. Patrick (1974). "Incineration," *Energy from Solid Waste*, Noyes Data Corporation, Park Ridge, NJ, 79–150.

Hahn, G. J., and W. Q. Meeker (1991). *Statistical Intervals: A Guide for Practitioners*, John Wiley & Sons, New York.

Hogg, R. V., and J. Ledolter (1987). *Engineering Statistics*, Macmillan, New York.

Neal, H. A., and J. R. Schubel (1987). *Solid Waste Management and the Environment*, Prentice Hall, Englewood Cliffs, NJ.

Thunberg, B., ed. (1987). *Acidification and Air Pollution, A Brief Guide*, National Swedish Environmental Protection Board, Solna, Sweden.

Chapter 8

Model Calibration

All models are wrong, but some are useful.

George E.P. Box (1991)

The procedure for model calibration consists of the following steps:

- Data collection
- Model identification
- Parameter estimation
- Model validation

Based on data from statistical experiments, we can form a basis for identifying a model. When the data set is of sufficient size, a histogram is useful, but theory, knowledge and experience with a system also provide useful information when choosing a model. In the validation stage, the model and its parameter estimates are tested to ensure that they are representative of the data.

The procedure is general and can be applied to a broad class of models. In this chapter, we consider calibration of probability models and regression models. Monte Carlo simulation is a procedure that generates a data set for a complex system by enumeration. These data can be used, among other things, for model calibration.

8.1 PROBABILITY MODELS

Assume that a data set, $y_1, y_2, \ldots, y_n$, has been collected where n = number of observations in the data set. The next task is to estimate a distribution function $f(y)$ for Y, the unknown parent distribution.

Knowledge of the system and the properties of common probability distributions can be used to select a model for Y. For example, an exponential function may be a good candidate for determining the failure distribution T of machine parts. The normal function, on the other hand, may prove to be a good candidate for a model of Y expected to have a point of central tendency. For example, when the data set is large enough, a histogram will give a good indicator if the initial model choice of $f(y)$ is a good one. The histogram of power output shown in Figure 4 of Chapter 7 gives little reason to doubt that a normal distribution will adequately describe this process. If there is a lack of observations, then a model must be chosen based primarily on knowledge of the process.

Assuming the model of choice $f(y)$ is the true distribution of the parent distribution Y, we can estimate the parameter or parameters of $f(y)$ with the sample data. Once the model is fit, the assumption that $f(y)$ is representative of the parent distribution will be challenged and tested in the model validation stage.

Model Parameter Estimates

Model parameters are estimated with sample statistics as discussed in Chapter 7. Consider the problem of estimating m unknown parameters of an assumed parent distribution $f(y)$ of Y. The *method of moments* is one parameter estimation method. A set of m equations are established by equating the kth order theoretical moment and the kth order sample moment for $k = 1, 2, \ldots m$, and then solving for the k unknown parameter values. Assuming a probability density function $f(y)$, the kth equation is

$$E[Y^k] = \int_{-\infty}^{\infty} y^k f(y)\,dy = \frac{\sum_{j=1}^{n} y_j^k}{n}$$

where

n = the number of observations in the sample

and

y_j = the jth observation of y

The first-order moment of Y, or $k = 1$, is the mean $\mu = E[Y]$. The theoretical and sample means form the equality equation,

$$\mu = \bar{y}$$

When $k = 2$, $E[Y^2] = \sigma^2 + \mu^2$. In lieu of using this relationship, the theoretical and sample variances may be employed in establishing the second equality equation,

$$\sigma^2 = s^2 = \frac{\sum_{j=1}^{n} (y_j - \bar{y})^2}{n-1}$$

The probability distributions introduced in this book have been limited to one or two parameter models, that is, $m = 1$ or $m = 2$. Thus, the first and second moments, or equivalently, the mean and variance equations are sufficient for estimating model parameter estimates.

If Y is a discrete random variable, then a probability function $f(y)$ is used and the moments are calculated with sums in lieu of a probability density function and integrals as shown here.

Consider the exponential and normal distributions. Since the exponential distribution $f(y) = \rho e^{-\rho y}$ has a single parameter ρ, with theoretical mean $E(Y) = \mu = \frac{1}{\rho}$. The parameter estimate is

$$\hat{\rho} = \frac{1}{\bar{y}}$$

Without drawing attention to it, the method of moments is used in Example 7.10. The parameter estimate is $\hat{\rho} = 0.063$ and $f_Y(y) = 0.063e^{-0.063y}$.

For the case where Y is assumed to have a normal distribution $N(\mu, \sigma^2)$, the estimates of μ and σ are

$$\hat{\mu} = \bar{y}$$

and

$$\hat{\sigma} = s$$

For example, consider the MSW energy data given in Table 1 of Chapter 7. The weighted mean and variance are calculated with the following formulas, $\bar{y} \pm s = 25.3 \pm 2.82$MW for $n = 101$ observations. The parameter estimates are

$$\hat{\mu} = \bar{y} = 25.3$$
$$\hat{\sigma} = s = 2.82$$

Thus

$$f(y) = \frac{1}{\sqrt{2\pi 2.82^2}} \exp\left(-\frac{1}{2}\left(\frac{y - 25.3}{2.82}\right)^2\right)$$

The method of moments is introduced because it is conceptually simple and numerically easy to calculate. The method does not usually yield minimum variance unbiased estimates. If the method of moments does not give satisfactory results, then the method of maximum likelihood, for example, may be used. This method maximizes the likelihood function in terms of the unknown parameters of $f(y)$. The likelihood function l for a sample of size n is defined to be the product

$$l = \prod_{j=1}^{n} f(y_j).$$

This method is not discussed further in this book.

Model Validation

If the parent distribution using estimated parameters and the histogram match, then the theoretical model is assumed to be acceptable. In the MSW energy, the frequency diagram for $N(25.3, 2.82^2)$ is shown in Figure 1. The $f(y)$ plot and histogram of Figure 4 of Chapter 7 are well matched.

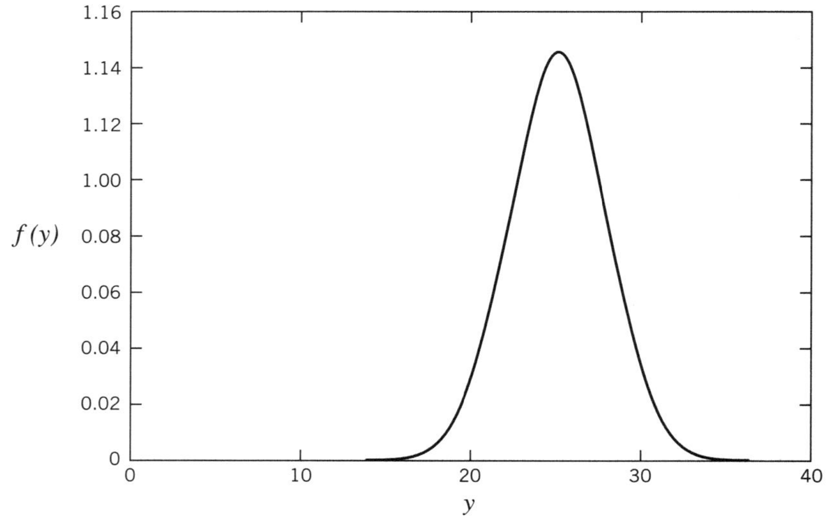

Figure 1 The $f(y)$ distribution for $N(25.3, 2.8^2)$.

Statistical tests are also used for model validation; however, the histogram or frequency diagram is preferred because the plot may suggest a modification to the model or the selection of a different model if the model appears to be unsatisfactory.

The Dot Plot

When a sample size is small, it is often difficult to judge if the fit is acceptable. The dot plot and the quantile–quantile plot are two graphical methods that are used for model validation. Often the dot plot, like the histogram, assists in model identification. These methods are illustrated by example.

Suppose that the 10 random samples of MSW density measurements in tons per m^3 are collected: 0.172, 0.289, 0.220, 0.245, 0.063, 0.073, 0.359, 0.250, 0.195, and 0.173. It is assumed that Y has a normal distribution.

Since there are only 10 observations, developing a frequency table and histogram will reveal little information about the parent distribution. A dot diagram, shown in Figure 2, uses a continuous scale; thus, a sense of the variability can be visually interpreted. There is no reason to reject the assumption that the parent distribution is a normal distribution. The data appear to be clustered in the 0.15 to 0.3 tons per m^3 range. We might speculate a mean equal to about 0.22.

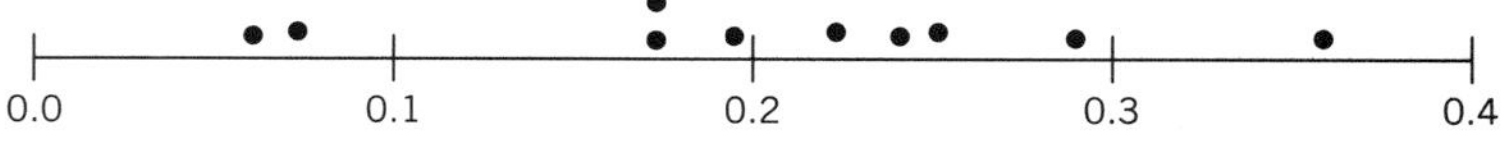

Figure 2 Dot diagram.

The sample average and standard deviation are calculated to be

$$\hat{\mu} \pm \hat{\sigma} = \bar{y} + s = 0.204 \pm 0.086$$

The probability model $N(0.204, 0.086^2)$ is investigated. In order to give added assurance that the normal distribution is an acceptable choice for Y, the frequency diagram, shown in Figure 3, and quantile-quantile plot are constructed. It is a less subjective method for evaluating goodness of fit.

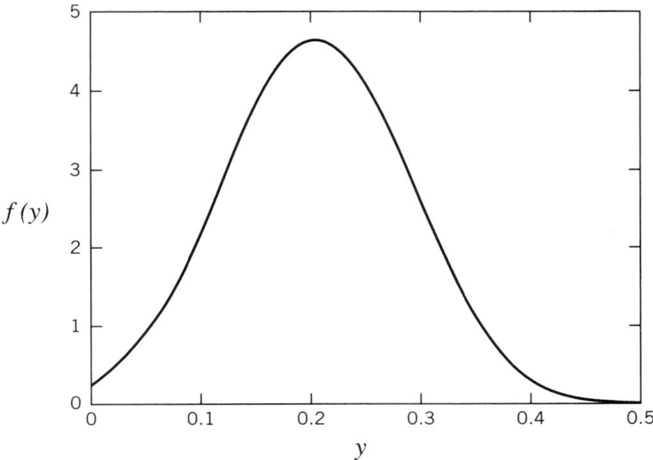

Figure 3 The frequency for $N(0.204, 0.086^2)$.

The Quantile–Quantile (q–q) Plot

Let $\hat{y}_j$ represent a model forecast corresponding to the sample point y_j. The difference $(\hat{y}_j - y_j)$ is a measure of model prediction error. If the model fit is perfect, than $(\hat{y}_j - y_j) = 0$ for all j. A graphical means to perform the same test is to construct a pairwise plot of $(\hat{y}_j, y_j)$ for all n observations in the data set. For a perfect model fit, all $(\hat{y}_j, y_j)$ points will lie on a 45° line passing through the origin. Deviation from the straight line is an indication that the distribution is an inappropriate choice.

A quantile–quantile or q–q plot uses this principle for testing the adequacy of an assumed $f(y)$ distribution of Y. Since $f(y)$ distributions are generally nonlinear, the data are ordered and transformed to a linear form so that they can be more easily interpreted on a graph. The ordered observations are called *order statistics* and are depicted as

$$y_{(1)} \leq y_{(2)} \leq \ldots \leq y_{(n)}$$

where the numbers in parentheses are the ranks of the observation. The quartile percentage or frequency corresponding to each of these order statistics is

$$\omega_j = \frac{j - 0.5}{n}$$

where

$$0 < \omega_1 < \omega_2 < \ldots \omega_n < 1$$
$$\text{for } j = 1, 2, \ldots, n.$$

and the cumulative distribution is used to calculate the corresponding forecasts. Let $\hat{y}_{(j)}$ = model forecast for observation j. Let

$$F\left(y_{(j)}\right) = \omega_j$$

Solving for $\hat{y}_{(j)}$ gives

$$\hat{y}_{(j)} = F^{-1}(\omega_j)$$

The q–q plot is constructed by using the following steps:

1. Arrange the sample data as order statistics, $y_{(j)}$ for $j = 1,2, \ldots, n$.
2. Calculate $\omega_j = \dfrac{j - 0.5}{n}$ for $j = 1,2, \ldots, n$.
3. Calculate $\hat{y}_{(j)} = F^{-1}(\omega_j)$ for $j = 1,2, \ldots, n$.
4. Plot $\hat{y}_{(j)}$ versus $y_{(j)}$.
5. Test for linearity.

The following table shows the steps for the MSW density data set. Step 3 is performed in two stages; first, the unit normal values of z_j are determined for each ω_j, and, second, the values of $\hat{y}_{(j)}$ are calculated. The plot of $y_{(j)}$ versus $\hat{y}_{(j)}$, shown in Figure 4, shows that the relationship is approximately linear; thus, the normal probability model $N(0.204, 0.086^2)$ is considered adequate.

	Step 1	Step 2	Step 3	
j	$y_{(j)}$	$\omega_j = \dfrac{j-0.5}{n}$	$z_j = F^{-1}(\omega_j)$	$\hat{y}_{(j)} = \hat{\mu} + z_j\hat{\sigma}$
1	0.063	0.05	–1.64	0.063
2	0.073	0.15	–1.03	0.115
3	0.172	0.25	–0.67	0.146
4	0.173	0.35	–0.38	0.171
5	0.195	0.45	–0.12	0.194
6	0.225	0.55	0.12	0.214
7	0.242	0.65	0.38	0.237
8	0.250	0.75	0.67	0.262
9	0.289	0.85	1.03	0.293
10	0.359	0.95	1.64	0.346

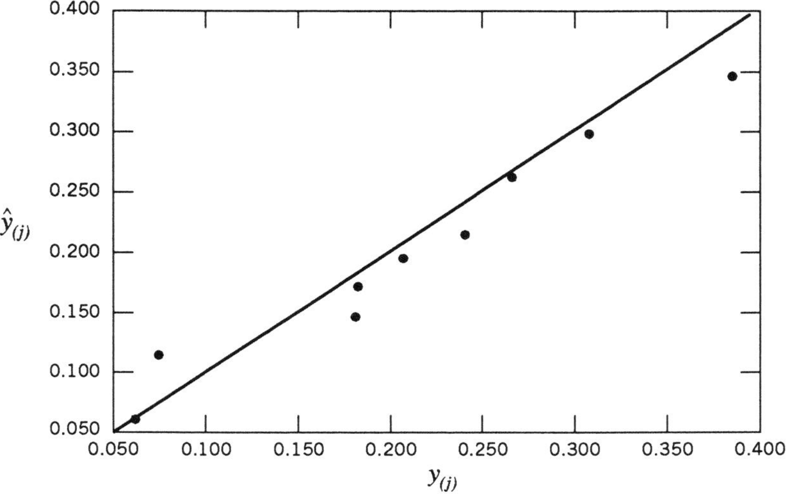

Figure 4 Quantile–quantile (q–q) plot.

EXAMPLE 8.1 *Instrument Failure*

Contamination causes an instrument to become fouled and the readings to drift. The time in hours to serious instrument drift for 30 independent trials is as follows:

6.76	0.55	3.08	7.92	8.48
36.55	14.85	1.91	7.95	11.57
8.12	1.81	2.17	2.19	39.62
2.36	0.59	8.78	10.84	0.38
1.33	9.01	8.13	4.46	7.06
6.86	0.62	10.48	4.54	2.27

Where T is assumed to be an exponential distribution, define

$$T = \{\text{time to instrument failure}\}$$

then estimate the model parameter and confirm this assumption.

SOLUTION

A frequency table using a time interval of 4 hours is

j	Interval	t_j, Interval Midpoint	ω_j, Frequency
1	0 – 4	2	12
2	4 – 8	6	7
3	8 – 12	10	8
4	12 – 16	14	1
5	16 – 20	18	0
6	20 – 24	22	1
7	24 – 28	26	0
8	28 – 32	30	0
9	32 – 36	34	0
10	36 – 40	38	1

The sample mean and standard deviation are $\bar{t} = 7.16$ hours and $s = 7.72$ hours. The histogram is

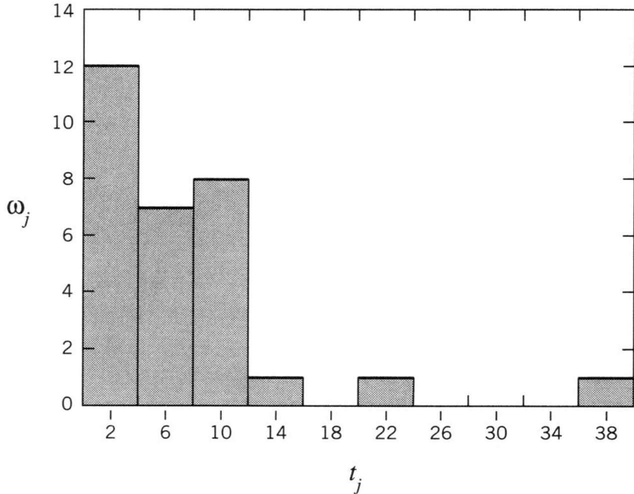

The histogram and sample statistics seem to confirm the assumption that the process has an exponential distribution. Since $\bar{t} \approx s$, it gives further evidence that an exponential distribution is a good choice.

Using the method of moments, we find that the estimate of ρ is

$$\hat{\rho} = \frac{1}{\bar{t}} = \frac{1}{7.16} = 0.14 \text{ failure per hour}$$

The q–q plot will be used in validating the model. The order statistics of $t_{(j)}$ and $\hat{t}_{(j)}$ are summarized in the following table. The model forecasts are determined by solving for $\hat{t}_{(j)}$.

$$F(t_{(j)}) = 1 - e^{-0.14 t_{(j)}} = \omega_j$$

where

$$\omega_j = \frac{j - 0.5}{n}$$

The result is

$$\hat{t}_{(j)} = \frac{1}{0.14} ln\left(\frac{1}{1 - \omega_j}\right)$$

j	$t_{(j)}$	ω_j	$\hat{t}_{(j)}$
1	0.381	0.017	0.120
2	0.548	0.050	0.367
3	0.594	0.083	0.623
4	0.618	0.117	0.888
5	1.326	0.150	1.163
6	1.812	0.183	1.449
7	1.912	0.217	1.748
8	2.175	0.250	2.059
9	2.185	0.283	2.384
10	2.275	0.317	2.725
11	2.362	0.350	3.083
12	3.083	0.383	3.460
13	4.463	0.417	3.858
14	4.537	0.450	4.279
15	6.765	0.483	4.726
16	6.861	0.517	5.203
17	7.057	0.550	5.715
18	7.917	0.583	6.266
19	7.954	0.617	6.862
20	8.121	0.650	7.514
21	8.134	0.683	8.230
22	8.478	0.717	9.026
23	8.775	0.750	9.922
24	9.011	0.783	10.946
25	10.479	0.817	12.141
26	10.835	0.850	13.578
27	11.566	0.883	15.376
28	14.848	0.917	17.784
29	20.020	0.950	21.440
30	39.617	0.983	29.303

The table shows reasonable predictions for $\hat{t}_{(j)}$, except for the distribution tail point at $t = 30$. The q–q plot suggests that the exponential probability distribution is an adequate model.

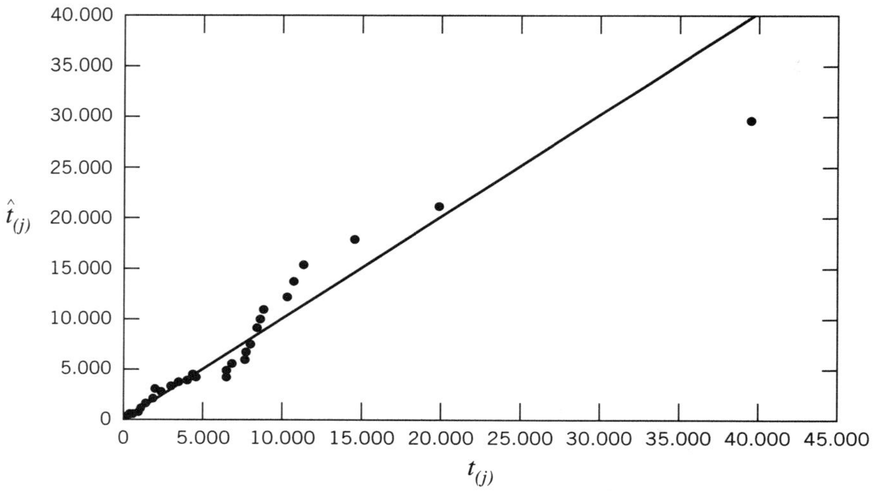

QUESTION FOR DISCUSSION

8.1.1 Define the following terms:

method of moments order statistics q–q plot

QUESTIONS FOR ANALYSIS

Instructions: The solutions to questions marked with an asterisk (*) are most easily obtained with the aid of a spreadsheet computer program.

8.1.2 Let

$$Y = \{\text{weight of discarded household waste in pounds per person-day}\}$$

The following data have been observed:

4.42	3.45	2.82	2.28	2.99	2.99	3.89	3.45	4.25	3.97
4.18	5.24	3.50	1.69	4.81	2.92	1.48	2.04	1.77	4.19

Assume that the parent distribution for Y is a normal distribution. The aim of this exercise is to confirm this assumption.

(a) Plot a dot plot for the data.

(b) Plot a histogram using intervals of 0.75 pound per person-day.

(c) Calculate the sample mean and standard deviation, $\bar{y} \pm s$, and assume that $\hat{\mu} = \bar{y}$ and $\hat{\sigma} = s$ are the best parameter estimate for the normal distribution, $N(\mu, \sigma^2)$.

(d) Draw a q–q plot to test the model found in (c).

(e) Use the dot, histogram, and q–q plots to evaluate the assumption that the parent distribution $f(y)$ is a normal distribution. Do you recommend or reject this distribution?

8.1.3 Let

$$T = \{\text{time to failure in months}\}$$

The following observations have been made:

10.3	5.82	0.06	0.44	0.58	2.41	5.24	1.53	14.9	10.1
2.83	4.63	1.10	2.53	4.42	10.7	5.64	4.85	2.89	6.24

Assume that the parent distribution for T is an exponential distribution. The aim of this exercise is to confirm this assumption.

(a) Plot a dot plot for the data.

(b) Plot a histogram using intervals of two months.

(c) Calculate the sample mean and standard deviation, $\bar{t} \pm s$ and assume that $\hat{\rho} = \frac{1}{\bar{t}}$ is the best estimate of ρ for an exponential distribution.

(d) Draw a q–q plot to test the model found in (c).

(e) Use the dot, histogram, and q–q plots to evaluate the assumption that the parent distribution $f(t)$ is an exponential distribution. Do you recommend or reject this distribution?

8.1.4* The number of pounds of newspaper discarded each week by 100 households is as follows:

13	6	15	7	13	10	8	14	14	5
9	17	10	5	9	12	12	16	5	10
13	10	8	11	16	20	14	16	10	13
7	10	7	16	12	14	12	8	16	16
15	17	14	7	4	16	9	7	11	19
16	11	12	15	10	19	18	13	6	7
15	18	11	13	19	14	15	11	11	13
11	17	10	13	13	13	4	16	16	16
12	17	13	20	18	12	15	6	15	9
18	11	8	7	12	6	15	11	13	12

A histogram of the data, Problem 7.3.2, shows it to have a bell-shaped distribution. From the solution to Problem 7.4.4, the mean and standard deviation for this data set is calculated to be $\bar{x} \pm s = 12.2 \pm 3.93$. It seems reasonable to assume that the parent distribution of this data set is $N(12.2, 3.93^2)$.

Evaluate this assumption by drawing a q–q plot to show that this mathematical model is a good fit of the data.

8.1.5* The number of pounds of aluminum discarded each week by 100 households is as follows:

0.1	0.4	0.5	0.7	0.1	1.5	0.0	0.7	0.0	0.0
0.0	0.0	0.0	0.8	0.0	0.5	0.8	0.8	0.0	1.3
1.1	0.0	1.2	0.4	0.0	0.6	0.6	1.1	0.7	0.0
0.0	0.0	1.4	0.4	0.0	0.6	1.4	0.5	0.0	1.1
1.0	0.0	0.0	1.4	0.5	0.0	0.0	0.0	1.3	0.0
0.1	0.5	0.7	0.3	0.0	0.0	1.2	1.2	0.7	0.0
0.0	1.0	1.2	0.8	0.9	0.5	0.0	0.0	1.3	0.0
0.0	0.9	0.0	1.0	0.4	0.8	1.3	0.8	0.0	0.1
0.1	0.0	1.3	0.0	0.0	0.0	1.5	1.2	0.6	0.0
0.0	0.6	0.9	0.6	0.1	0.4	0.7	0.0	0.1	0.1

A histogram of the data, Problem 7.3.3, indicates that a piece-wise distribution may be most appropriate for the range. The cumulative probability function is assumed to be

$$F(x) = \begin{cases} 0.4x \text{ for } 0 \leq x < 0.1 \\ 0.4 + 0.6\,\dfrac{x - 0.1}{1.4} \text{ for } 0.1 \leq x \leq 1.5 \end{cases}$$

(a) Plot $F(x) - x$.

(b) Evaluate the assumption by drawing a q–q plot to show that $F(x)$ is a good fit of the data.

8.2 MONTE CARLO SIMULATION

A Monte Carlo simulation uses numerical methods to gain insight into the behavior of a complex random process. The method proves useful when the direct calculation of probability functions is complicated. Statistical summaries, frequency distributions, graphs, and other statistical tools are used to evaluate the data and understand the behavior of the system. One of the major advantages of a Monte Carlo simulation is that large data sets can be generated without conducting physical experiments. Consequently, a Monte Carlo simulation could prove to be a cost-effective design tool.

The Method

In order to conduct a Monte Carlo simulation, certain assumptions have to be made about the system under study. The rules that govern the process must be specified. Consider the relationship

$$Y = g(X)$$

The input variable is specified to be a random variable X with known probability distribution $f(x)$. A set of values of the random number X, $x_1, x_2, \ldots, x_m$, is generated, and the responses, $y_1, y_2, \ldots, y_m$, are calculated from the relationship $Y = g(X)$. The y data set can be analyzed to determine the nature of the process under study.

The Monte Carlo simulation procedure can be extended to a multivariate function of Y, where $Y = g(X_1, X_2, \ldots, X_n)$ with known frequency distributions $f(x_1), f(x_2), \ldots,$ and $f(x_n)$. This will be demonstrated with an example. First, we will introduce basic principles.

A Monte Carlo Experiment

Consider the generation of an observation for the random variable X. A simulation run uses a cumulative distribution function of X and a uniform probability density function for a random number generator R. A simple spinner, found in some games of chance, is a mechanical random number generator. The spinner is assumed to be perfectly balanced and to have equally spaced markings labeled between zero and one. In some statistics and probability books, tables of random numbers are provided. Many computer languages and application programs include built-in pseudo-random generator routines. In this book, since computer solutions are advocated, random numbers are generated by computer.

The cumulative distribution for R is

$$P(R \leq r) = \int_0^r dt = r, \quad \text{for} \quad 0 \leq r \leq 1$$

In this special situation where r is restricted in the range of 0 and 1, there is a one-to-one correspondence between r and $P(R \leq r)$.

A value of x is determined by solving the equality,

$$P(R \leq r) = r = F(x)$$

where r is a generated random number. The value of x is determined by the inverse function,

$$x = F^{-1}(r)$$

The solution is depicted in Figure 5.

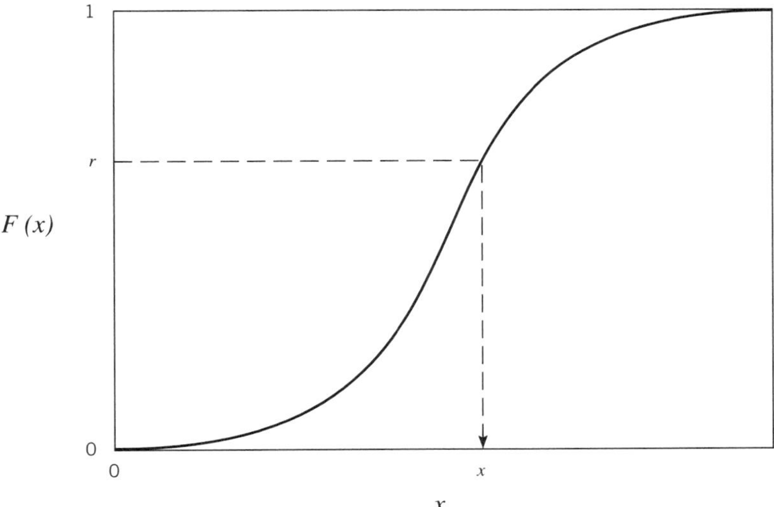

Figure 5 The determination of x given r.

Use of the cumulative continuous distribution of X guarantees that a single observation of x for each r exists because, by definition, $F(x)$ is strictly increasing. For a discrete probability function, the cumulative probability will be increasing but as a step function.

A Simulation

In order to calculate a set of m independent observations, this basic step is conducted m times for m independent observations of R. The set for X and Y are determined by performing the following steps:

1. Generate an independent random number, r_i, where R is a uniform distribution with $0 \le r \le 1$.
2. Solve $x_i = F^{-1}(r_i)$.
3. Solve $y_i = g(x_i)$.
4. Repeat steps 1 through 3 for $i = 1, 2, \ldots, m$.

The Exponential Distribution

The governing equation for an exponential random variable T with known parameter ρ is easily derived. Let

$$F(t) = 1 - e^{-\rho t} = r$$

where r is assumed to be known. Solving for t gives

$$t = \frac{1}{\rho} ln\left(\frac{1}{1-r}\right)$$

If t is the time failure and ρ is the failure rate, then t is the time to failure at probability r.

The Normal Distribution

The governing equation for $X = N(\mu, \sigma^2)$ with known parameters μ and σ requires the use of standard unit normal distribution $Z = N(0, 1)$. Assuming r is known, establish the equality,

$$\Phi(z) = r$$

and solve for z,

$$z = \Phi^{-1}(r)$$

with the aid of a standard normal cumulative distribution table. Since $z = \dfrac{x-\mu}{\sigma}$, it follows that

$$x = \mu + z\sigma$$

Writing x as a function of r, gives

$$x = \mu + \Phi^{-1}(r)\sigma$$

EXAMPLE 8.2 ***Availability of a System with a Uniformly Distributed Repair Rate***

A system is expected to have a failure rate of $\rho = 1$ per year and a repair rate of $\eta = 6$ repairs per year. The exponential and uniform distributions are used to describe the failure and repair processes, respectively. Since the availability function for a single repairable unit system, given in Section 6.4, is restricted to failure and repair processes that are exponential distributions, it is not applicable here.

(a) Use a Monte Carlo simulation to determine the availability $a(t)$ at times $t = 0.25$, 0.5, 0.75, 1.0, 1.25, 1.5, 1.75, and 2 years.

(b) Compare the availability values from the Monte Carlo simulation with the limiting state availability $a(\infty)$ for exponential failure and repair distributions.

SOLUTION

The availability $a(t)$ is defined at time t as the ratio of the number of working units to the number of total units, or as a probability as

$$a(t) = \frac{n(t)}{n(0)}$$

For this problem, the function is evaluated at times $t = 0.25, 0.5, 0.75, 1.0, 1.25, 1.5, 1.75$, and 2 years. In order to estimate $a(t)$ at one of these specified times, we will employ the following scheme. First, we determine the times of failure T_f and the times of repair T_r by conducting 20 Monte Carlo experiments, $m = 20$. In other words, we are evaluating 20 systems; thus, $n(0) = m = 20$. Since the system is repairable, we will determine the times of failure and repair for four failure and repair sequences, $p = 1, 2, 3, 4$. (The assignment of 4 failure-repair systems is arbitrary. In retrospect, it proved satisfactory for this problem.) After these values are determined, estimates of $a(t)$ are made at each of the specified times of t. Define the sample space $Z = \{S, S'\}$ where

$$S = \{\text{system unavailable at time } t\} = \{Y = 0\}$$

and

$$S' = \{\text{system available at time } t\} = \{Y = 1\}$$

Also, define

$$T_f = \{\text{time to failure}\} \text{ with } F(t_f) = 1 - e^{-t_f}$$

$$T_r = \{\text{time to repair}\} \text{ with } F(t_r) = \frac{t_r - 0}{1/6 - 0} = 6t_r$$

The governing equations for calculating t_f and t_r by Monte Carlo simulation are

$$t_f(p) = \left[\frac{1}{1 - r_1(p)}\right] \quad \text{for } p = 1, 2, 3, 4$$

and

$$t_r(p) = \frac{r_2(p)}{6} \quad \text{for } p = 1, 2, 3, 4$$

where $r_1(p)$ and $r_2(p)$ are two independent, uniformly distributed random variables with range between 0 and 1, and p = counter indicating the first, second, third, and fourth incident of system failure and repair.

A series of $m = 20$ Monte Carlo experiments where $j = 1, 2, \ldots, 20$ and $p = 1, 2, 3, 4$ are made and are summarized in the following table. For each experiment it is assumed that at time $t = 0$ that all 20 systems are available for service. The times to failure $t_f(p)$ and repair $t_r(p)$ are

Experiment j	$t_f(1)$	$t_r(1)$	$t_f(2)$	$t_r(2)$	$t_f(3)$	$t_r(3)$	$t_f(4)$	$t_r(4)$
1	1.02	0.08	1.94	0.00	0.76	0.08	0.48	0.16
2	1.31	0.16	2.03	0.16	0.18	0.02	0.94	0.09
3	0.67	0.03	0.26	0.14	0.12	0.08	0.54	0.09
4	0.30	0.07	2.56	0.06	1.03	0.14	0.05	0.12
5	0.20	0.11	1.94	0.08	0.05	0.03	0.59	0.04
6	1.34	0.01	0.28	0.08	0.20	0.11	1.40	0.01
7	2.35	0.04	1.81	0.10	0.07	0.15	0.15	0.13
8	0.71	0.05	0.64	0.09	0.09	0.05	0.47	0.12
9	0.73	0.13	1.16	0.13	0.03	0.06	0.29	0.13
10	0.85	0.11	0.61	0.06	0.68	0.05	0.87	0.16
11	1.67	0.16	1.09	0.04	0.34	0.07	1.71	0.10
12	2.51	0.04	1.78	0.14	0.24	0.10	0.65	0.06
13	1.85	0.10	0.13	0.00	0.65	0.03	1.10	0.10
14	0.02	0.15	0.08	0.03	1.56	0.07	2.72	0.01
15	1.95	0.12	0.93	0.05	1.41	0.06	3.40	0.02
16	0.42	0.10	2.40	0.13	1.50	0.07	0.72	0.04
17	0.43	0.08	0.58	0.13	1.05	0.00	0.43	0.14
18	0.66	0.02	0.35	0.08	0.12	0.06	2.21	0.12
19	0.08	0.06	0.03	0.00	0.92	0.04	1.67	0.06
20	0.25	0.08	1.09	0.06	0.20	0.01	0.12	0.11

The availability $a(t)$, defined as a probability in terms of Y, is

$$a(t) = P(Y = 1) = \frac{n(t)}{n(0)}$$

For our estimates of $a(t)$, it is assumed, as previously stated, that the number of units is equal to the number of runs, or $n(0) = m = 20$. Time t are designated at discrete times; whereas the time to failure $t_f(p)$ and the time to repair $t_r(p)$ are given as time intervals. In order to determine when the system is available and unavailable, the time intervals of $t_f(p)$ and $t_r(p)$ are analyzed to the same discrete time scale as t. Define

$$u(p) = \text{time } t \text{ when the pth failure occurs,}$$

where

$$u(p)=\begin{cases}\sum_{p=1}^{q} t_f(p)+t_r(p-1) \text{ where } t_r(0)=0\\ \sum_{p=1}^{q} t_f(p)+t_r(p)\end{cases}$$

$w(p)$ = time t when the system is repaired and returned to service, where

$$w(p) = \sum_{p=1}^{q} t_f(p) + t_r(p)$$

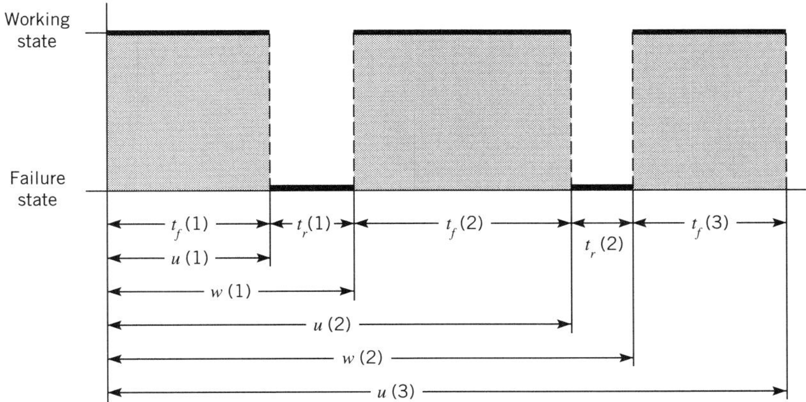

For example, for run j = 1, the system becomes unavailable at times.

$$u(1) = t_f(1) = 1.02$$

$$u(2) = t_f(1) + t_r(1) + t_f(2) = 1.02 + 0.08 + 1.94 = 3.04$$

$$u(3) = t_f(1) + t_r(1) + t_f(2) + t_r(2) + t_f(3) = 1.02 + 0.08 + 1.94 + 0.00 + 0.76 = 3.80$$

$$u(4) = t_f(1) + t_r(1) + t_f(2) + t_r(2) + t_f(3) + t_r(3) + t_f(4) = \\ 1.02 + 0.08 + 1.94 + 0.00 + 0.76 + 0.08 + 0.48 = 4.36$$

The system is put back in service at times.

$$w(1) = t_f(1) + t_r(1) = 1.02 + 0.08 = 1.10$$

$$w(2) = t_f(1) + t_r(1) + t_f(2) + t_r(2) = 1.02 + 0.08 + 1.94 + 0.00 = 3.04$$

$$w(3) = t_f(1) + t_r(1) + t_f(2) + t_r(2) + t_f(3) + t_r(3) = 1.02 + 0.08 + 1.94 + 0.00 + 0.76 = 3.88$$

$$w(4) = t_f(1) + t_r(1) + t_f(2) + t_r(2) + t_f(3) + t_r(3) + t_f(4) + t_r(4) =$$
$$1.02 + 0.08 + 1.94 + 0.00 + 0.76 + 0.08 + 0.48 = 4.36.$$

The system is unavailable at time t and $Y = 0$ if the inequality is satisfied,

$$u(p) \le t < w(p) \text{ for } p = 1, 2, 3, 4$$

Otherwise, the system is available at time t and $Y = 1$.

The experiments when the system is unavailable at time t are:

j	$u(1)$	$\le t <$	$s(1)$	$u(2)$	$\le t <$	$s(2)$	$u(3)$	$\le t <$	$s(3)$	$u(4)$	$\le t <$	$s(4)$
1	1.02	$\le t <$	1.10	3.04	$\le t <$	3.04	3.80	$\le t <$	3.88	4.36	$\le t <$	4.52
2	1.31	$\le t <$	1.47	3.50	$\le t <$	3.66	3.84	$\le t <$	3.86	4.80	$\le t <$	4.89
3	0.67	$\le t <$	0.70	0.96	$\le t <$	1.10	1.22	$\le t <$	1.30	1.84	$\le t <$	1.93
4	0.3	$\le t <$	0.37	2.93	$\le t <$	2.99	4.02	$\le t <$	4.16	4.21	$\le t <$	4.33
5	0.2	$\le t <$	0.31	2.25	$\le t <$	2.33	2.38	$\le t <$	2.41	3.00	$\le t <$	3.04
6	1.34	$\le t <$	1.35	1.63	$\le t <$	1.71	1.91	$\le t <$	2.02	3.42	$\le t <$	3.43
7	2.35	$\le t <$	2.39	4.20	$\le t <$	4.30	4.37	$\le t <$	4.52	4.67	$\le t <$	4.80
8	0.71	$\le t <$	0.76	1.40	$\le t <$	1.49	1.58	$\le t <$	1.63	2.10	$\le t <$	2.22
9	0.73	$\le t <$	0.86	2.02	$\le t <$	2.15	2.18	$\le t <$	2.24	2.53	$\le t <$	2.66
10	0.85	$\le t <$	0.96	1.57	$\le t <$	1.63	2.31	$\le t <$	2.36	3.23	$\le t <$	3.39
11	1.67	$\le t <$	1.83	2.92	$\le t <$	2.96	3.30	$\le t <$	3.37	5.08	$\le t <$	5.18
12	2.51	$\le t <$	2.55	4.33	$\le t <$	4.47	4.71	$\le t <$	4.81	5.46	$\le t <$	5.52
13	1.85	$\le t <$	1.95	2.08	$\le t <$	2.08	2.73	$\le t <$	2.76	3.86	$\le t <$	3.96
14	0.02	$\le t <$	0.17	0.25	$\le t <$	0.28	1.84	$\le t <$	1.91	4.63	$\le t <$	4.64
15	1.95	$\le t <$	2.07	3.00	$\le t <$	3.05	4.46	$\le t <$	4.52	7.92	$\le t <$	7.94
16	0.42	$\le t <$	0.52	2.92	$\le t <$	3.05	4.55	$\le t <$	4.62	5.34	$\le t <$	5.38
17	0.43	$\le t <$	0.51	1.09	$\le t <$	1.22	2.27	$\le t <$	2.27	2.70	$\le t <$	2.84
18	0.66	$\le t <$	0.68	1.03	$\le t <$	1.11	1.23	$\le t <$	1.29	3.50	$\le t <$	3.62
19	0.08	$\le t <$	0.14	0.17	$\le t <$	0.17	1.09	$\le t <$	1.13	2.80	$\le t <$	2.86
20	0.25	$\le t <$	0.33	1.42	$\le t <$	1.48	1.68	$\le t <$	1.69	1.81	$\le t <$	1.92

For example, given $t = 0.25$ year, the system is unavailable, $Y = 0$, for experiments 5, 14, and 20 as indicated in the above table. Since the number of times the system is found to be unavailable is 3 of 20 experiments, the system is available $n(0.25) = 17$ of 20 experiments. The availability for $t = 0.25$ year is calculated to be

$$\hat{a}(0.25) = \frac{17}{20} = 0.85$$

The availabilities at times $t = 0.25, 0.5, \ldots, 2.00$ years are determined in a similar manner and summarized in the following table.

t	0.25	0.50	0.75	1.00	1.25	1.50	1.75	2.00
$a(t)$	0.85	0.90	0.90	0.95	0.90	1.00	0.95	0.90

The estimate of the limit state availability using an exponential distribution for times to failure and repair is $a(\infty) = 0.92$. It appears that there is little or no difference in the two models for this set of random numbers. More Monte Carlo computer runs would clarify this conjecture.

This problem illustrates the basic principles and procedures used in performing a Monte Carlo simulation. In this case, the principles of statistics were used to determine a probability estimate of $a(t)$.

EXAMPLE 8.3 *Migration Time Model*

In Example 6.8, the migration times for groundwater plume containing a toxic material is assumed to be exponentially distributed as

$$P(T \leq t) = 1 - e^{-\rho t}, \text{ for } t > 0$$

where $T = \{\text{migration time in years}\}$. The parameter $\rho = 1/\mu$ where μ = average time for the plume to travel 200 feet in silt is 238 years.

Justify the assumption that the exponential distribution with $\mu = 238$ is reasonable.

SOLUTION

The overall scheme to solve this problem is as follows. First, the fluid flow through a porous medium and the soil properties, which are not known with certainty and are introduced to the problem as random variables, are described. Then a Monte Carlo simulation is used to generate the times for water to travel $x = 200$ feet in a silty soil. These data are used to plot a histogram of the simulated travel times, and the method of moments is used to estimate the rate ρ of the exponential distribution. Finally, a q–q plot is used to evaluate the validity of the model.

Plan View

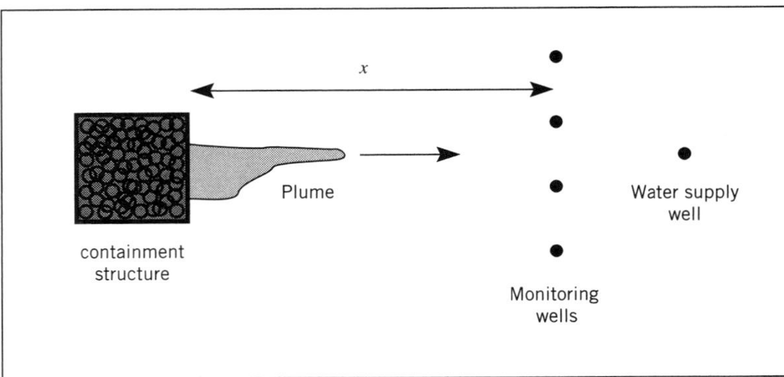

Cross-section view

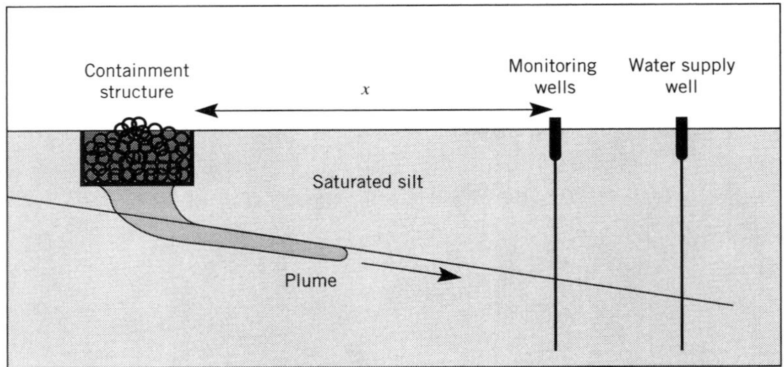

Source: Massman and Freeze (1987)

Darcy's Law is used to describe fluid flow through porous media. The pore water velocity v (meters per year) is

$$v = \frac{ki}{\phi}$$

where

k = hydraulic conductivity
ϕ = porosity
i = hydraulic gradient of the groundwater table = 0.002

The migration time t for a plume to reach the compliance surface is calculated as distance $x = 200$ meters divided by velocity v,

$$t = \frac{x}{v} = \frac{x\phi}{ki} = \frac{200\,\phi}{0.002\,k} = \frac{\phi}{k} \times 10^5$$

The soil properties for silt have been observed to have a wide range of values for k and ϕ.

$$0.003 \le k \le 224 \text{ meters per year}$$
$$0.34 \le \phi \le 0.61$$

In order to evaluate this variability, the hydraulic conductivity and porosity are assumed to be random variables with uniform probability density function. Let

$$T = \frac{\Phi}{K} x 10^5$$

where T is a bivariate function of F and K, or $T(\Phi, K)$ and

$$K = \{\text{hydraulic conductivity}\}$$

with

$$F(k) = \frac{k - 0.003}{224 - 0.003} = \frac{k - 0.003}{224}$$

and

$$\mathrm{F} = \{\text{soil porosity}\}$$

with

$$F(\phi) = \frac{\phi - 0.34}{0.61 - 0.34} = \frac{\phi - 0.34}{0.27}$$

The Monte Carlo simulation employs the following steps:

1. Generate two independent, uniformly distributed random numbers r_1 and r_2 with ranges between zero and one. Calculate k and ϕ as

   $$F(k) = r_1 \rightarrow k = 0.003 + 224 r_1$$

   and

   $$F(\phi) = r_2 \rightarrow \phi = 0.34 + 0.27 r_2$$

2. Calculate $t = \frac{f}{k} x 10^5$.
3. Repeat steps 1 and 2 until a representative sample of t is obtained. The following results are for $m = 200$ runs.
4. Plot a histogram to help identify distribution T. The result is

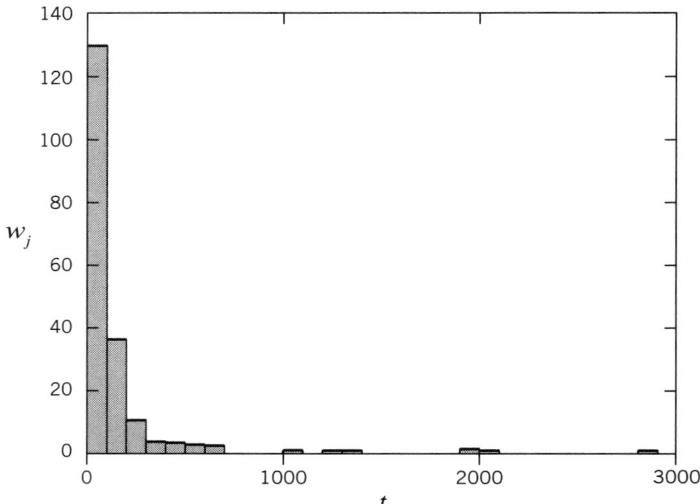

5. Since the histogram shows that the underlying distribution T may be an exponential, the method of moments is applied. The average migration time for the m = 200 runs is

$$\hat{u} = \bar{y} = 238 \text{ years}$$

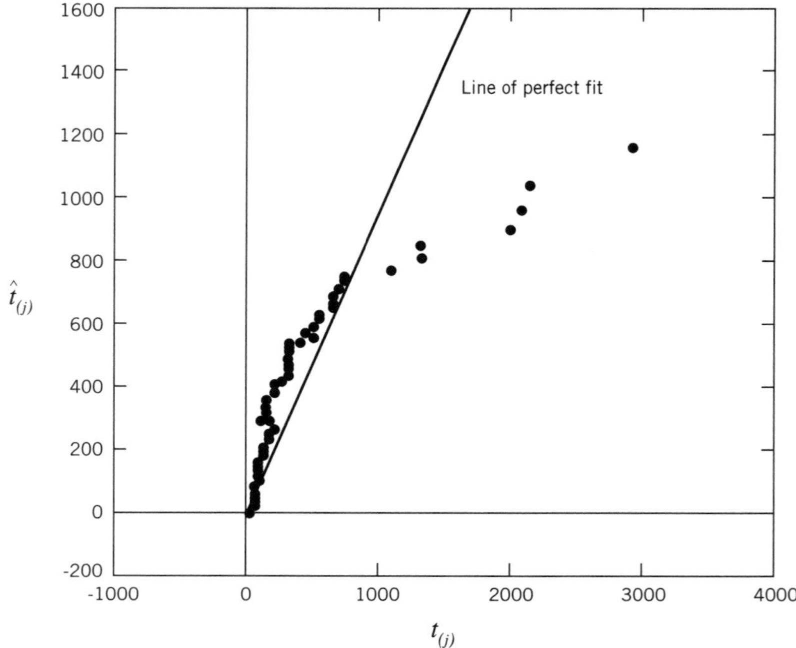

This is truly an impervious material. The fitted exponential distribution is

$$F(t) = 1 - e^{-t/238}$$

where

$$\hat{\rho} = \frac{1}{\hat{\mu}} = \frac{1}{238}$$

6. In order to test the validity of the model, a q–q plot is constructed.

The model tends to overestimate the observed mitigation times for t between 300 and 1000 years. The exponential model for silt with $\mu = 238$ years appears to be satisfactory in the time range of interest from 0 to 25 years.

QUESTIONS FOR ANALYSIS

Instructions: The following problems are most easily solved with the aid of a computer. In order to perform sensitivity analysis, it is often advantageous to use the same random numbers for Monte Carlo experiments. Data sets 1 and 2 are each a set of 50 random numbers.

Data Set 1:

0.708	0.943	0.459	0.633	0.031
0.041	0.315	0.031	0.889	0.221
0.697	0.873	0.535	0.045	0.209
0.371	0.413	0.652	0.913	0.254
0.921	0.114	0.706	0.595	0.772
0.333	0.040	0.866	0.497	0.628
0.475	0.373	0.453	0.426	0.644
0.888	0.234	0.386	0.133	0.919
0.114	0.071	0.185	0.508	0.834
0.070	0.462	0.071	0.803	0.889

Data Set 2:

0.049	0.445	0.335	0.527	0.625
0.496	0.253	0.454	0.677	0.444
0.803	0.384	0.601	0.791	0.066
0.852	0.680	0.190	0.173	0.530
0.619	0.948	0.191	0.942	0.053
0.463	0.764	0.537	0.954	0.988
0.128	0.289	0.221	0.672	0.006
0.980	0.089	0.736	0.680	0.030
0.844	0.440	0.473	0.146	0.964
0.327	0.295	0.716	0.094	0.410

8.2.1* Let T = {time to failure}, where T is a Weibull distribution with an average failure rate ρ and shape factor γ,

$$F(t) = 1 - e^{-\rho t^{\gamma}} \quad \text{for } t \geq 0$$

(a) Solve for governing equation for $t = F^{-1}(r)$, where R is a uniform distribution with $0 \leq r \leq 1$.

(b) Perform a sensitivity analysis by using a Monte Carlo simulation to determine the effect of the shape parameter γ. Compare the histograms of two Weibull distributions, where $\rho = 1$, $\gamma = 1$ (the exponential distribution) for one distribution and $\gamma = 2$ for the other distribution.

In order to make a fair comparison, use the same random numbers, data set 1, for plotting the histograms of the two distributions. Use an interval width of 0.5 unit for the histogram.

(c) In order to analyze the effect of random numbers on the results, repeat (b) using data set 2.

8.2.2* Let T = {time to failure}, where $T = T_1 + T_2$ is a derived distribution for units 1 and 2. The average failure times are $\mu_1 = 1$ and $\mu_2 = 3$ units per year. The probability density function for T is

$$f_T(t) = \frac{\rho_1 \rho_2}{\rho_1 - \rho_2}\left(e^{-\rho_2 t} - e^{-\rho_1 t}\right)$$

for $t \geq 0$.

(a) Derive $F(t)$ and plot the $F(t)$–t diagram.

(b) Perform a Monte Carlo simulation and plot a histogram with an interval width of 2 units for the random numbers of data set 1. Calculate $t = F^{-1}(r)$ as

$$t = t_1 + t_2 = F_1^{-1}(r) + F_2^{-1}(r)$$

the sum of the inverses of two exponential distributions. Note that a closed-form solution of the inverse function, $t = F^{-1}(r)$, cannot be obtained; therefore, the values of t are found as the summation.

(c) Repeat (b) for the random numbers of data set 2.

8.2.3* In Example 8.3, Darcy's Law and Monte Carlo simulation are used to calculate migration times for a groundwater plume containing a toxic material through $x = 200$ meters of a silt media. The conclusion drawn from this exercise is that an exponential distribution is adequate for describing T, where T = {migration time for silt}.

The purpose of this exercise is to determine if the exponential distribution for T is also suitable for a sand medium. The hydraulic conductivity and porosity for fine sand have the following ranges:

$$6.3 \leq k \leq 5960 \text{ meters per year}$$
$$0.26 \leq \phi \leq 0.53$$

Assume k and ϕ are random variables with uniform distributions, K and Φ, respectively.

(a) In order to determine the adequacy of the exponential distribution for T, perform a Monte Carlo simulation and plot a histogram for the random variables of T with an interval width of 20 years. Use the random numbers of data set 1. Does an exponential distribution for T seem to be a reasonable assumption?

(b) Reverse the data set assignments of random numbers for K and Φ and repeat (a).

(c) The theoretical mean and standard deviation for an exponential distribution are equal. Calculate the sample means and standard deviations from (a) and (b). With this criterion used as a decision-making aid, does an exponential distribution for T seem to be a reasonable assumption?

8.2.4* Determine the sample mean and standard deviation time to failure for this pumping system. System failure occurs when either pump 1 or 2 fails to operate. Both pumps are identical.

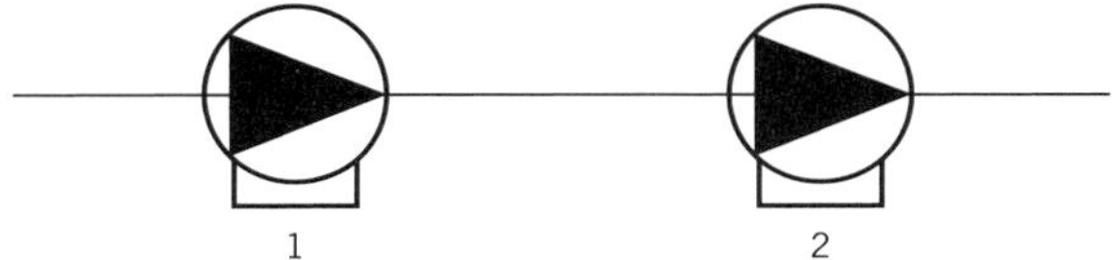

Let T = {time to pump failure}, where T is a Weibull distribution with an average failure rate ρ = 1 failure per year and shape factor $\gamma = 4$,

$$F(t) = 1 - e^{-t^4} \quad \text{for } t \geq 0$$

Conduct a Monte Carlo simulation. Use the random numbers of data set 1.

8.3 SIMPLE LINEAR REGRESSION MODELS

Linear regression analysis plays an important role in decision making. One extremely important application is to determine a relationship between a response variable y and an explanatory variable x. We begin our discussion by assuming that a cause-effect relationship exists between x and y and that the relationship between them is linear. These assumptions are expressed as a simple linear regression model, or for simple regression:

$$y = b_0 + b_1 x$$

where

$$b_0 = \text{intercept} \quad \text{and} \quad b_1 = \text{slope of the linear equation}$$

The same model calibration principles and procedures that apply to a simple regression model are applicable to other linear models, such as

No intercept model: $y = bx$

Quadratic model: $y = b_0 + b_1 x + b_2 x^2$

The quadratic model, a nonlinear function in terms of x^2, is a linear regression model because it is a linear function in terms of the model parameters b_0, b_1, and b_2. This model can be written as a function of two explanatory variables of $x_1 = x$ and $x_2 = x^2$. The quadratic model can be analyzed as a multiple linear regression model, or for multiple regression:

$$y = b_0 + b_1x_1 + b_2x_2$$

or, using matrix notation, the model is

$$y = \mathbf{xb}$$

where

$$\mathbf{x} = [1\ x_1\ x_2] \text{ and } \mathbf{b} = \begin{bmatrix} b_0 \\ b_1 \\ b_2 \end{bmatrix}$$

Basic principles and procedures for simple linear regression model analysis are introduced in this section in the context of the four steps of model calibration given in the introduction to this chapter. In Section 8.4, these same principles are used to explore multiple regression models and their application.

The Scatter Plot

Suppose a cause-effect relationship $y = f(x)$ is suspected where x is an explanatory variable and y is a response variable. For purposes of illustration, let $y = f(x)$ be a dose-response relationship for some toxic chemical, where y is the probability that an animal dies at chemical dose x. We hypothesize that $f(x)$ is the linear relationship, or

$$y = b_0 + b_1x$$

We will justify our hypothesis by conducting an experiment and exploring the data set.

The test animals are divided into n groups, and the animals of each group j, where $j = 1, 2, \ldots, n$, are fed the same diet containing a specified concentration of toxic chemical. Let

$$x_j = \text{concentration of a toxic chemical fed to group } j$$

and

$$y_j = \text{a probability estimate that an animal dies in group } j$$

where

$$y_j = \frac{\text{number of animal deaths in group } j}{\text{number of animals in group } j}$$

The data set of $[(x_1, y_1), (x_2, y_2), \ldots, (x_n, y_n)]$ is obtained.

A scatter plot, pair-wise plot of raw data, (x_j, y_j), is a graphical tool to test the hypothesis that a higher concentration x will cause a greater fraction of animals to die and $y = f(x) = b_0 + b_1x$ can adequately describe the process. In this example, the intercept b_0 is the background probability. It is the probability that an animal, which receives a zero chemical dose, will die. The probability rate is b_1, measured as the probability of an animal's death per unit concentration.

This completes the data collection and model identification steps of the calibration procedure. Before turning our attention to the model parameter estimation and model validation, we discuss the theoretical properties of the regression model.

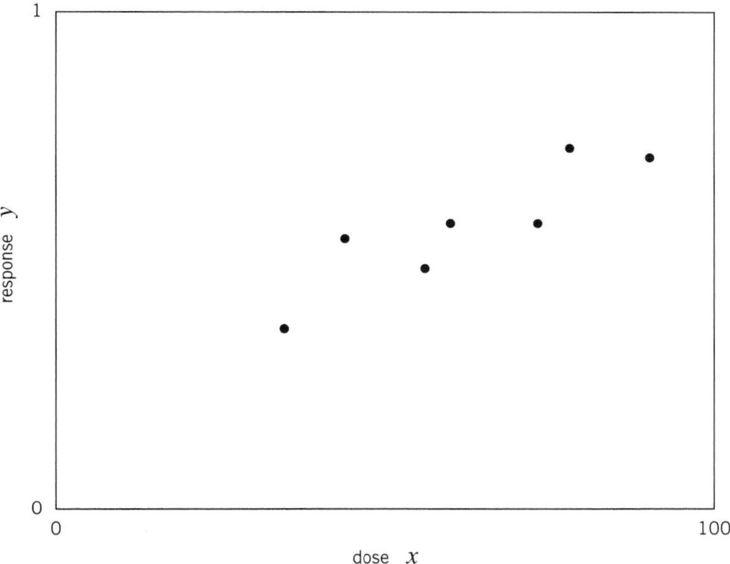

Figure 6 The scatter plot of (x_j, y_j).

The Line of Best Fit

In Figure 7, a response function has been labeled as a "line of best fit." The line, which is subjectively drawn, appears to be indicative of the overall behavior that an increase in

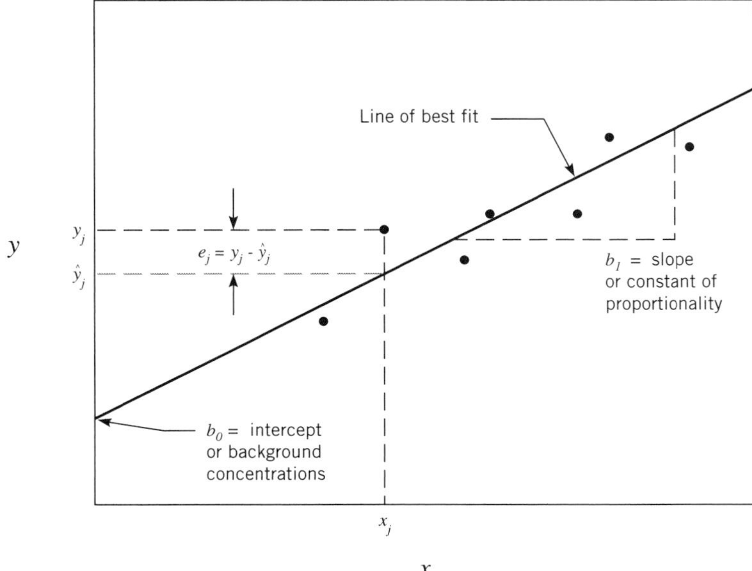

Figure 7 The Hypothesized model.

chemical dose results in a larger probability of an animal dying. The line has 3 data points above the line and 4 data points below it. Presently, the parameters of the model b_0 and b_1 will be estimated using an objective analysis called the method of least squares.

Now, consider the "line of best fit" as a prediction function. At dosage x_j the value of $\hat{y}_j$ is a forecast. The difference $e_j = \hat{y}_j - y_j$ is the *prediction error*. Suppose that $\hat{y}_j = 0.75$ and $y_j = 0.6$. The prediction error is 0.15 deaths. Given the variability in an animal test, this may not be an unexpected result.

Given that prediction errors are expected, let us treat the response y as the random variable Y where Y is defined to be

$$Y = \beta_0 + \beta_1 x + \varepsilon$$

It is the sum of the "line of best fit," a deterministic term, plus a random disturbance term ε. Given the properties of the errors associated with the "line of best fit," it seems reasonable to define ε to be a normal distributed random variable with mean $\mu = 0$ and constant variance σ^2, $N(0, \sigma^2)$. The expected value of Y is the "line of best fit,"

$$\hat{y} = E(Y) = E(b_0 + b_1 x + \varepsilon) = b_0 + b_1 x$$

The variance of Y is

$$\text{var}(Y) = E[(Y - E[Y])^2] = E(\varepsilon^2) = \sigma^2$$

In summary, the "line of best fit," $\hat{y} = b_0 + b_1 x$, is simply a *expection function*, where b_0 and b_1 are estimates of β_0 and β_1.

Assumptions for Regression Models

The assumptions for regression models are stated here and shown in Figure 8:

1. The response Y at x is the expected value of Y at x plus random error.
2. Each error has a normal distribution.
3. Each error has a zero mean.
4. The errors have equal variances.
5. The errors are distributed independently.

These assumptions are checked in the model validation step.

Model Parameter Estimation

The *method of least squares* is used to estimate the unknown parameters of a regression model. The parameter estimation problem is an unconstrained optimization problem. The objective function is to minimize the sum of square error, or

$$\text{minimize } S = \sum_{j=1}^{n} e_j^2$$

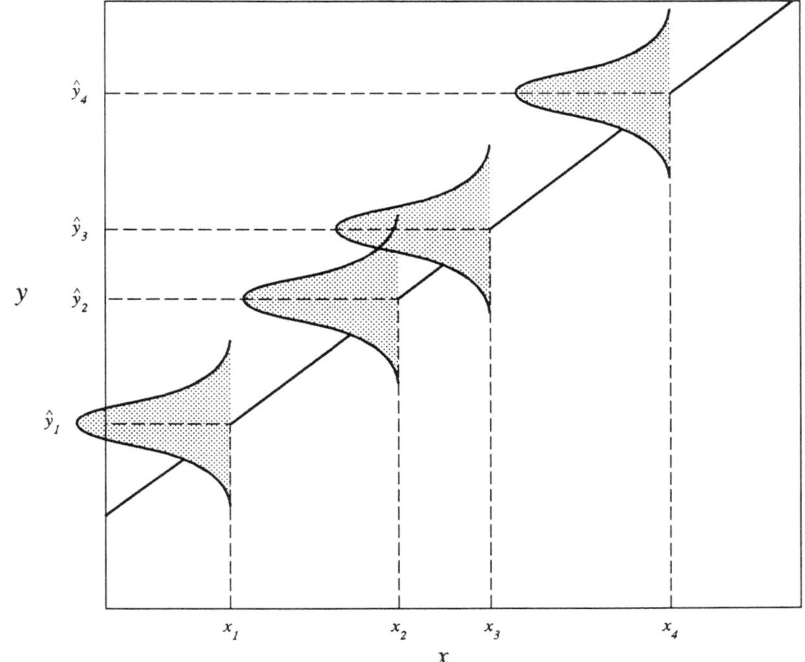

Figure 8 A simple regression model.

where $e_j = y_j - \hat{y}_j$. Since the functional form of $\hat{y}$ is specified, e_j written in terms of the unknown parameters b_0 and b, or

$$e_j = y_j - b_0 - b_1 x_j$$

Consequently, the objective is to find estimates of b_0 and b_1 that

$$\text{minimize } S = \sum_{j=1}^{n} (y_j - b_0 - b_1 x_j)^2$$

The minimum of S occurs at the critical point or point of zero slope.

$$\frac{\partial S}{\partial b_0} = 2 \sum_{j=1}^{n} (y_j - b_0 - b_1 x_j)(-1) = 0$$

$$\frac{\partial S}{\partial b_1} = 2 \sum_{j=1}^{n} (y_j - b_0 - b_1 x_j)(-x_j) = 0$$

Expanding and simplifying, we obtain

$$\frac{\partial S}{\partial b_0} = \sum_{j=1}^{n} y_j - \sum_{j=1}^{n} b_0 - b_1 \sum_{j=1}^{n} x_j = 0$$

$$\frac{\partial S}{\partial b_1} = \sum_{j=1}^{n} x_j y_j - b_0 \sum_{j=1}^{n} x_j - b_1 \sum_{j=1}^{n} x_j^2 = 0$$

Introducing

$$\sum_{j=1}^{n} x_j = n\bar{x} \text{ and } \sum_{j=1}^{n} y_j = n\bar{y}$$

where $\bar{x}$ and $\bar{y}$ are samples means of x and y, respectively, and then rearranging the first equation, we obtain

$$b_0 = \bar{y} - b_1 \bar{x}$$

Substituting the expressions for b_0 and $\bar{x}$ into the second expression, we obtain

$$\frac{\partial S}{\partial b_1} = \sum_{j=1}^{n} x_j y_j - n\bar{x}\left(\sum_{j=1}^{n} y_j - b_1 \bar{x}\right) - b_1 \sum_{j=1}^{n} x_j^2 = 0$$

or

$$b_1 = \frac{\sum_{j=1}^{n} (x_j - \bar{x}) y_j}{\sum_{j=1}^{n} x_j^2 - n\bar{x}^2} = \frac{\sum_{j=1}^{n} (x_i - \bar{x})(y_j - \bar{y})}{\sum_{j=1}^{n} (x_j - \bar{x})^2}$$

Dividing the numerator and denominator by $n - 1$, we can rewrite b_1 in terms of the sample covariance of x and y and variance of x or in terms of the correlation coefficient and the sample variances of x and y.

$$b_1 = \frac{\text{cov}(x, y)}{s_x^2} = \frac{\rho_{XY} s_Y}{s_X}$$

The correlation coefficient is a measure of linear dependency and has the limits of $-1 \leq \rho_{XY} \leq 1$.

The fitted model is $\hat{y} = b_0 + b_1 x$. The estimate of σ^2 or s^2 is an average of the sum of the squared error S,

$$s^2 = \frac{\sum_{j=1}^{n} e_j^2}{n-2} = \frac{\sum_{j=1}^{n} (y_j - b_0 - b_1 x_j)^2}{n-2}$$

where n = number of observations, p = number of model parameters = 2, and $n - p$ = degrees of freedom = $n - 2$. The value of s^2 is an unbiased estimate of σ^2.

In Section 8.4, matrix algebra is introduced and used for parameter estimation and determining other statistics. For the present, the above relationships serve to introduce basic concepts of model calibration.

Model Validation

One of the most effective means of determining the adequacy of a model is by plotting the error-response plots, that is, $(e_j, \hat{y}_j)$ and (e_j, x_j) plots. The purpose of these plots is to determine if assumptions for regression models are satisfied.

Since ε is assumed to have an $N(0, \sigma^2)$ distribution, the e_j points should be scattered around the y and x axes in a random pattern resembling a normal distribution. Along the y and x axes for the $(e_j, \hat{y}_j)$ and (e_j, x_j) plots, the points should lack any regular patterns or trends. Patterns or trends usually indicate that the model is inadequate in describing the process. The model should be rejected and an another one identified, calibrated, and tested.

For example, if an explanatory variable is x and the (e_j, x_j) plot shows a regular pattern, then it indicates that e and x are dependent variables. This is a violation of the assumption that errors are independent. This model should be rejected and another one calibrated and tested. In many instances, the original model can be modified. This will be demonstrated by example.

The following examples demonstrate the use of fundamental principles for model calibration, particularly issues dealing with model identification and model validation. For the present, scatter and error response plots will be used. In Section 8.4 where multiple regression and matrix methods are introduced, other validation tests are presented.

EXAMPLE 8.4 ***An HCl-Lime Dose Model***

A series of eight experiments at different lime dosages have been conducted to determine the relationship between HCl flue gas concentration and lime dose rate. The results of the study are as follows:

j	x, Lime Dose (kg per hour)	y, HCL Concentration (ppmv)
1	70	304
2	90	326
3	110	332
4	130	219
5	150	274
6	170	156
7	190	91
8	210	101

Source: Based on J. Bergström and J. Lundqvist (1983).

Use these data for model identification, calibration, and validation.

SOLUTION

A scatter plot of the data indicates that a simple linear regression model should be adequate. The hypothesized model is

$$Y = \beta_0 + \beta_1 x + \varepsilon$$

where the slope term is expected to be negative, $b_1 < 0$.

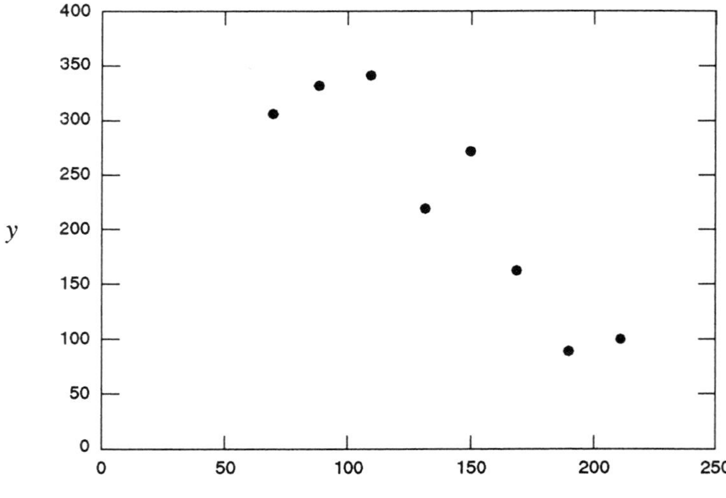

The parameters are estimated using the method of least squares. The estimates of parameters are $b_0 = 481$ and $b_1 = -1.83$. Details of the model calibration procedure using matrix methods are given in Example 8.6. The fitted model or expectation function is

$$\hat{y} = 481 - 1.83x$$

and

$$n - p = \text{degrees of freedom} = 6$$

The residual plots of e–x and e– $\hat{y}$ show that the five key assumptions for regression models (summarized in Figure 8) are satisfied. The error appears to be normally distributed, and it has a zero mean and an equal variance. Since there are no apparent trends or patterns in either the $(e_j, \hat{y}_j)$ or (e_j, x_j) plots, the error appears to satisfy the assumption that it is distributed independently.

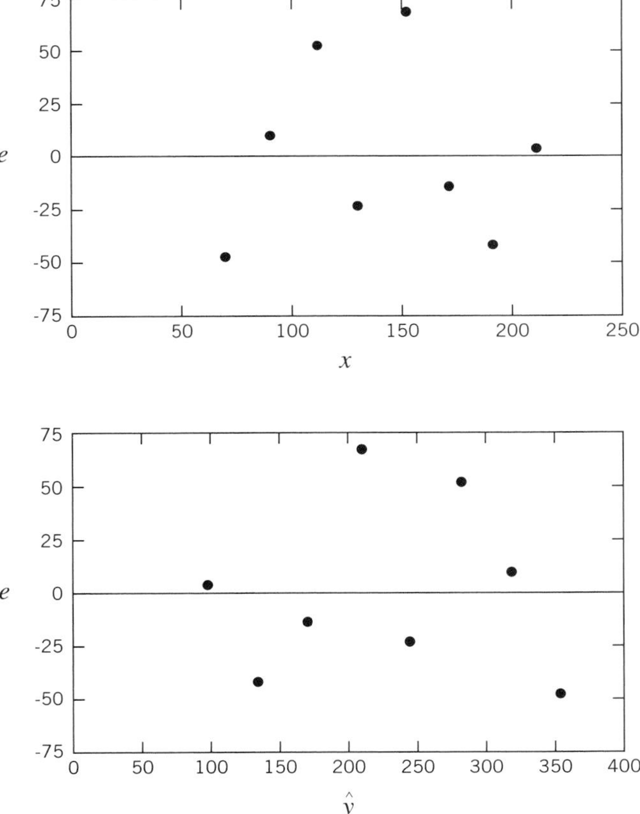

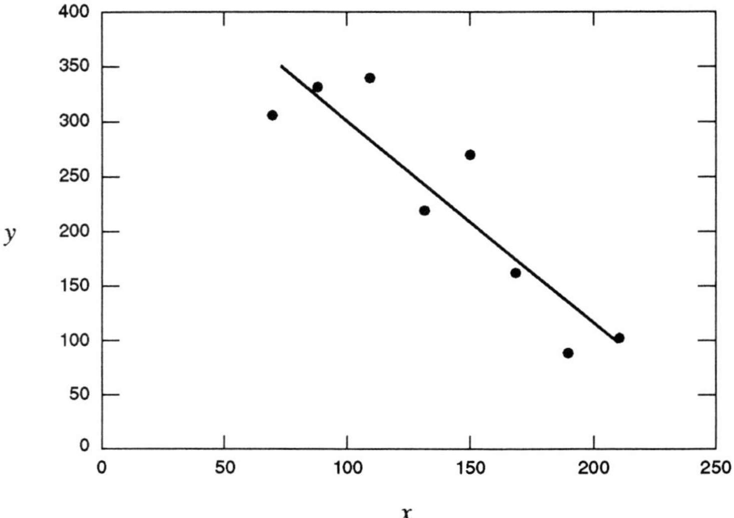

The fitted line imposed on the scatter plot shows a reasonable fit of all data points.

Model Transformations

Through linear regression analysis, it is possible to transform a nonlinear model into a linear form and estimate the parameters. For example, the decay model,

$$z = c_0 e^{-\rho t}$$

where

$$c_0 = \text{initial concentration}$$
$$\rho = \text{decay rate}$$
$$t = \text{time}$$

is a nonlinear model in terms both of its model parameters c_0 and ρ, and of its response variable z and explanatory variable t. It can be transformed into a linear model by taking the natural logarithm of each side of the equation. The result is

$$ln(z) = ln(c_0) - \rho t$$
$$y = b_0 + b_1 t$$

where

$$y = ln(z)$$
$$b_0 = ln(c_0)$$
$$b_1 = -\rho$$

After estimates of b_0 and b_1 are made, the estimates c_0 and ρ can be determined by solving these equality relationships. The use of this technique is shown by example.

EXAMPLE 8.5 ***A DDT Dose-Response Model as a Weibull Model***

Risk assessment studies subject animals to a lifetime exposure to a fixed amount of a toxic chemical. The number of animals of a population of $n(d)$ showing a response $r(d)$ at dose d are counted. Response means the animal dies or develops a cancerous tumor. The probability of a response is estimated as a frequency, $\hat{\theta} = r/n$, at chemical dose d.

The following table show the liver cancer responses of animals fed a daily dose of DDT.

Group j	Dose d_j	Response r_j	Number of Animals n_j	Lifetime Risk Probability Estimate, $\hat{\theta}_j = \frac{r_j}{n_j}$
1	0	4	111	0.036
2	2	4	105	0.038
3	10	11	124	0.089
4	50	13	104	0.125
5	250	60	90	0.667

Source: Van Ryzin (1980).

Calibrate a Weibull model of the form, $\theta = 1 - e^{-\beta d^{\gamma}}$.

SOLUTION

The Weibull model can be transformed into a linear equation by taking natural logarithms of both sides of the equation two times. First, rearrange the equation as

$$1 - \theta = e^{-\beta d^{\gamma}}$$

Taking the natural logarithm of the equation gives

$$-ln(1 - \theta) = \beta d^{\gamma}$$

This model is nonlinear in terms of its parameters β and γ. Taking the natural logarithm once again gives

$$ln[-ln(1 - \theta)] = ln(\beta) + \gamma ln(d)$$

This equation has a linear form

$$y = b_0 + b_1 x$$

where

$$\begin{aligned} y &= ln[-ln(1-\theta)] \\ b_0 &= ln(\beta) \\ b_1 &= \gamma \\ x &= ln(d) \end{aligned}$$

Since the transformed model is linear in terms of its parameters, a simple linear regression model is hypothesized,

$$Y = \beta_0 + \beta_1 x + \varepsilon$$

After making the logarithmic transformations of the data, a scatter plot of y–x shows that a simple linear regression line appears to be an appropriate choice. Since $\ln(0) = -\infty$ for $d = 0$, the first data point is dropped from the data set. This data point will be introduced into the Weibull model as a background probability $\theta(0) = 0.036$.

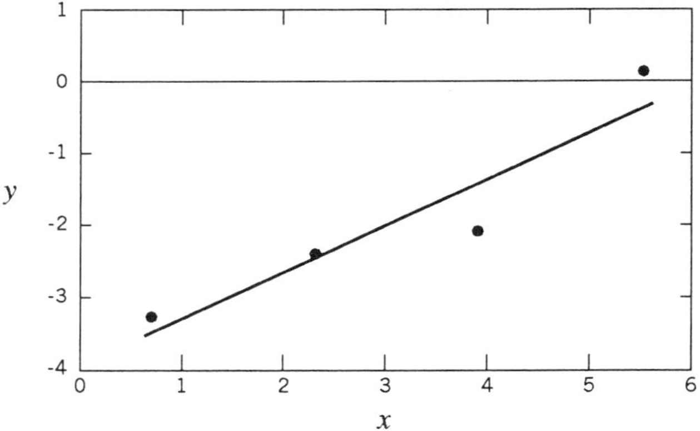

The method of least squares gives

$$\hat{y} = -3.89 + 0.646x$$

with $s^2 = 0.564^2$. The x–y plot shows that this line fits the data well.

Validation tests are performed on the linear model to assure that the fundamental assumptions for a linear regression model are not violated. The error plot shows that no violation of assumptions has occurred.

The parameters of β and γ are easily determined:

$$\begin{aligned} \beta &= e^{b_0} = e^{-3.89} = 0.00525 \\ \gamma &= b_1 = 0.646 \end{aligned}$$

The fitted Weibull response function is

$$\theta = 1 - \exp(-0.0204d^{0.646})$$

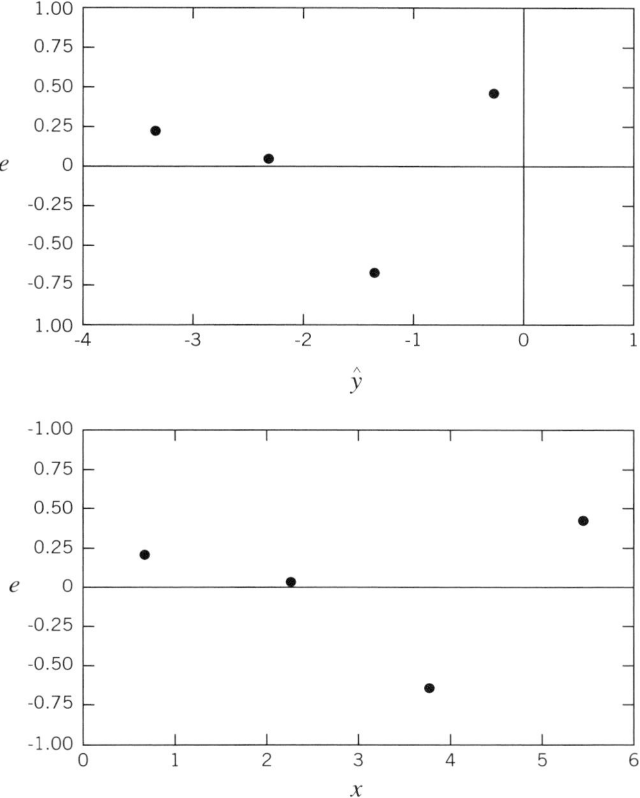

The background probability is introduced into the model,

$$\theta'(d) = \theta(0) + [1 - \theta(0)]\theta(d)$$
$$\theta'(d) = 0.036 + 0.964[1 - \exp(-0.0204d^{0.646})].$$

The following plot shows the two models and the original data set on a single graph.

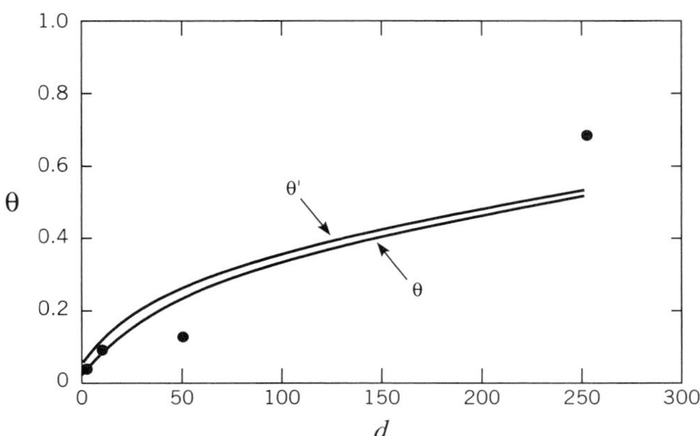

DISCUSSION

The use of model transformation and linear regression analysis gives no guarantee that the nonlinear model will fit the data well. In this example, we should question the use of these models at higher dosages. The above plot suggests that the model seriously underestimates the observed response at $d = 250$.

Nonlinear regression analysis uses the same basic principles for calibrating nonlinear models as given in this text for linear models. The method of least squares criterion for parameter estimation is the same for linear and nonlinear models. That is, to minimize the sum of the square error, or

$$\text{minimize } S = \sum_{j=1}^{n} e_j^2$$

The most fundamental and important difference lies in defining the error terms.

For linear regression: $e_j = y_j - b_0 - b_1 x_j$.

where x_j and y_j represent the transformed observations for d_j and θ_j.

For nonlinear regression: $e_j = \theta_j - [1 - \exp(-\beta d_j^{\gamma})]$.

In the nonlinear case, e_j is a direct measure of the prediction error. In the linear case, since y_j and x_j are transforms of the observed data, e_j is an indirect measure of the prediction error.

The calibrated model using nonlinear regression analysis is

$$\theta = 1 - \exp(-0.00163 d^{1.17})$$

The following diagram compares the goodness of fit for the dose response model using nonlinear and linear regression analyses. Clearly, the model calibrated as a nonlinear regression model has a better overall fit.

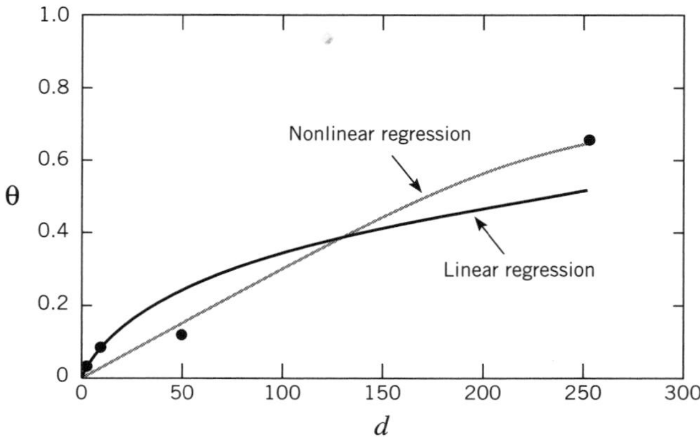

The model transformation and linear regression analysis technique can often give good results. Owing to its mathematical simplicity, this approach is commonly used in practice.

This illustration serves as a warning that the model transform approach must be used with caution. Nonlinear regression analysis is beyond the scope of this book, and the method will not be discussed further. It is introduced in this example simply for completeness. It also shows that it is sometimes necessary to resort to more complex methods of analysis to reach desired results.

In Example 8.7, a linear DDT dose-response model is calibrated and compared to the results from this example. As this example and upcoming examples will show, feedback analysis is an important part of model calibration. Different model structures must be analyzed critically before one is recommended for application. This problem will be addressed again in Example 8.7.

QUESTION FOR DISCUSSION

8.3.1 Give the following definitions:

expectation function | prediction error
method of least squares

QUESTIONS FOR ANALYSIS

8.3.2 (a) Use the method of least squares to show that the governing equation for estimating the model parameter b in a no-intercept model, $Y = \beta x + \varepsilon$, is

$$b = \frac{\sum_{j=1}^{n} y_j x_j}{\sum_{j=1}^{n} x_j^2}$$

where b is the estimate of β.

(b) What is the equation to estimate the variance σ^2 of $Y = \beta x + \varepsilon$?

8.3.3* A new-mass burn MSW plant with electrical power generation is to be constructed. The following data, which are also given in Problem 7.4.2, are representative of the plant being considered for construction. An expectation function for construction cost, $\hat{k} = ub$, is desired, where the variable b = design capacity of the plant, and parameter u = unit cost of construction.

i	b, Design Capacity (tons/day)	k, Capital Cost ($M)
1	3000	$253
2	2250	$200
3	400	$40
4	3000	$290
5	400	$30
6	2250	$250
7	518	$40
8	1650	$113

Source: Neal and Schubel (1987)

The following statistics have been calculated:

$$\bar{k} \pm s_{\mathrm{K}} = \$152\mathrm{M} \pm 109\mathrm{M}$$
$$\bar{b} \pm s_{\mathrm{B}} = 1684 \pm 1118 \text{ tons per day,}$$

and

$$\rho_{BK} = 0.948$$

(a) Consider a no-intercept model, $K = uB + \varepsilon$. Plot a scatter plot of the data in order to determine if the selected model form is an acceptable choice. Estimate the unknown model parameter using the method of least squares. See Problem 8.3.2.

(b) Use error-response plots to test model validity. If the no-intercept model fails one or more of the assumptions for linear regression, formulate a new model and determine if it is a satisfactory model.

(c) Superimpose the acceptable expected value function model found in (b) on a scatter plot of the data. Based on the results from (a) and (b), can the hypothesized model be justified?

8.3.4* Temperature control is considered to be an effective means to minimize the public's exposure to toxic air emissions. The following data have been collected:

x, Furnace Temperature (degrees C)	y, Air Emission Concentration of PCDD (ppmv)
573	3.9
587	5.0
650	1.7
655	2.3
670	1.8
553	5.8
670	1.1
650	1.1

Source: Hasselriis (1987).

The following statistics have been calculated:

$$\bar{x} \pm s_X = 626 \pm 49.1$$
$$\bar{y} \pm s_Y = 2.84 \pm 1.82 \text{ ppm},$$

and

$$\rho_{XY} = -0.938$$

(a) Plot a scatter plot of the data to determine if a simple linear regression model is an acceptable choice. Estimate the unknown model parameters using the method of least squares.

(b) Use error-response plots to test model validity. If the model fails one or more of the assumptions for linear regression, formulate a new model and determine if it is a satisfactory model.

(c) Superimpose the acceptable expected value function model found in (b) on a scatter plot of the data. Based on the results from (a) and (b), can the model be justified?

8.3.5* Owing to the expense and difficulty in measuring PCDD concentration, CO concentration is being considered as its surrogate measure. The following data have been collected:

x Carbon Monoxide (CO) Emissions, (ppm)	y Air Emission Concentration of PCDD (ppmv)
850	3.9
1020	5.0
390	1.7
510	2.3
490	1.8
500	5.8
450	1.1
310	1.1

Source: Hasselriis (1987).

A no-interpret model, $Y = \beta x + \varepsilon$, is hypothesized.

(a) Construct scatter plots and determine if a linear or some nonlinear model will be best suited for the data. Use the method of least squares to estimate the model parameters. See Problem 8.3.2.

(b) Use error-response plots to test model validity.

(c) Based on the results from (a) and (b), can the model be justified?

8.3.6* The following data show the number of rats that developed a liver tumor due to the exposure of dimethylnitrosamine (DMN).

Group j	Dose d_j (ppm)	Number of Animals, n_j	Number Responses r_j
1	0	29	0
2	2	18	0
3	5	62	4
4	10	5	2
5	20	23	15
6	50	12	10

Source: Van Ryzin (1980).

(a) Calibrate and validate a one-hit model:

$$\theta = 1 - e^{-\beta d}$$

where θ = probability that a rat will develop liver tumor at dose d of DMN. Transform the model into a linear model and use the method of least squares for linear regression analysis.

(b) Plot $\theta - d$ relationship on a scatter plot.

8.4 MULTIPLE LINEAR REGRESSION MODELS

The same theories and principles that apply to the calibration of simple regression models apply to the calibration of multiple regression models. Since there are more parameters, the mathematical complexity is greater. Matrix methods are most convenient and compact in expressing complex relations; therefore, its use is advocated. In addition, some spreadsheet programs have built-in matrix functions, and commercially available statistics and linear regression computer programs make use of matrix notation for computation and for conveniently inputting and outputting data.

Assumptions for Regression Models

Simple and multiple regression models, written in matrix notation, can be expressed as

$$\text{Model: } \mathbf{Y} = x\beta + \varepsilon$$

where

$\mathbf{Y}$ = response vector for n observations = $[y_1\ y_2\ \ldots\ y_n]^T$

$\mathbf{X}$ = $n \times p$ matrix of explanatory variables,
where n = number of observations and
p = number of model parameters

or

$$\mathbf{X} = \begin{bmatrix} x_{11} & x_{12} & \cdot\ \cdot & x_{1p} \\ x_{21} & x_{22} & \cdot\ \cdot & x_{2p} \\ \cdot & \cdot & \cdot\ \cdot & \cdot \\ \cdot & \cdot & \cdot\ \cdot & \cdot \\ x_{n1} & x_{n2} & \cdot\ \cdot & x_{np} \end{bmatrix}$$

$$\beta = [\beta_1\ \beta_2 \ldots \beta_p]^T$$
$$\varepsilon = [\varepsilon_1\ \varepsilon_2 \ldots \varepsilon_n]^T$$

where the error ε is normally distributed with $E(\varepsilon) = 0$ and its covariance matrix is $\text{cov}(\varepsilon) = \sigma^2(\mathbf{X}^T\mathbf{X})^{-1}$.

Matrix notation is compact; thus, key relationships can be neatly summarized.

Response or expectation value function: $\hat{\mathbf{y}} = E(\mathbf{Y}) = \mathbf{Xb}$

Parameter estimates: $\mathbf{b} = (\mathbf{X}^T\mathbf{X})^{-1}\mathbf{X}^T\mathbf{y}$

Error function: $\mathbf{e} = \mathbf{y} - \mathbf{Xb}$

Residual sum of squares: $S(\mathbf{b}) = \mathbf{e}^T\mathbf{e} = (\mathbf{y} - \mathbf{Xb})^T(\mathbf{y} - \mathbf{Xb})$

Variance estimate: $s^2 = \dfrac{S(\mathbf{b})}{n-p}$

Standard error for parameter estimate q:

$$se(b_q) = s\sqrt{[(\mathbf{X}^T\mathbf{X})^{-1}]_{qq}}$$

where

$$[(\mathbf{X}^T\mathbf{X})^{-1}]_{qq} = \text{the } q\text{th diagonal term of } (\mathbf{X}^T\mathbf{X})^{-1}$$

The Significance of Model Parameter Estimates

Confidence intervals for estimates of model parameter **b** can be established and used to determine their significance. A $100(1 - \alpha)\%$ confidence interval for b_q is

$$[L_{b_q}, U_{b_q}] = b_q \pm se(b_q)\, t_{(1-\alpha/2,\, n-p)}$$

where

$$se(b_q) = \text{standard error for a parameter estimate } q$$

and

$$n - p = \text{degrees of freedom}$$

If the interior of the confidence interval for parameter estimate b_q is exclusive of zero, then the parameter is considered to be significantly different than zero at the $100(1 - \alpha)\%$ level.

On the other hand, if the interior of the confidence interval contains zero, the parameter is not considered to be significantly different from zero. The model is rejected, and a new model is fit and tested again. The new model will not contain b_q.

The use of matrix algebra and the importance of testing for the significance of a model parameter are best shown by example. Examples 8.7 through 8.9 also show the need for feedback analysis.

EXAMPLE 8.6 *A HCl-Lime Dose Model*

In Example 8.4, the following linear regression model is calibrated,

$$\hat{y} = 481 - 1.83x$$

with $s = 45.4$ for the following data set:

j	x, Lime Dose (kg per hour)	y, HCL Concentration (ppmv)
1	70	304
2	90	326
3	110	332
4	130	219
5	150	274
6	170	156
7	190	91
8	210	101

(a) Use matrix algebra to estimate the parameters of a sample linear regression model.

(b) Test the significance of the model parameters at the 95% confidence level.

SOLUTION

(a) The parameters are estimated using $\mathbf{b} = (\mathbf{X}^T\mathbf{X})^{-1}\mathbf{X}^T\mathbf{y}$, where

$$\mathbf{y} = \begin{bmatrix} 304 \\ 326 \\ 323 \\ 219 \\ 274 \\ 156 \\ 91 \\ 101 \end{bmatrix} \quad \text{and} \quad \mathbf{X} = \begin{bmatrix} 1 & 70 \\ 1 & 90 \\ 1 & 110 \\ 1 & 130 \\ 1 & 150 \\ 1 & 170 \\ 1 & 190 \\ 1 & 210 \end{bmatrix}$$

The vector of parameter estimates is

$$\mathbf{b} = \begin{bmatrix} b_0 \\ b_1 \end{bmatrix} = \begin{bmatrix} 481 \\ -1.83 \end{bmatrix}$$

and the fitted model is

$$\hat{y} = 481 - 1.83x$$

with a variance estimate of $s^2 = 45.4^2$ and

$$(\mathbf{X}^T\mathbf{X})^{-1} = \begin{bmatrix} 1.29 & -0.00833 \\ -0.00833 & 0.0000595 \end{bmatrix}$$

(b) The 95% confidence level for the intercept term is

$$[L_{b0}, U_{b0}] = b_0 \pm se(b_0)t_{(1-\alpha/2,\, n-p)}$$

$$[L_{b0}, U_{b0}] = 481 \pm (51.10)(2.447) = [355, 607]$$

where

$$t_{(0.975,\, 6)} = 2.447$$

and

$$se(b_0) = s\sqrt{\left[\left(\mathbf{X}^T\mathbf{X}\right)^{-1}\right]_{11}} = 45\sqrt{1.29} = 51.10$$

The 95% confidence level for the slope term is

$$[L_{b1}, U_{b1}] = b_1 \pm se(b_1)t_{(1-\alpha/2,\, n-p)}$$

$$[L_{b1}, U_{b1}] = -1.83 \pm (0.347)(2.447) = [-2.68, -0.970]$$

where

$$se(b_1) = s\sqrt{\left[\left(\mathbf{X}^T\mathbf{X}\right)^{-1}\right]_{22}} = 45\sqrt{0.0000595} = 0.347$$

Since the confidence intervals $[L_{b0}, U_{b0}]$ and $[L_{b1}, U_{b1}]$ are exclusive of zero, the intercept and slope terms are considered to be significantly different from zero.

This validation test and the error plot test, presented in Example 8.4, give assurances that the model, $\hat{y} = 481 - 1.83x$, is adequate.

EXAMPLE 8.7 *A DDT Dose-Response Model as a Linear Model*

In Example 8.5, a Weibull model was calibrated. Use the same data from that example to calibrate a linear dose-response model,

$$\theta = b_0 + b_1 d$$

where

b_0 = background probability

and

b_1 = probability rate, measured as the probability of an animal's death per unit concentration

The data are

Group j	Dose d_j	Lifetime Risk Estimate, $\hat{\theta}_j = \frac{r_j}{n_j}$
1	0	0.036
2	2	0.038
3	10	0.089
4	50	0.125
5	250	0.667

(a) Use matrix algebra to estimate the parameters.

(b) Validate the model by using error plots and determining the significance of the model parameters at the 95% confidence level.

SOLUTION

(a) The parameters are estimated using $\mathbf{b} = (\mathbf{X}^T\mathbf{X})^{-1}\mathbf{X}^T\mathbf{y}$, where $y = \theta$, $x = d$ and

$$\mathbf{y} = \begin{bmatrix} 0.036 \\ 0.038 \\ 0.089 \\ 0.125 \\ 0.667 \end{bmatrix} \text{ and } \mathbf{X} = \begin{bmatrix} 1 & 0 \\ 1 & 2 \\ 1 & 10 \\ 1 & 30 \\ 1 & 250 \end{bmatrix}$$

The vector of parameter estimates is

$$\mathbf{b} = \begin{bmatrix} b_0 \\ b_1 \end{bmatrix} = \begin{bmatrix} 0.034 \\ 0.00250 \end{bmatrix}$$

and the fitted model is

$$\hat{y} = 0.034 + 0.00250x$$

with a variance estimate of $s^2 = 0.0264^2$ and

$$(\mathbf{X}^T\mathbf{X})^{-1} = \begin{bmatrix} 0.285 & -0.0014 \\ -0.0014 & 0.000022 \end{bmatrix}$$

(b) The 95% confidence level for the intercept term is

$$[L_{b0}, U_{b0}] = b_0 \pm se(b_0)t_{(1-\alpha/2,\, n-p)}$$

$$[L_{b0}, U_{b0}] = 0.034 \pm (0.014)(3.182)$$

$$[L_{b0}, U_{b0}] = [-0.0104, 0.0795]$$

where

$$t_{(0.975,\, 3)} = 3.182$$

and

$$se(b_0) = s\sqrt{\left[\left(\mathbf{X}^T\mathbf{X}\right)^{-1}\right]_{11}} = 0.0264\sqrt{0.285} = 0.014$$

This parameter is not significantly different from zero.

The 95% confidence level for the slope term is

$$[L_{b1}, U_{b1}] = b_1 \pm se(b_1)t_{(1-\alpha/2,\, n-p)}$$

$$[L_{b1}, U_{b1}] = 0.00250 \pm (0.0012)(3.182)$$

$$[L_{b1}, U_{b1}] = [0.00211, 0.00289]$$

where

$$se(b_1) = s\sqrt{\left[\left(\mathbf{X}^T\mathbf{X}\right)^{-1}\right]_{22}} = 0.0264\sqrt{0.000022} = 0.0012$$

The error plots are

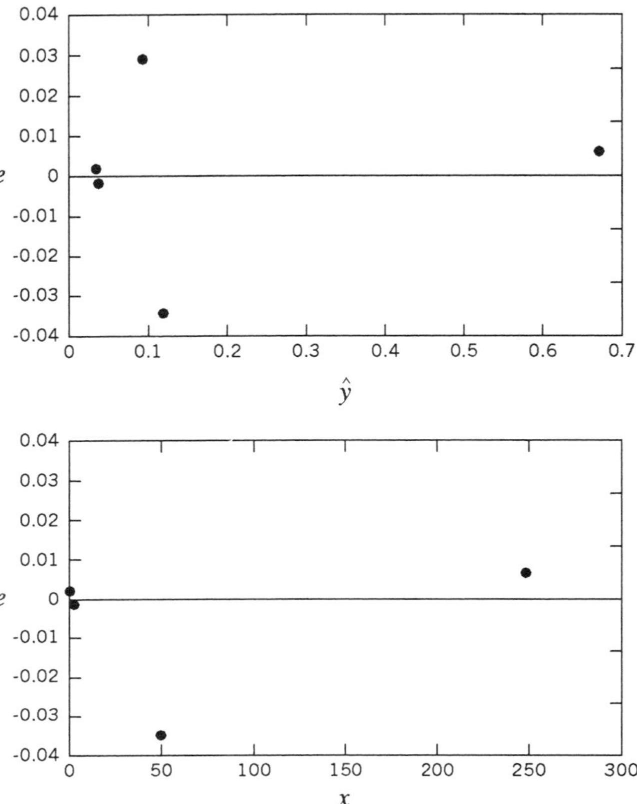

The error plots show no apparent violations in the assumptions for a linear regression model. However, since $[L_{b0}, U_{b0}]$ contains zero, the intercept term is not considered to be significantly different from zero at the 95% level. The model of $y = 0.034 + 0.00250x$ is rejected.

A no intercept model, $y = bx$, is considered. All model parameter and validation procedures must be performed once again on this model.

The parameter b is estimated using $b = (\mathbf{X}^T\mathbf{X})^{-1}\mathbf{X}^T\mathbf{y}$, where

$$\mathbf{y} = \begin{bmatrix} 0.036 \\ 0.038 \\ 0.089 \\ 0.125 \\ 0.667 \end{bmatrix} \quad \text{and} \quad \mathbf{X} = \begin{bmatrix} 0 \\ 2 \\ 10 \\ 50 \\ 250 \end{bmatrix}$$

The parameter estimate is $b = 0.00267$, and the fitted model is

$$y = 0.00267x$$

with a variance estimate of $s^2 = 0.0397^2$ and $(\mathbf{X}^T\mathbf{X})^{-1} = 2.42 \times 10^{-8}$.

The 95% confidence level for the slope term is

$$[L_{b_1}, U_{b_1}] = b_1 \pm se(b_1)t_{(1-\alpha/2,\, n-p)}$$

$$[L_{b_1}, U_{b_1}] = 0.00267 \pm \left(0.0397\sqrt{2.42 \times 10^{-8}}\right)(3.182) = [0.00218, 0.00317]$$

where

$$t_{(0.0975,\, 4)} = 2.776.$$

The intercept term is significantly different from zero.

The error plots are

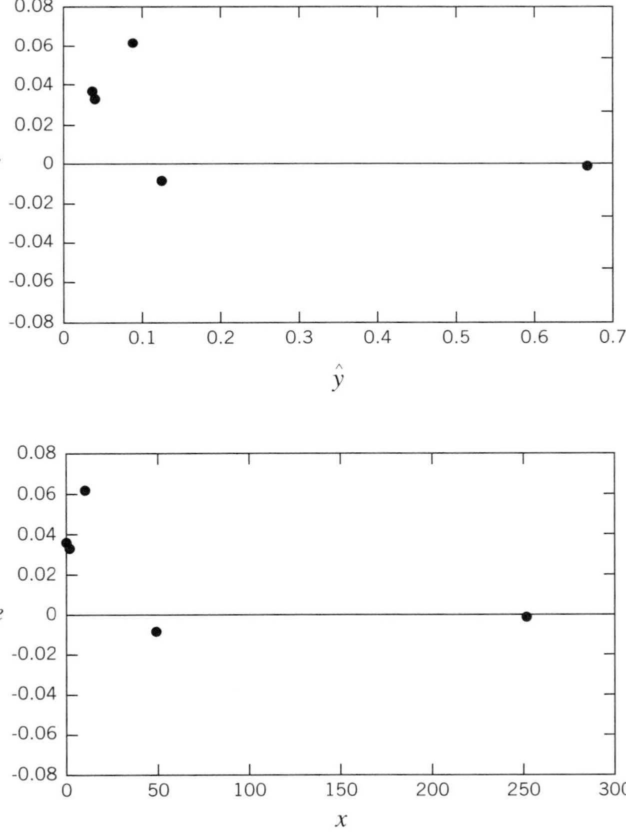

The error plots show a lack of fit at low dosages at and near $d = 0$, and a possible violation of the assumption that the model error mean estimate ε is zero and its variance is equal. It is difficult to recommend either the intercept (background) or the no-intercept (no-background) models. Neither model completely satisfies all validation criteria.

DISCUSSION

The example shows that in real-world applications an analysis can lead to conflicting conclusions and uncertainty. Factors, other than statistics, will be introduced in this decision-making process for selecting a model. Consider the results from this example and the discussion from Example 8.5. The following no-background models have been considered:

$$\text{Weibull: (LR)}\ \ \theta = 1 - \exp(-0.0204d^{0.646})$$

calibrated using model transformation and linear regression (LR) analysis,

$$\text{Weibull: (NLR)}\ \ \theta = 1 - \exp(-0.00163d^{1.17})$$

calibrated using non-linear regression (NLR) analysis,

$$\text{No intercept: } \theta = 0.00267d$$

presented in (a) as $y = 0.00267x$.

The Weibull models do not use the first observation of the data set of $\hat{\theta}_1 = 0.036$ at $d_1 = 0$. The no-intercept model uses all five data points in its model calibration. Nevertheless, all models at $d = 0$ have the same expected value of $\theta = 0$. The background probability is introduced into all three models with the following model,

$$\theta'(d) = \theta(0) + [1 - \theta(0)]\theta(d) = 0.036 + 0.964\theta(d)$$

The following plots show how well each model fits the data.

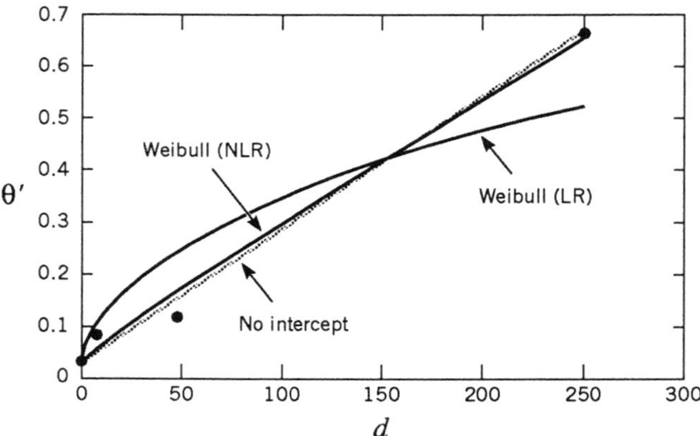

The Weibull (NLR) and no intercept models appear to fit the data equally well. This result shows that the no-intercept model can give as good results as a more complex Weibull model. Therefore, one is inclined to use the no-intercept model.

Modeling offers a method of summarizing complex relationships, but it cannot always eliminate and explain all uncertainties associated with the process under study. As mentioned in Chapter 5, when dose-response models are used for establishing a regulatory standard, a safety factor sf is employed. For example, the maximum contamination standard for a carcinogen using the no-intercept model is

$$\text{MCL} = \frac{w\theta}{r\beta sf}$$

where β is the slope from a no-intercept model and w and r are the standard weight and intake rates for humans. See Chapter 5 for further discussion.

Experimental Design

An important and primary objective for designing an experiment is to choose values of the design X matrix that will lead to good parameters with small variances or standard errors of $se(b_q)$. That is,

$$\text{minimize } se(b_q) = s\sqrt{\left[\left(\mathbf{X}^T\mathbf{X}\right)^{-1}\right]_{qq}}$$

The objective is achieved by selecting the values of $\mathbf{X}$ such that the determinant of $D = |\mathbf{X}^T\mathbf{X}|$ is maximized. This method is called the *D-optimality criterion.*

It application can be most easily illustrated with a no intercept model.

$$y = bx$$

where the matrix $\mathbf{X}$ is the vector

$$\mathbf{X} = [x_1\ x_2 \ldots x_n]^T$$

and n = the number of observations in the experiment. In this case, the standard error of the intercept term is simply,

$$se(b) = \frac{s}{\sqrt{\mathbf{X}^T\mathbf{X}}}$$

The value of $se(b)$ is minimized when the vector $\mathbf{X}^T\mathbf{X}$ is made long. Since the square root of the product $\mathbf{X}^T\mathbf{X}$ is a measure of distance; a long vector simply means, in this case, that the $\mathbf{X}$ vector contains values of x of large magnitude. The same conclusion is drawn if the *D*-optimality condition is used where $D = |\mathbf{X}^T\mathbf{X}|$.

In theory, since there is one parameter in the no intercept model, the experiment can consist of only one run, $n = p = 1$.

According to the *D*-optimality criterion where $\mathbf{X} = [x]$, $D = x^2$ is maximum when the x approaches infinity or some large value of x. While there are serious drawbacks to using the data from a single experimental run, the important point is that the size of experiment is not as important as selecting a well chosen value of x. In practice, the $\mathbf{x}$ vector should contain a few well chosen values of $\mathbf{x}$ over a wide range of x values including values of x near zero. The information from this experimental design can be used to test the assumption that the no intercept model is indeed a correct model specification. In addition, the experiment can include repeat observations at certain values of x so that the variability of y can be studied in more detail.

Now, consider a simple linear regression model,

$$y = b_0 + b_1 x$$

where

$$\mathbf{X} = [x_1\ x_2\ \ldots\ x_n]^T$$

with

$$\mathbf{x}_1 = [1\ \mathbf{x}_1],\ \mathbf{x}_2 = [1\ x_2],\ \ldots,\ x_n = [1\ x_n]$$

Since the model contains only two model parameters, assume for the moment that the experiment will consist of only two runs, $n = p = 2$; thus

$$\mathbf{X} = \begin{bmatrix} 1 & x_1 \\ 1 & x_2 \end{bmatrix}$$

and

$$D = x_1{}^2 - 2x_1x_2 + x_2{}^2$$

The value of D is a maximum when $x_1 = 0$ and x_2 approaches infinity or a large value. By making $\mathbf{X}$ long, $|\mathbf{X}^T\mathbf{X}|$ is maximized and the standard errors of b_0 and b_1 are both minimized. Increasing the size of the experiment to be greater than $n = 2$ with a wide range of values in $\mathbf{X}$ including repeat runs can be used to determine if the simple regression model is a correct specification and to study the variability of y. These are the same conclusions that we obtained for the no intercept model.

For multiple regression models, parameter correlation must be considered in designing an experiment. Failure to do so can lead to strong linear dependencies between two or more model parameters and misleading conclusions about the process under study. The extent of the dependency between two parameters can be measured with the covariance matrix $cov(\varepsilon)$. The covariance of b_q and b_r is the off diagonal term of $cov(\varepsilon)$, or $cov(b_q, b_r) = cov(\varepsilon)_{qr}$. We will employ the D-optimality criteria and for purposes of illustration consider the following multiple regression model with $p = 2$ parameters.

$$y = b_1x_1 + b_2x_2$$

For simplicity, let $n = p = 2$ and consider the values of $\mathbf{X}$ where

$$\mathbf{X} = \begin{bmatrix} x_{11} & x_{21} \\ x_{21} & x_{22} \end{bmatrix}$$

and

$$D = x_{12}^2 x_{21}^2 - 2x_{11}x_{12}x_{21}x_{22} + x_{11}^2 x_{22}^2$$

The value of D is a maximum when $x_{12} = x_{21} = 0$ and magnitudes of x_{11} and x_{22} are made large. For this assignment, the design matrix **X** is long and it leads no correlation between parameters, $cov(b_1, b_2) = 0$ because the correlation between x_1 and x_2 for this design equals zero, $\rho_{12} = 0$.

Consider another design where $x_{11} = x_{12} = x_{21} = x_{22}$. Here, $D = 0$. The D-optimality criteria is not satisfied because D is minimized, but more importantly, there is a perfect correlation between model parameters, $(\mathbf{X}^T\mathbf{X})^{-1}$ does not exist, and no parameter estimates of b_1 and b_2 can be found because $(\mathbf{X}^T\mathbf{X})^{-1}$ is a singular matrix. The data contained in **X** form an ill-conditioned matrix. The correlation between x_1 and x_2 for this design is perfect, $[\rho_{12}] = 1$. Clearly, steps must be taken to minimize the correlation between explanatory variables.

Coefficient of Determination

The *coefficient of determination*, R^2, is a summary statistic for measuring how well the model fits the data.

$$R^2 = \frac{ssr}{ssto} = \frac{\sum_{j=1}^{m} (\hat{y}_j - \bar{y})^2}{\sum_{j=1}^{m} (y_j - \bar{y})^2}$$

where

$$0 \le R^2 \le 1$$

where

$$ssr = \text{sum of squares error explained by the regression}$$

and

$$ssto = \text{total sum of square error}$$

In matrix form,

$$R^2 = \frac{(\mathbf{Xb} - \bar{y})^T (\mathbf{Xb} - \bar{y})}{(n-2) s_Y^2}$$

where s_Y^2 = sample variance of the observed response variable y. This notation is used to distinguish it from the variance estimate for the model, $s^2 = \dfrac{S(\mathbf{b})}{n-p}$.

If $R^2 = 0$, the model explains no variation observed in *ssto*, and if $R^2 = 1$, the fit is perfect because it explains all the variation explained in *ssto*. The residual plot is a preferred model validation tool because all data are evaluated and it can be helpful in suggesting reformulated models if there is a lack of fit. The coefficient of variation complements the error plots. For simple regression, the absolute value of the correlation coefficient ρ_{XY} is equal to R^2.

EXAMPLE 8.8 A CO-O_2 Incinerator Performance Model

Oxygen (O_2) is considered an important control variable in minimizing carbon monoxide (CO) emissions. Results for an experiment conducted at an MSW incinerator are as follows:

x	3.5	4.0	6.6	6.9	7.0	7.0	7.6	7.7	8.0	8.0	8.3	10.2	10.8	11.3	13.2	13.3
y	2	11	3	2	8	5	1	1	2	2	8	15	2	25	120	73

Source: Hasselriis (1987).

where

y = response variable and CO concentration in ppmv
x = O_2 concentration as a % dry weight basis

Calibrate and test a regression model. Assume $\alpha = 5\%$.

SOLUTION

The following scatter plot suggests that a nonlinear function may be adequate. For academic reasons, let us first fit a linear model and investigate the residual error. This exercise will demonstrate the need for a quadratic model and the use of error plots in suggesting new model structures.

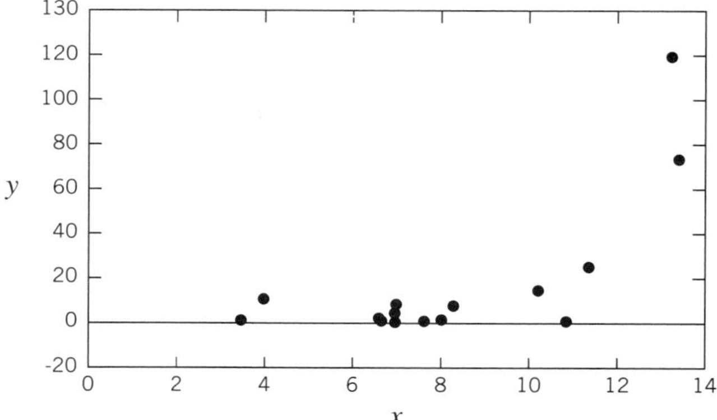

The Linear Model: The hypothesized model is $Y = b_0 + b_1 x + \varepsilon$. The parameters are estimated using $\mathbf{b} = (\mathbf{X}^T\mathbf{X})^{-1}\mathbf{X}^T\mathbf{y}$, where

$$\mathbf{y} = \begin{bmatrix} 2 \\ 11 \\ 3 \\ 2 \\ 8 \\ 5 \\ 1 \\ 1 \\ 2 \\ 2 \\ 8 \\ 15 \\ 2 \\ 25 \\ 120 \\ 73 \end{bmatrix} \quad \text{and} \quad \mathbf{X} = \begin{bmatrix} 1 & 3.5 \\ 1 & 4 \\ 1 & 6.6 \\ 1 & 6.9 \\ 1 & 7 \\ 1 & 7 \\ 1 & 7.6 \\ 1 & 7.7 \\ 1 & 8 \\ 1 & 8 \\ 1 & 8.3 \\ 1 & 10.2 \\ 1 & 10.8 \\ 1 & 11.3 \\ 1 & 13.2 \\ 1 & 13.3 \end{bmatrix}$$

The vector of parameter estimates is

$$\mathbf{b} = \begin{bmatrix} b_0 \\ b_1 \end{bmatrix} = \begin{bmatrix} -50.4 \\ 8.15 \end{bmatrix}$$

and the fitted model is

$$\hat{y} = -50.4 + 8.15x$$

with a variance estimate of $s^2 = 24.1^2$. The coefficient of determination is $R^2 = 0.491$.

The error plots show a nonlinear pattern, suggesting a nonlinear term is necessary.

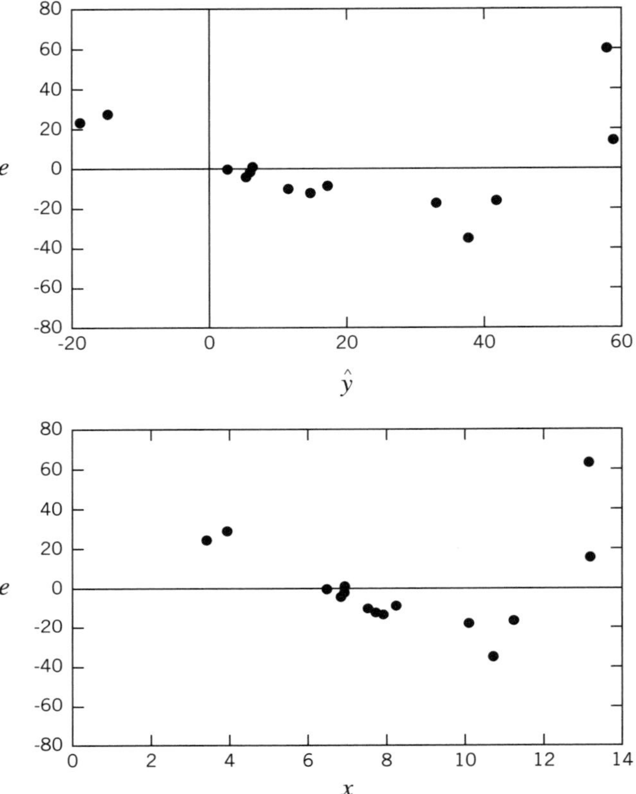

The 95% confidence intervals for the model parameters for $t_{(14,\,0.975)} = 2.145$ are

$$[L_{b0}, U_{b0}] = [-92.0, -8.88]$$
$$[L_{b1}, U_{b1}] = [3.41, 12.9]$$

The model parameter estimates are considered to be statistically significant; however, the response function y fails to capture the nonlinear behavior exhibited by the data. The following diagram shows that the linear model poorly fits the data.

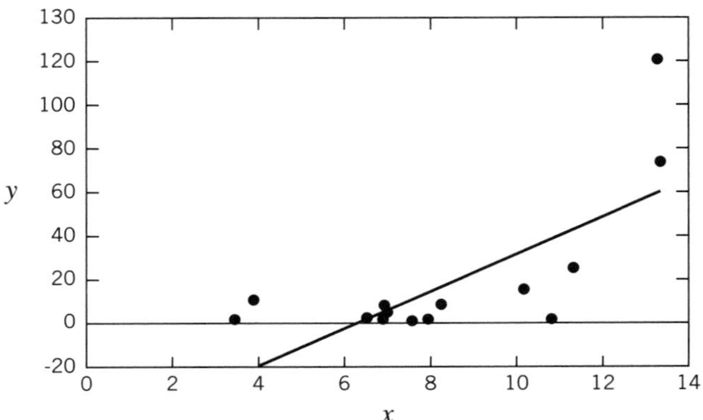

The Quadratic Model: The hypothesized model becomes $Y = b_0 + b_1x + b_2x^2 + \varepsilon$. The parameters are estimated using $\mathbf{b} = (\mathbf{X}^T\mathbf{X})^{-1}\mathbf{X}^T\mathbf{y}$, where

$$\mathbf{y} = \begin{bmatrix} 2 \\ 11 \\ 3 \\ 2 \\ 8 \\ 5 \\ 1 \\ 1 \\ 2 \\ 2 \\ 8 \\ 15 \\ 2 \\ 25 \\ 120 \\ 73^\circ \end{bmatrix} \quad \text{and} \quad \mathbf{X} = \begin{bmatrix} 1 & 3.5 & 3.5^2 \\ 1 & 4.0 & 4.0^2 \\ 1 & 6.6 & 6.6^2 \\ 1 & 6.9 & 6.9^2 \\ 1 & 7.0 & 7.0^2 \\ 1 & 7.0 & 7.0^2 \\ 1 & 7.6 & 7.6^2 \\ 1 & 7.7 & 7.7^2 \\ 1 & 8.0 & 8.0^2 \\ 1 & 8.0 & 8.0^2 \\ 1 & 8.3 & 8.3^2 \\ 1 & 10.2 & 10.2^2 \\ 1 & 10.8 & 10.8^2 \\ 1 & 11.3 & 11.3^2 \\ 1 & 13.2 & 13.2^2 \\ 1 & 3.3 & 13.3^2 \end{bmatrix}$$

The vector of parameter estimates is

$$\mathbf{b} = \begin{bmatrix} b_0 \\ b_1 \\ b_2 \end{bmatrix} = \begin{bmatrix} 81.2 \\ -25.7 \\ 1.96 \end{bmatrix}$$

The fitted model is

$$\hat{y} = 81.6 - 25.7x + 1.96x^2$$

and the variance estimate is $s^2 = 15.7^2$. The coefficient of determination is $R^2 = 0.80$.

The 95% confidence intervals for the model parameters for $t_{(13,\ 0.975)} = 2.160$ are

$$[L_{b0}, U_{b0}] = [12.4, 151]$$
$$[L_{b1}, U_{b1}] = [-42.4, -9.11]$$
$$[L_{b2}, U_{b2}] = [1.01, 2.90]$$

All model parameter estimates are statistically significant.
The error plots are

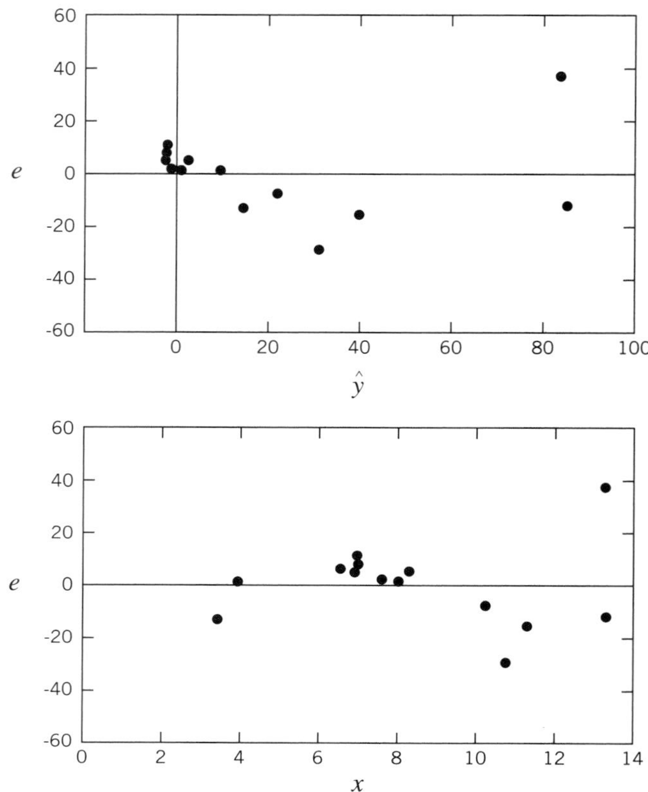

Even though the error plot of $(e_j,\ \hat{y}_j)$ shows that a pattern still exists, the function shows a reasonable fit. Owing to scarcity of data, it seems inappropriate to suggest another model.

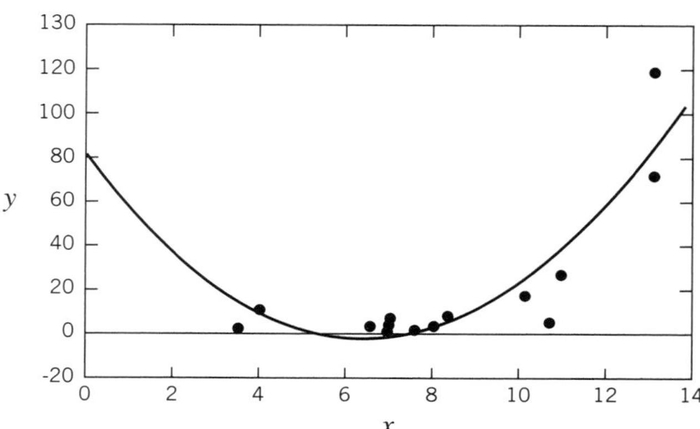

A 100(1 – α)% Confidence Interval for the Conditional Mean

The expected value of simple regression model $Y = b_0 + b_1x + \varepsilon$, given a known value of $x = x_0$, is a conditional mean of Y and denoted as Y_0, compact notation for $Y|x = x_0$. Here, it is assumed that x_0 is known with certainty. When discussing the effects of variability on regression, it is convenient to express the conditional expectation function of $\hat{y}_0 = b_0 + b_1x_0$

$$\hat{y}_0 = E(Y_0) = E(Y \mid x = x_0) = \bar{y} + b_1(x_0 - \bar{x})$$

where x_0 is given.

The slope b_1 and the distance $(x_0 - \bar{x})$ are emphasized. The variance of $\hat{y}_0$ is

$$\sigma^2_{\hat{y}_0} = \text{Var}\left[\bar{y} + b_1(x_0 - \bar{x})\right] = \frac{\sigma^2}{n} + (x_0 - \bar{x})^2 \sigma^2_{b_1}$$

It can be shown that $\sigma^2_{b_1} = \dfrac{\sigma^2}{s_x^2 (n-1)}$, where s_x^2 = sample variance of the observed explanatory variable x. Using s^2 as an estimate of σ^2, we estimate the value of $\sigma^2_{\hat{y}_0}$ to be

$$s^2_{\hat{y}_0} = \left(\frac{1}{n} + \frac{(x_0 - \bar{x})^2}{s_x^2 (n-1)}\right) s^2$$

where

$$s^2 = \frac{S(\mathbf{b})}{n-p}$$

Using the assumption that $\dfrac{Y_0 - \hat{y}_0}{s_{\hat{y}_0}} \sim N(0,1)$, it can be shown that the two-sided 100(1 – α)% confidence interval is

$$\left[L_{\hat{y}_c}, U_{\hat{y}_c}\right] = \bar{y} + \hat{b}_1(x_0 - \bar{x}) \pm t_{(1-\alpha, n-2)} \left(\frac{1}{n} + \frac{(x_0 - \bar{x})^2}{s_x^2 (n-1)}\right)^{1/2} s$$

The lower and upper one-sided 100(1 – α)% confidence bounds for a simple linear regression model are

$$L_{\hat{y}_c} = \bar{y} + \hat{b}_1(x_0 - \bar{x}) - t_{(1-\alpha, n-2)} \left(\frac{1}{n} + \frac{(x_0 - \bar{x})^2}{s_x^2 (n-1)}\right)^{1/2} s$$

and

$$L_{\hat{y}_c} = \bar{y} + \hat{b}_1(x_0 - \bar{x}) - t_{(1-\alpha, n-2)} \left(\frac{1}{n} + \frac{(x_0 - \bar{x})^2}{s_x^2 (n-1)}\right)^{1/2} s$$

Note that the lengths of a confidence intervals for one- and two-sided confidence intervals are minimum when $x_0 = \bar{x}$.

For a multiple regression model, the two-sided 100(1 – α)% confidence interval for the conditional mean:

$$[L_{\hat{y}_c}, U_{\hat{y}_c}] = \mathbf{x}_0\mathbf{b} \pm t_{(1-\alpha/2,\, n-p)} s \sqrt{\mathbf{x}_0^T (\mathbf{X}^T\mathbf{X})^{-1} \mathbf{x}_0^2}$$

given $\mathbf{x} = \mathbf{x}_0$ where $\mathbf{x}_0$ is assumed to be known with certainty. The one-sided lower and upper 100(1 – α)% confidence bounds for the conditional mean are

$$L_{\hat{y}_c} = \mathbf{x}_0^T\mathbf{b} - t_{(1-\alpha,\, n-p)} s \sqrt{\mathbf{x}_0^T (\mathbf{X}^T\mathbf{X})^{-1} \mathbf{x}_0}$$

and

$$U_{\hat{y}_c} = \mathbf{x}_0^T\hat{\mathbf{b}} + t_{(1-\alpha,\, n-p)} s \sqrt{\mathbf{x}_0^T (\mathbf{X}^T\mathbf{X})^{-1} \mathbf{x}_0}$$

A 100(1 – α)% Confidence for the Prediction Interval

In the preceding discussion on the conditional mean of Y_0, it is assumed that $\mathbf{x}_0$ is known with certainty. When forecasting a future event, the value of $\mathbf{x}_0$ is a predicted value and subject to forecast uncertainty. In order to compensate for this uncertainty, the inherent variability σ^2 of ε is introduced into the analysis. Consider the simple linear regression model, where $x = x_0$,

$$Y_0 = \bar{y} + b_1(x_0 - \bar{x}) + \varepsilon$$

Its mean and variance are

$$\hat{y}_0 = \bar{y} + b_1(x_0 - \bar{x})$$

and

$$\sigma_{\hat{y}_0}^2 = \frac{\sigma^2}{n} + (x_0 - \bar{x})^2\sigma_{b_1}^2 + \sigma^2$$

The two-sided 100(1 – α)% prediction bound is

$$[L_{\hat{y}p}, U_{\hat{y}p}] = \bar{y} + b_1(x_0 - \bar{x}) \pm t_{(1-\alpha/2,\, n-2)} \left(1 + \frac{1}{n} + \frac{(x_0 - \bar{x})^2}{s_x^2\ (n-1)}\right)^{1/2} s$$

The lower and upper one-sided 100(1 – α)% prediction bounds are

$$L_{\hat{y}p} = \bar{y} + b_1(\hat{x}_0 - \bar{x}) - t_{(1-\alpha,\, n-2)} \left(1 + \frac{1}{n} + \frac{(x_0 - \bar{x})^2}{s_x^2\ (n-1)}\right)^{1/2} s$$

$$U_{\hat{y}p} = \bar{y} + b_1(x_0 - \bar{x}) + -t_{(1-\alpha,\, n-2)} \left(1 + \frac{1}{n} + \frac{(x_0 - \bar{x})^2}{s_x^2\ (n-1)}\right)^{1/2} s$$

For multiple regression models, the two-sided 100(1 – α)% prediction bounds are:

$$[L_{\hat{y}_p}, U_{\hat{y}_p}] = \mathbf{x}_0 \mathbf{b} \pm t_{(1-\alpha/2, n-p)}\, s \sqrt{1 + \mathbf{x}_0^T (\mathbf{X}^T \mathbf{X})^{-1} \ \mathbf{x}_0}$$

given $\mathbf{x} = \mathbf{x}_0$, where $\mathbf{x}_0$ is assumed to be a forecast vector. The one-sided lower and upper 100(1 – α)% prediction bounds are

$$L_{\hat{y}_p} = \mathbf{x}_0 \mathbf{b} - t_{(1-\alpha, n-p)}\, s \sqrt{1 + \mathbf{x}_0^T (\mathbf{X}^T \mathbf{X})^{-1} \ \mathbf{x}_0}$$

and

$$U_{\hat{y}_p} = \mathbf{x}_0 \mathbf{b} + t_{(1-\alpha, n-p)}\, s \sqrt{1 + \mathbf{x}_0^T \ (\mathbf{X}^T \mathbf{X})^{-1} \mathbf{x}_0}$$

EXAMPLE 8.9 ***A CO–O_2 Incinerator Performance Model***

In Example 8.8, a performance model is calibrated to be

$$\hat{y} = 81.6 - 25.7x + 1.96x^2$$

with $s^2 = 15.7^2$ where

$\hat{y}$ = response variable and CO concentration in ppmv
x = O_2 concentration as a % dry weight basis

(a) Determine the O_2 control setting to minimize CO emissions.

(b) Determine the 99% upper confidence bound for CO emissions at the optimal control setting for O_2.

SOLUTION

(a) For optimum control x^* occurs when $\hat{y}$ is a minimum value,

$$\text{minimize } \hat{y} = 81.6 - 25.7x + 1.96x^2$$

or when

$$\frac{d\hat{y}}{dx} = -25.7 + 2(1.96)x = 0$$

or

$$x^* = 6.56\%$$

and

$$\hat{y}^* = \mathbf{x}^*\mathbf{b} = -3.08 \text{ ppmv}$$

Clearly, $\hat{y}^* < 0$ is an impossibility; $\hat{y}$ has a lower limit of 0.

(b) The upper one-sided 99% confidence bound, assuming x* is known with certainty, is estimated as

$$U_{\hat{y}_c} = \mathbf{x}^*\mathbf{b} + t_{(0.99,\,13)} s \sqrt{\mathbf{x}^{*T} (\mathbf{X}^T\mathbf{X})^{-1} \mathbf{x}^*} = 9.81$$

where $n - p = 16 - 3 = 13$ and $t_{(0.99,13)} = 2.650$.

The following diagram shows the upper one-sided confidence bound of the conditional mean over a range of x values.

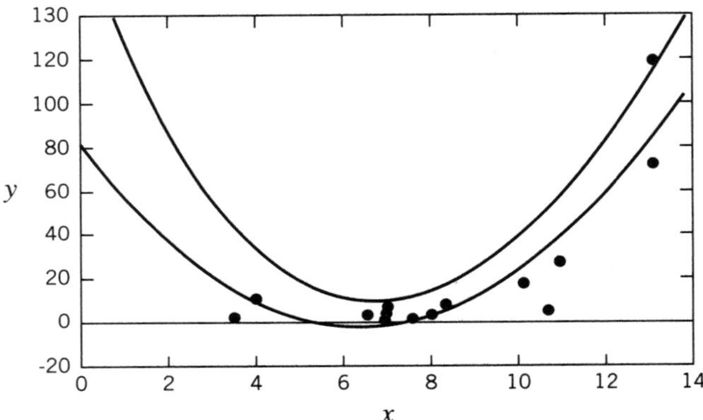

DISCUSSION

If it is assumed that x^* is a forecast value and is not known with certainty, the upper one-sided 99% prediction bound is

$$U_{\hat{y}_0} = \mathbf{x}^*\mathbf{b} + t_{(0.99,13)} s \sqrt{1 + \mathbf{x}^{*T} (\mathbf{X}^T\mathbf{X})^{-1} \mathbf{x}^*} = 40.3 \text{ ppmv CO}$$

Since linear regression analysis plays such an important role in decision making, the following example is presented to describe how linear regression analysis is used in MSW plant control for the purpose of minimizing exposure to dioxins and furans. The example shows the importance of data collection and how, in practice, to deal with uncertainty.

EXAMPLE 8.10 *Minimizing the Toxic Emissions by Plant Control*

Since dioxins and furans are so highly toxic, research has been conducted to determine if the combustion process can be controlled to reduce their levels to acceptable values. It is well known that these trace organics are formed during incomplete combustion of organics in the presence of chlorine. The reaction is quite complex and is not completely understood. It is known, however, that these products are virtually insoluble in water. Thus, there is little danger that the ash containing dioxins and furans will leach from landfills and contaminate water supplies. Most research has centered on air emission exposure.

Obtaining trace levels of PCDD is expensive, in both time and money; therefore, it is an unsuitable measure for plant control. A typical analysis for PCDDs and PCDFs in fly ash consists of sample extraction, extract cleanup steps to remove interferences, and analysis by gas chromatography. The sample extraction step requires that a sample be shaken for three hours with HCl and then be washed in distilled water.

The problem is to justify the use of CO concentrations as a suitable surrogate measure of dioxins and furans. Justification will be based on

(a) Survey data

(b) Data from controlled experiments for studying the relationship between PCDD and CO

(c) Data from controlled experiments for studying the relationship between PCDD and temperature

SOLUTION

(a) *Survey Results:* The scatter plot shows the results obtained from a survey of 14 plants in the United States and Canada. The quantities measured from these plants differed in some instances by three orders of magnitude; the total PCDD and PCDF stack emissions ranged from 5.7 to 8700 ppmv. In some plants, however, the emissions were below acceptable levels and were so low that they were below the detection level of the instrument and not reported in the scatter plot.

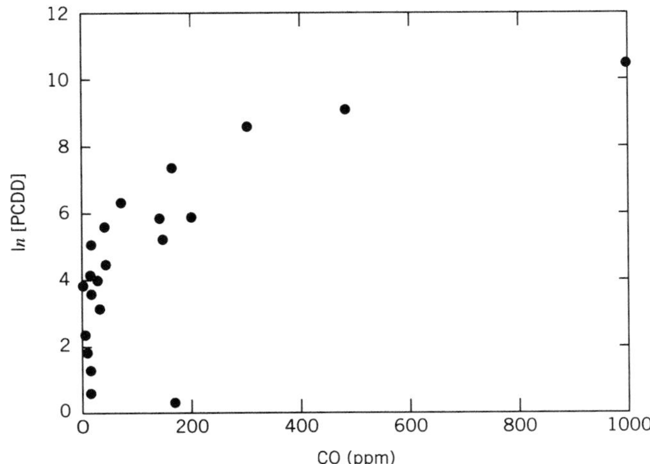

Source: Hasselriis (1987).

These survey data are used primarily to record plant performance. Controlled experiments to derive cause-effect models for CO-PCDD require careful design and evaluation. CO concentration has traditionally been the standard measure of combustion efficiency. Before MSW incinerator and coal-fired emissions became an issue, the operator's goal was to minimize CO emission levels. Thus, it is not surprising that so many observations were reported near zero or were below the monitor detection levels and not recorded.

While these data are considered unsuitable for cause-effect modeling, the scatter plot shows a relationship between CO and PCDD. In other words, it seems reasonable from these data that operators could continue to minimize CO emissions and be assured that dioxin and furan emissions are being minimized. Similar findings have been reported by others.

(b) *PCDD–CO Model*: The survey results served as an impetus for designing statistical experiments to explain the variations observed in the PCDD response variable. A series of 13 runs were conducted at a plant in Hamilton, Ontario, to determine the effect of changes in load, furnace temperature, combustion air distribution, oxygen concentration, and the composition and moisture content of the waste on dioxin and furan production.

It may be surprising that so few runs were conducted. The cost of obtaining data is often important in designing statistical experiments. Therefore, care must be used to obtain quality information. In the Hamilton study, the plant was operated for a minimum of four hours for a given set of conditions to make sure the detection limit of these trace organics was exceeded. The scatter plot shows the results of the study.

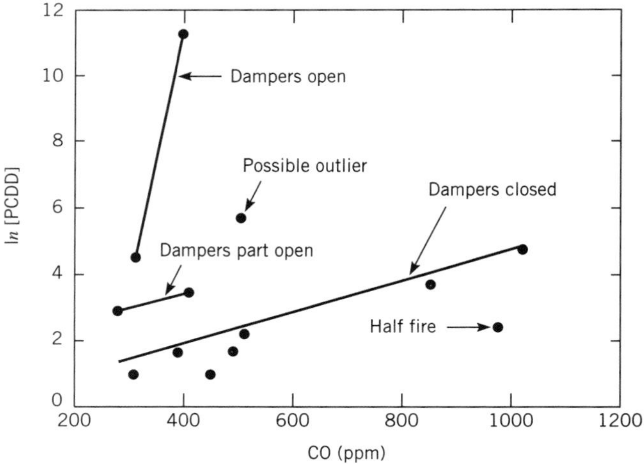

Source: Hasselriis (1987).

These data indicate that linear relationships exist between PCDD and CO concentrations. The implication that if control strategies can be found to minimize CO, a minimum level of dioxins and furans will be emitted from the plant.

The regression line for all the dampers closed data is shown on the scatter plot. If the possible outlier is eliminated from the data set, the regression line will fit the data better. Statistical experiments can increase our knowledge and confidence, but they certainly do not eliminate all uncertainty. The possible outlier remained in the data because there is insufficient information to remove it.

(c) *A PCDD-Temperature Model:* A similar approach is used to determine if there is a relationship between PCDD concentration and furnace temperature. Survey are analyzed and then followed by a controlled experiment. The survey data show that PCDD decreases as the temperature increases up to about 850°C, and then increases at temperatures in the vicinity of 1000°C.

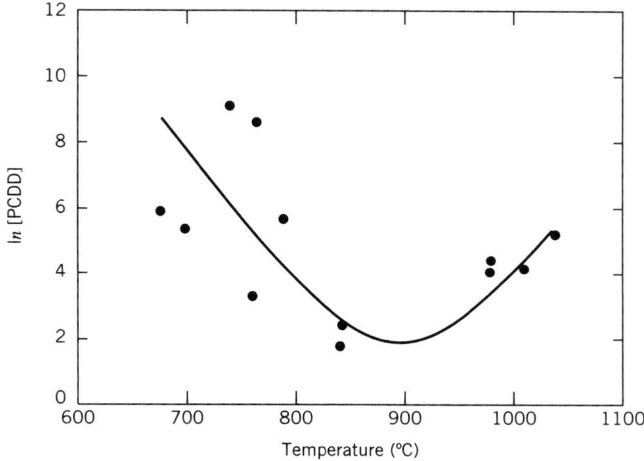

Source: Hasselriis (1987).

The line drawn on the figure is not a regression line. It is sketched on the figure to suggest that a PCDD and furnace temperature relationship may exist.

Data from the controlled experiment are shown in the following diagram.

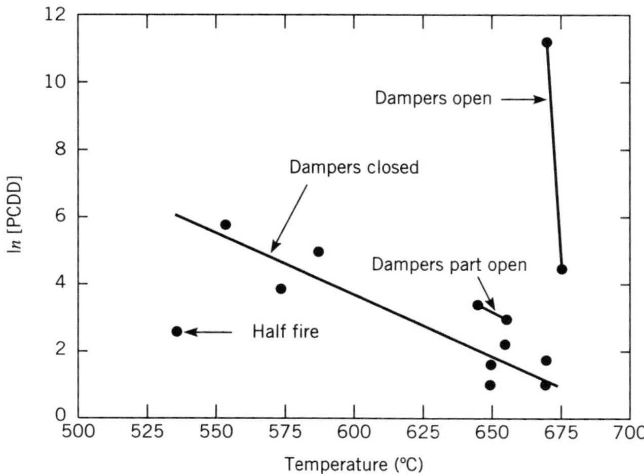

Source: Hasselriis (1987).

The data indicate that low PCDD concentrations are obtained with all dampers closed and relatively high furnace temperatures. The regression equation for all dampers closed data is shown.

Results from an independent MSW plant study conducted in Sweden confirm these findings and the premise that CO concentration is an acceptable measure of PCDD concentration. When the oxygen flow is less than 7%, then there is an oxygen deficiency causing incomplete combustion. When the oxygen flow is greater than 7%, the excess air has a cooling effect, causing incomplete combustion.

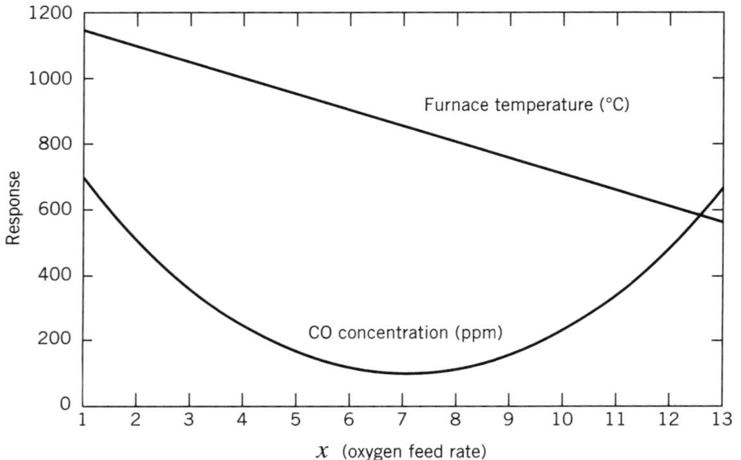

Source: Bergvall and Hult (1985).

In summary, the scientific approach to experimentation and observation coupled with mathematical modeling is extremely valuable in drawing conclusions about dependency between PCDD and CO and between PCDD and furnace temperature.

QUESTION FOR DISCUSSION

8.4.1 Define the following terms:

D-optimality criterion   Coefficient of determination

QUESTIONS FOR ANALYSIS

8.4.2 (a) Use the method of least squares and matrix algebra to estimate the parameters of the expectation function, $\hat{k} = k_f + cb$, where the variable b = design capacity of the plant, the parameter k_f = fixed cost of construction, and parameter c = unit cost of construction.

The data, which are the same for Problem 8.3.3, are

i	b, Design Capacity (tons/day)	k, Capital Cost ($M)
1	3000	$253
2	2250	$200
3	400	$40
4	3000	$290
5	400	$30
6	2250	$250
7	518	$40
8	1650	$113

Source: Neal and Schubel (1987).

(b) Determine the 95% confidence level for k_f and c.

(c) Use the results of (b) for determining the validity of the model by testing the significance of the model parameters at a 95% confidence level.

8.4.3* (a) Repeat question (a) of Problem 8.4.2 for a no-intercept model, $\hat{k} = cb$, where the variable b = design capacity of the plant and parameter c = unit cost of construction.

(b) Determine the 95% confidence level for c.

(c) Plot the expected value function and 95% construction cost confidence bounds for a plant capacity range of 0 to 3000 tons per day on a scatter plot of the data.

8.4.4* Use the results given in Example 8.6 to answer this question.

(a) Plot the expected value function and 99% HCl upper confidence bound for a lime dose range of 50 to 220 kg per hour on a scatter plot of the data. Assume that the lime dose concentration is known with certainty.

(b) Based on these data and a lime dose of 210 kg per hour, what percentage of the time will the plant meet a compliance of 175 ppmv HCl? State your assumptions.

8.4.5 Problems 8.3.5 and 8.3.6 analyze the linear relationships between furnace temperature and air emission concentration of PCDD and carbon monoxide and air emission concentration.

Since carbon monoxide is used as a surrogate measure of combustion efficiency and furnace temperature is an important emissions control, the following model is hypothesized:

$$\hat{y} = b_1 x_1 + b_2 x_2$$

where

$\hat{y}$ = PCDD concentration in ppmv
x_1 = furnace temperature in ^{0}C
x_2 = CO concentration in ppm

The data are

x_1 Furnace Temperature (degrees C)	x_2 Carbon Monoxide (CO) emissions, (ppm)	y Air Emission Concentration of PCDD (ppmv)
573	850	3.9
587	1020	5.0
650	390	1.7
655	510	2.3
670	490	1.8
553	500	5.8
670	450	1.1
650	310	1.1

Source: Hasselriis (1987).

(a) Use the method of least squares and matrix algebra to estimate the parameters of the hypothesized expectation function.

(b) Determine the 95% confidence levels for b_1 and b_2.

(c) Use the results of (b) for determining the validity of the model.

8.4.6 The following model is specified:

$$y = b_1x_1 + b_2x_2$$

where $0 \leq x_1 \leq 1$ and $0 \leq x_2 \leq 1$.

Design an experiment using the *D*-optimality criterion. Assume the size of the experiment is $n = 2$.

(a) Determine $\text{cov}(b_1, b_2)$.

(b) Assume that x_1 and x_2 are perfectly correlated, $\rho_{12} = 1$. Show that the parameters of the specified model are impossible to estimate under this condition.

(c) Since $\rho_{12} = 1$ from (b), no information is gleaned from both x_1 and x_2. Show that a model for y is improperly specified and that a correct model specification of y is a simple linear regression equation of x_1 or x_2 exclusively. Use the following procedure to show this: (1) specify a functional relationship between x_1 and x_2 employing the fact that $\rho_{12} =$ 1, and then (2) substitute this relationship into $y = b_1x_1 + b_2x_2$ and simplify.

8.4.7 This problem is concerned with the design of experiments for multiple linear regression. In this academic exercise, the parent distribution of Y is known to be

$$Y = 100 + 10x_1 + 20x_2 + \varepsilon$$

where

$$\varepsilon = N(0, 5^2)$$

A Monte Carlo simulation is used to generate two data sets for this model. The data are:

Data set 1

x_1	x_2	y
1	11	333
2	10	320
3	9	312
4	8	304
5	7	294
6	6	280
7	5	270
8	4	65
9	3	250
10	2	240
11	1	231
12	0	223

Data set 2

x_1	x_2	y
3	2	174
3	4	215
3	6	253
3	8	291
6	2	201
6	4	243
6	6	283
6	8	320
9	2	233
9	4	274
9	6	310
9	8	350

(a) Use the method of least squares and matrix algebra to estimate the parameters of the expectation function, $\hat{y} = b_0 + b_1x_1 + b_2x_2$. Why is it impossible to estimate the model parameters for data set 1. Calculate ρ_{12} = correlation coeffiicient for X_1 and X_2.

(b) Repeat (a) for the expectation function, $\hat{y} = b_0 + b_1x_1 + b_2x_2$, for data set 2.

(c) Repeat (a) for the expectation function, $\hat{y} = b_0 + b_1x_1$, for data set 1.

(d) Repeat (a) for the expectation function, $\hat{y} = b_0 + b_2x_2$, for data set 1.

(e) Compare the models from (b), (c), and (d) with the intention of selecting the best cause-effect model. Which model do you find the best one? Explain your answer.

REFERENCES

Bates, D. G., and D. G. Watts (1988). *Nonlinear Regression Analysis and Its Applications*, Wiley Interscience, New York.

Beck, M. B. (1986). "Identification, Estimation and Control of Biological Wastewater Treatment Processes," *IEEE Proceedings*, **133**, (Pt. D, No. 5), 254–264.

Cargo, D. B. (1978). *Solid Wastes: Factors Influencing Generation Rates*, The University of Chicago, Department of Geography, Research Paper No. 174, Chicago.

Draper, N. R., and H. Smith (1981). *Applied Regression Analysis*, John Wiley & Sons, New York.

Hasselriis, F. (1987). "Optimization of Combustions to Minimize Dioxin Emissions," *Waste Management & Research* **5**, 311–326.

Hickman, H. L., W. D. Turner, R. Hopper, F. Hasselriis, J. L. Kuester, and G. L. Trezak (1984). *Thermal Conversion Systems for Municipal Solid Waste*, Noyes Publications, Park Ridge, NJ.

Neal, H. A., and J. R. Schubel (1987). *Solid Waste Management and the Environment*, Prentice Hall, Englewood Cliffs, NJ.

Swedish Association of Public Cleansing and Solid Waste Management (1988). *Solid Waste Management in Sweden*, S-211 22, Malmo, Sweden.

Van Ryzin, J. (1980). "Quantitative Risk Assessment," *Journal of Occupational Medicine*, **22**, 321–326.

Appendices

A. PROBABILITY CONCEPTS

The laws of probability are:

1. $0 \leq P(S) \leq 1$ where S is any event.
2. $P(\phi) = 0$ and $P(Z) = 1$ where ϕ represents the null event and Z represents the entire sample space.
3. If $S_1, S_2, \ldots, S_k$ is a collection of mutually exclusive events, then

$$P(S_1 \cup S_2 \cup \ldots \cup S_k) = P(S_1) + P(S_2) + \cdots + P(S_k).$$

Events S_i and S_j are mutually exclusive if $S_i \cap S_j = \phi$. If S_i is observed, then it is impossible to observe the event S_j. A collection is mutually exclusive if all pairs of the events are mutually exclusive.

4. If the complementary event of S in Z is designated as S', then

$$P(S') = 1 - P(S).$$

5. If S_1 and S_2 are any events, then

$$P(S_1 \cup S_2) = P(S_1) + P(S_2) - P(S_1 \cap S_2).$$

Note that if S_1 and S_2 are mutually exclusive events, then this reduces to $P(S_1 \cup S_2) = P(S_1) + P(S_2)$.

Discrete Random Variables

The following definitions apply to a discrete random variable X with a probability function $f(x)$ and sample space Z.

1. $P(X = x) = f(x)$ where $0 \leq f(x) \leq 1$ and $x \in Z$

2. $\sum_{x \in Z} f(x) = 1$

3. $P(X \in A) = \sum_{x \in A} f(x)$ where A is a subset of Z

4. $F(x) = P(X \le x) = \sum_{t \le x} f(t)$ where $A = \{t \in Z, t \le x\}$

Continuous Random Variables

The following definitions apply to a continuous random variable X with a probability density function $f(x)$ and sample space Z.

1. $f(x) \ge 0$ for all x
2. $P(X = c) = 0$ where c is a constant

and

$$P(X \in Z) = P(-\infty < X < \infty) = \int_{-\infty}^{\infty} f(x)\, dx = 1$$

3. The complement of $X \le x$ is $X > x$, and the respective probabilities are

$$P(X \le x) = \int_{-\infty}^{x} f(t)\, dt$$

and

$$P(X > x) = 1 - P(X \le x)$$

4. The cumulative distribution function of X, $F(x)$, is defined to be the probability that X is in the interval from $-\infty$ to x.

$$F(x) = P(X \le x) = \int_{-\infty}^{x} f(t)\, dt$$

It follows that as $x \to \infty$, the entire range of x is spanned and $F(x)$ must approach one.

Expectation

The expected value of a function $g(X)$ is a weighted average of the possible values of the function. If X is a discrete random variable, the expected value is expressed as the sum,

$$E[g(X)] = \sum_{x \in Z} g(x)\, f(x)$$

where

$$f(x) = P(X = x) = \text{probability function of } X$$

and, if X is a continuous variable, it is

$$E[g(X)] = \int_{-\infty}^{\infty} g(x) f(x)\, dx$$

where

$$f(x) = \text{probability density function of } X$$

The mean μ, variance σ^2, and standard deviation $\sigma = \sqrt{\sigma^2}$ are common measures that describe a random variable. The mean of X is a measure of its central tendency, and the variance measures its dispersion or variability. When X is a discrete random variable, its mean and variance are

$$\mu = E(X) = \sum_{x \in Z} x f(x)$$

$$\sigma^2 = E\left[(X - \mu)^2 \right] = \sum_{x \in Z} (x - \mu)^2 f(x) = \sum_{-\infty}^{\infty} x^2 f(x) - \mu^2$$

and when X is a continuous random variable,

$$\mu = E(X) = \int_{-\infty}^{\infty} x f(x)\, dx$$

$$\sigma^2 = E\left[(X - \mu)^2 \right] = \int_{-\infty}^{\infty} (x - \mu)^2 f(x)\, dx = \int_{-\infty}^{\infty} x^2 f(x)\, dx - \mu^2$$

The expected value of $g(X) = c$, where c is a constant, is

$$E[g(X)] = E(c) = c$$

and when $g(X) = c_1 g_1(X) + c_2 g_2(X)$, then

$$E[g(X)] = c_1 E[g_1(X)] + c_2 E[g_2(X)]$$

B. TABLES

TABLE I Proximate and Ultimate Analyses and Heating Value of Waste Components

	Proximate Analysis (as-received) Weight %				Ultimate Analysis (dry) Weight %						Higher Heating Value (kcal/kg)		
Waste Component	Moisture	Volatile Matter	Fixed Carbon	Non-Comb.	C	H	O	N	S	Non-Comb.	As Received	Dry	Moisture and Ash Free
Paper and Paper Products													
Paper, Mixed	10.24	75.94	8.44	5.38	43.41	5.82	44.32	0.25	0.20	6.00	3778	4207	4475
Newsprint	5.97	81.12	11.48	1.43	49.14	6.10	43.03	0.05	0.16	1.52	4430	4711	4778
Brown Paper	5.83	83.92	9.24	1.01	44.90	6.08	47.34	0.00	0.11	1.07	4031	4281	4333
Trade Magazine	4.11	66.39	7.03	22.47	32.91	4.95	38.55	0.07	0.09	23.43	2919	3044	3972
Corrugated Boxes	5.20	77.47	12.27	5.06	43.73	5.70	44.93	0.09	0.21	5.34	3913	4127	4361
Plastic-Coated Paper	4.71	84.20	8.45	2.64	45.30	6.17	45.50	0.18	0.08	2.77	4078	4279	4411
Waxed Milk Cartons	3.45	90.92	4.46	1.17	59.18	9.25	30.13	0.12	0.10	1.22	6293	6518	6606
Paper Food Cartons	6.11	75.59	11.80	6.50	44.74	6.10	41.92	0.15	0.16	6.93	4032	4294	4583
Junk Mail	4.56	73.32	9.03	13.09	37.87	5.41	42.74	0.17	0.09	13.72	3382	3543	4111
Food and Food Waste													
Vegetable Food Waste	78.29	17.10	3.55	1.06	49.06	6.62	37.55	1.68	0.20	4.89	997	4594	4833
Citrus Rinds and Seeds	78.70	16.55	4.01	0.74	47.96	5.68	41.67	1.11	0.12	3.46	948	4453	4611
Meat Scraps (cooked)	38.74	56.34	1.81	3.11	59.59	9.47	24.65	1.02	0.19	5.08	4235	6913	7283
Fried Fats	0.00	97.64	2.36	0.00	73.14	11.54	14.82	0.43	0.07	0.00	9148	9148	9148
Mixed Garbage I	72.00	20.26	3.26	4.48	44.99	6.43	28.76	3.30	0.52	16.00	1317	4713	5611
Mixed Garbage II	—	—	—	—	41.72	5.75	27.62	2.97	0.25	21.87	—	4026	5144
Trees, Wood, Brush, Plants													
Green Logs	50.00	42.25	7.25	0.50	50.12	6.40	42.26	0.14	0.08	1.00	1168	2336	2361
Rotten Timbers	26.80	55.01	16.13	2.06	52.30	5.5	39.0	0.2	1.2	2.8	2617	3538	3644
Demolition Softwood	7.70	77.62	13.93	0.75	51.0	6.2	41.8	0.1	<.1	0.8	4056	4398	4442
Waste Hardwood	12.00	75.05	12.41	0.54	49.4	6.1	43.7	0.1	<.1	0.6	3572	4056	4078
Furniture Wood	6.00	80.92	11.74	1.34	49.7	6.1	42.6	0.1	<.1	1.4	4083	4341	4411
Evergreen Shrubs	69.00	25.18	5.01	0.81	48.51	6.54	40.44	1.71	0.19	2.61	1504	4853	4978
Balsam Spruce	74.35	20.70	4.13	0.82	53.30	6.66	35.17	1.49	0.20	3.18	1359	5301	5472
Flowering Plants	53.94	35.64	8.08	2.34	46.65	6.61	40.18	1.21	0.26	5.09	2054	4459	4700

Lawn Grass I	75.24	18.64	4.50	1.62	46.18	5.96	36.43	4.46	0.42	6.55	1143	4618	4944
Lawn Grass II	65.00	—	—	2.37	43.33	6.04	41.68	2.15	0.05	6.75	1494	4274	4583
Ripe Leaves I	9.97	66.92	19.29	3.82	52.15	6.11	30.34	6.99	0.16	4.25	4436	4927	5150
Ripe Leaves II	50.00	—	—	4.10	40.50	5.95	45.10	0.20	0.05	8.20	1964	3927	4278
Wood and Bark	20.00	67.89	11.31	0.80	50.46	5.97	42.37	0.15	0.05	1.00	3833	4785	4833
Brush	40.00	—	—	5.00	42.52	5.90	41.20	2.00	0.05	8.33	2636	4389	4778
Mixed Greens	62.00	26.74	6.32	4.94	40.31	5.64	39.00	2.00	0.05	13.00	1494	3932	4519
Grass, Dirt, Leaves	21–62	—	—	—	36.20	4.75	26.61	2.10	0.26	30.08	—	3491	4994
Domestic Wastes													
Upholstery	6.9	75.96	14.52	2.62	47.1	6.1	43.6	0.3	.1	2.8	3867	4155	4272
Tires	1.02	64.92	27.51	6.55	79.1	6.8	5.9	0.1	1.5	6.6	7667	7726	8278
Leather	10.00	68.46	12.49	9.10	60.00	8.00	11.50	10.00	0.40	10.10	4422	4917	5472
Leather Shoe	7.46	57.12	14.26	21.16	42.01	5.32	22.83	5.98	1.00	22.86	4024	4348	5639
Shoe, Heel & Sole	1.15	67.03	2.08	29.74	53.22	7.09	7.76	0.50	1.34	30.09	6055	6126	8772
Rubber	1.20	83.98	4.94	9.88	77.65	10.35	—	—	2.00	10.00	6222	6294	7000
Mixed Plastics	2.0	—	—	10.00	60.00	7.20	22.60	—	—	10.20	7833	7982	8889
Plastic Film	3–20	—	—	—	67.21	9.72	15.82	0.46	0.07	6.72	—	7692	8261
Polyethylene	0.20	98.54	0.07	1.19	84.54	14.18	0.00	0.06	0.03	1.19	10,932	10,961	11,111
Polystyrene	0.20	98.67	0.68	0.45	87.10	8.45	3.96	0.21	0.02	0.45	9122	9139	9172
Polyurethane	0.20	87.12	8.30	4.38	63.27	6.26	17.65	5.99	0.02	4.38†	6224	6236	6517
Polyvinyl Chloride	0.20	86.89	10.85	2.06	45.14	5.61	1.56	0.08	0.14	2.06‡	5419	5431	5556
Linoleum	2.10	64.50	6.60	26.80	48.06	5.34	18.70	0.10	0.40	27.40	4528	4617	6361
Rags	10.00	84.34	3.46	2.20	55.00	6.60	31.20	4.12	0.13	2.45	3833	4251	4358
Textiles	15–31	—	—	—	46.19	6.41	41.85	2.18	0.20	3.17	—	4464	4611
Oils, Paints	0	—	—	16.30	66.85	9.63	5.20	2.00	—	16.30	7444	7444	8889
Vacuum Cleaner Dirt	5.47	55.68	8.51	30.34	35.69	4.73	20.08	6.26	1.15	32.09	3548	3753	5533
Household Dirt	3.20	20.54	6.26	70.00	20.62	2.57	4.00	0.50	0.01	72.30	2039	2106	7583
Municipal Wastes													
Street Sweepings	20.00	54.00	6.00	20.00	34.70	4.76	35.20	0.14	0.20	25.00	2667	3333	4444
Mineral§	2–6	—	—	—	0.52	0.07	0.36	0.03	0.00	99.02	—	47	—
Metallic§	3–11	—	—	—	4.54	0.63	4.28	0.05	0.01	90.49	—	412	4333
Ashes	10.00	2.68	24.12	63.2	28.0	0.5	0.8	—	0.5	70.2	2089	2318	7778

Source: Walter R. Neissen. *Combustion and Incineration Process; Applications in Environmental Engineering.* Marcel Dekker Inc., New York (1978).

† Remaining 2.42% is chlorine.

‡ Remaining 45.41% is chlorine.

§ Heat and organic content from labels, coatings, and remains of contents of containers.

TABLE II Price Indices

The following tables give annual averages of PPI, CPI, CCI, and BCI indices for selected sectors of the U. S. economy. With the exception of processed fuels and lubricants, there has been an increase in all prices over the period shown.

PRODUCER PRICE INDEX (PPI) FOR SELECTED ITEMS

1979	1980	1981	1982	1983	1984	1985	1986	1987	1988	1989	1990
FINISHED GOODS											
77.6	88.0	96.1	100.0	101.6	103.7	104.7	103.2	105.4	108.0	113.6	119.2
CONSUMER GOODS											
77.5	88.6	96.6	100.0	101.3	103.3	103.8	101.4	103.6	106.2	112.1	118.2
CAPITAL EQUIPMENT											
77.5	85.8	94.6	100.0	102.8	105.2	107.5	109.7	111.7	114.3	118.8	122.9
INTERMEDIATE MANUFACTURING MATERIALS											
78.4	90.3	98.6	100.0	100.6	103.1	102.7	99.1	101.5	107.1	112.0	114.5
MATERIALS AND COMPONENTS FOR MANUFACTURING											
80.9	91.7	98.7	100.0	101.2	104.1	103.3	102.2	105.3	113.2	118.1	118.7
MATERIALS AND COMPONENTS FOR CONSTRUCTION											
84.2	98.3	97.9	100.0	102.8	105.6	107.3	108.1	109.8	116.1	121.3	122.9
PROCESSED FUELS & LUBRICANTS											
61.6	85.0	106.6	100.0	95.4	95.7	92.8	72.7	73.3	71.2	76.4	85.9

CONSUMER PRICE INDEX (CPI), U. S. CITY AVERAGE (ALL ITEMS)

1979	1980	1981	1982	1983	1984	1985	1986	1987	1988	1989	1990
72.6	82.4	90.9	96.5	99.6	103.9	107.6	109.6	113.6	118.3	124.0	130.7

Source: U.S. Department of Labor, *Monthly Labor Review*, (December 1991).

BUILDING COST INDEX (BCI) (ALL ITEMS)

1979	1980	1981	1982	1983	1984	1985	1986	1987	1988	1989	1990	1991
1819	1941	2097	2225	2384	2417	2428	2483	2541	2598	2634	2702	2751

CONSTRUCTION COST INDEX (CCI), U. S. CITY AVERAGE (ALL ITEMS)

1979	1980	1981	1982	1983	1984	1985	1986	1987	1988	1989	1990	1991
3003	3237	3535	3825	4066	4146	4195	4295	4406	4519	4615	4732	4835

Source: Engineering News Record (March 30, 1992).

TABLE III Standard Normal Cumulative Distribution Probabilities: $P(Z \le z) = \Phi(z)$

z	.00	.01	.02	.03	.04	.05	.06	.07	.08	.09
–0.0	.5000	.4960	.4920	.4880	.4840	.4801	.4761	.4721	.4681	.4641
–0.1	.4602	.4562	.4522	.4483	.4443	.4404	.4364	.4325	.4286	.4247
–0.2	.4207	.4168	.4129	.4090	.4052	.4013	.3974	.3936	.3897	.3859
–0.3	.3821	.3783	.3745	.3707	.3669	.3632	.3594	.3557	.3520	.3483
–0.4	.3446	.3409	.3372	.3336	.3300	.3264	.3228	.3192	.3156	.3121
–0.5	.3085	.3050	.3015	.2981	.2946	.2912	.2877	.2843	.2810	.2776
–0.6	.2743	.2709	.2676	.2643	.2611	.2578	.2546	.2514	.2483	.2451
–0.7	.2420	.2389	.2358	.2327	.2296	.2266	.2236	.2206	.2177	.2148
–0.8	.2119	.2090	.2061	.2033	.2005	.1977	.1949	.1922	.1894	.1867
–0.9	.1841	.1814	.1788	.1762	.1736	.1711	.1685	.1660	.1635	.1611
–1.0	.1587	.1562	.1539	.1515	.1492	.1469	.1446	.1423	.1401	.1379
–1.1	.1357	.1335	.1314	.1292	.1271	.1251	.1230	.1210	.1190	.1170
–1.2	.1151	.1131	.1112	.1093	.1075	.1056	.1038	.1020	.1003	.0985
–1.3	.0968	.0951	.0934	.0918	.0901	.0885	.0869	.0853	.0838	.0823
–1.4	.0808	.0793	.0778	.0764	.0749	.0735	.0721	.0708	.0694	.0681
–1.5	.0668	.0655	.0643	.0630	.0618	.0606	.0594	.0582	.0571	.0559
–1.6	.0548	.0537	.0526	.0516	.0505	.0495	.0485	.0475	.0465	.0455
–1.7	.0446	.0436	.0427	.0418	.0409	.0401	.0392	.0384	.0375	.0367
–1.8	.0359	.0351	.0344	.0336	.0329	.0322	.0314	.0307	.0301	.0294
–1.9	.0287	.0281	.0274	.0268	.0262	.0256	.0250	.0244	.0239	.0233
–2.0	.0228	.0222	.0217	.0212	.0207	.0202	.0197	.0192	.0188	.0183
–2.1	.0179	.0174	.0170	.0166	.0162	.0158	.0154	.0150	.0146	.0143
–2.2	.0139	.0136	.0132	.0129	.0125	.0122	.0119	.0116	.0113	.0110
–2.3	.0107	.0104	.0102	$.0^{2}990$	$.0^{2}964$	$.0^{2}939$	$.0^{2}914$	$.0^{2}889$	$.0^{2}866$	$.0^{2}842$
–2.4	$.0^{2}820$	$.0^{2}798$	$.0^{2}776$	$.0^{2}755$	$.0^{2}734$	$.0^{2}714$	$.0^{2}695$	$.0^{2}676$	$.0^{2}657$	$.0^{2}639$
–2.5	$.0^{2}621$	$.0^{2}604$	$.0^{2}587$	$.0^{2}570$	$.0^{2}554$	$.0^{2}539$	$.0^{2}523$	$.0^{2}508$	$.0^{2}494$	$.0^{2}480$
–2.6	$.0^{2}466$	$.0^{2}453$	$.0^{2}440$	$.0^{2}427$	$.0^{2}415$	$.0^{2}402$	$.0^{2}391$	$.0^{2}379$	$.0^{2}368$	$.0^{2}357$
–2.7	$.0^{2}347$	$.0^{2}336$	$.0^{2}326$	$.0^{2}317$	$.0^{2}307$	$.0^{2}298$	$.0^{2}289$	$.0^{2}280$	$.0^{2}272$	$.0^{2}264$
–2.8	$.0^{2}256$	$.0^{2}248$	$.0^{2}240$	$.0^{2}233$	$.0^{2}226$	$.0^{2}219$	$.0^{2}212$	$.0^{2}205$	$.0^{2}199$	$.0^{2}193$
–2.9	$.0^{2}187$	$.0^{2}181$	$.0^{2}175$	$.0^{2}169$	$.0^{2}164$	$.0^{2}159$	$.0^{2}154$	$.0^{2}149$	$.0^{2}144$	$.0^{2}139$
–3.0	$.0^{2}135$	$.0^{2}131$	$.0^{2}126$	$.0^{2}122$	$.0^{2}118$	$.0^{2}114$	$.0^{2}111$	$.0^{2}107$	$.0^{2}104$	$.0^{2}100$
–3.1	$.0^{3}968$	$.0^{3}935$	$.0^{3}904$	$.0^{3}874$	$.0^{3}845$	$.0^{3}816$	$.0^{3}789$	$.0^{3}762$	$.0^{3}736$	$.0^{3}711$
–3.2	$.0^{3}687$	$.0^{3}664$	$.0^{3}641$	$.0^{3}619$	$.0^{3}598$	$.0^{3}577$	$.0^{3}557$	$.0^{3}538$	$.0^{3}519$	$.0^{3}501$
–3.3	$.0^{3}483$	$.0^{3}466$	$.0^{3}450$	$.0^{3}434$	$.0^{3}419$	$.0^{3}404$	$.0^{3}390$	$.0^{3}376$	$.0^{3}362$	$.0^{3}349$
–3.4	$.0^{3}337$	$.0^{3}325$	$.0^{3}313$	$.0^{3}302$	$.0^{3}291$	$.0^{3}280$	$.0^{3}270$	$.0^{3}260$	$.0^{3}251$	$.0^{3}242$
–3.5	$.0^{3}233$	$.0^{3}224$	$.0^{3}216$	$.0^{3}208$	$.0^{3}200$	$.0^{3}193$	$.0^{3}185$	$.0^{3}178$	$.0^{3}172$	$.0^{3}165$
–3.6	$.0^{3}159$	$.0^{3}153$	$.0^{3}147$	$.0^{3}142$	$.0^{3}136$	$.0^{3}131$	$.0^{3}126$	$.0^{3}121$	$.0^{3}117$	$.0^{3}112$
–3.7	$.0^{3}108$	$.0^{3}104$	$.0^{4}996$	$.0^{4}957$	$.0^{4}920$	$.0^{4}884$	$.0^{4}850$	$.0^{4}816$	$.0^{4}784$	$.0^{4}753$
–3.8	$.0^{4}723$	$.0^{4}695$	$.0^{4}667$	$.0^{4}641$	$.0^{4}615$	$.0^{4}591$	$.0^{4}567$	$.0^{4}544$	$.0^{4}522$	$.0^{4}501$
–3.9	$.0^{4}481$	$.0^{4}461$	$.0^{4}443$	$.0^{4}425$	$.0^{4}407$	$.0^{4}391$	$.0^{4}375$	$.0^{4}359$	$.0^{4}345$	$.0^{4}330$
–4.0	$.0^{4}317$	$.0^{4}304$	$.0^{4}291$	$.0^{4}279$	$.0^{4}267$	$.0^{4}256$	$.0^{4}245$	$.0^{4}235$	$.0^{4}225$	$.0^{4}216$

TABLE III (*continued*)

z	.00	.01	.02	.03	.04	.05	.06	.07	.08	.09
0.0	.5000	.5040	.5080	.5120	.5160	.5199	.5239	.5279	.5319	.5359
0.1	.5398	.5438	.5478	.5517	.5557	.5596	.5636	.5675	.5714	.5753
0.2	.5793	.5832	.5871	.5910	.5948	.5987	.6026	.6064	.6103	.6141
0.3	.6179	.6217	.6255	.6293	.6331	.6368	.6406	.6443	.6480	.6517
0.4	.6554	.6591	.6628	.6664	.6700	.6736	.6772	.6808	.6844	.6879
0.5	.6915	.6950	.6985	.7019	.7054	.7088	.7123	.7157	.7190	.7224
0.6	.7257	.7291	.7324	.7357	.7389	.7422	.7454	.7486	.7517	.7549
0.7	.7580	.7611	.7642	.7673	.7704	.7734	.7764	.7794	.7823	.7852
0.8	.7881	.7910	.7939	.7967	.7995	.8023	.8051	.8078	.8106	.8133
0.9	.8159	.8186	.8212	.8238	.8264	.8289	.8315	.8340	.8365	.8389
1.0	.8413	.8438	.8461	.8485	.8508	.8531	.8554	.8577	.8599	.8621
1.1	.8643	.8665	.8686	.8708	.8729	.8749	.8770	.8790	.8810	.8830
1.2	.8849	.8869	.8888	.8907	.8925	.8944	.8962	.8980	.8997	.9015
1.3	.9032	.9049	.9066	.9082	.9099	.9115	.9131	.9147	.9162	.9177
1.4	.9192	.9207	.9222	.9236	.9251	.9265	.9279	.9292	.9306	.9319
1.5	.9332	.9345	.9357	.9370	.9382	.9394	.9406	.9418	.9429	.9441
1.6	.9452	.9463	.9474	.9484	.9495	.9505	.9515	.9525	.9535	.9545
1.7	.9554	.9564	.9573	.9582	.9591	.9599	.9608	.9616	.9625	.9633
1.8	.9641	.9649	.9656	.9664	.9671	.9678	.9686	.9693	.9699	.9706
1.9	.9713	.9719	.9726	.9732	.9738	.9744	.9750	.9756	.9761	.9767
2.0	.9772	.9778	.9783	.9788	.9793	.9798	.9803	.9808	.9812	.9817
2.1	.9821	.9826	.9830	.9834	.9838	.9842	.9846	.9850	.9854	.9857
2.2	.9861	.9864	.9868	.9871	.9875	.9878	.9881	.9884	.9887	.9890
2.3	.9893	.9896	.9898	$.9^{2}010$	$.9^{2}036$	$.9^{2}061$	$.9^{2}086$	$.9^{2}111$	$.9^{2}134$	$.9^{2}158$
2.4	$.9^{2}180$	$.9^{2}202$	$.9^{2}224$	$.9^{2}245$	$.9^{2}266$	$.9^{2}286$	$.9^{2}305$	$.9^{2}324$	$.9^{2}343$	$.9^{2}361$
2.5	$.9^{2}379$	$.9^{2}396$	$.9^{2}413$	$.9^{2}430$	$.9^{2}446$	$.9^{2}461$	$.9^{2}477$	$.9^{2}492$	$.9^{2}506$	$.9^{2}520$
2.6	$.9^{2}534$	$.9^{2}547$	$.9^{2}560$	$.9^{2}573$	$.9^{2}585$	$.9^{2}598$	$.9^{2}609$	$.9^{2}621$	$.9^{2}632$	$.9^{2}643$
2.7	$.9^{2}653$	$.9^{2}664$	$.9^{2}674$	$.9^{2}683$	$.9^{2}693$	$.9^{2}702$	$.9^{2}711$	$.9^{2}720$	$.9^{2}728$	$.9^{2}736$
2.8	$.9^{2}744$	$.9^{2}752$	$.9^{2}760$	$.9^{2}767$	$.9^{2}774$	$.9^{2}781$	$.9^{2}788$	$.9^{2}795$	$.9^{2}801$	$.9^{2}807$
2.9	$.9^{2}813$	$.9^{2}819$	$.9^{2}825$	$.9^{2}831$	$.9^{2}836$	$.9^{2}841$	$.9^{2}846$	$.9^{2}851$	$.9^{2}856$	$.9^{2}861$
3.0	$.9^{2}865$	$.9^{2}869$	$.9^{2}874$	$.9^{2}878$	$.9^{2}882$	$.9^{2}886$	$.9^{2}889$	$.9^{2}893$	$.9^{2}896$	$.9^{2}900$
3.1	$.9^{3}032$	$.9^{3}065$	$.9^{3}096$	$.9^{3}126$	$.9^{3}155$	$.9^{3}184$	$.9^{3}211$	$.9^{3}238$	$.9^{3}264$	$.9^{3}289$
3.2	$.9^{3}313$	$.9^{3}336$	$.9^{3}359$	$.9^{3}381$	$.9^{3}402$	$.9^{3}423$	$.9^{3}443$	$.9^{3}462$	$.9^{3}481$	$.9^{3}499$
3.3	$.9^{3}517$	$.9^{3}534$	$.9^{3}550$	$.9^{3}566$	$.9^{3}581$	$.9^{3}596$	$.9^{3}610$	$.9^{3}624$	$.9^{3}638$	$.9^{3}651$
3.4	$.9^{3}663$	$.9^{3}675$	$.9^{3}687$	$.9^{3}698$	$.9^{3}709$	$.9^{3}720$	$.9^{3}730$	$.9^{3}740$	$.9^{3}749$	$.9^{3}758$
3.5	$.9^{3}767$	$.9^{3}776$	$.9^{3}784$	$.9^{3}792$	$.9^{3}800$	$.9^{3}807$	$.9^{3}815$	$.9^{3}822$	$.9^{3}828$	$.9^{3}835$
3.6	$.9^{3}841$	$.9^{3}847$	$.9^{3}853$	$.9^{3}858$	$.9^{3}864$	$.9^{3}869$	$.9^{3}874$	$.9^{3}879$	$.9^{3}883$	$.9^{3}888$
3.7	$.9^{3}892$	$.9^{3}896$	$.9^{4}004$	$.9^{4}043$	$.9^{4}080$	$.9^{4}116$	$.9^{4}150$	$.9^{4}184$	$.9^{4}216$	$.9^{4}247$
3.8	$.9^{4}277$	$.9^{4}305$	$.9^{4}333$	$.9^{4}359$	$.9^{4}385$	$.9^{4}409$	$.9^{4}433$	$.9^{4}456$	$.9^{4}478$	$.9^{4}499$
3.9	$.9^{4}519$	$.9^{4}539$	$.9^{4}557$	$.9^{4}575$	$.9^{4}593$	$.9^{4}609$	$.9^{4}625$	$.9^{4}641$	$.9^{4}655$	$.9^{4}670$
4.0	$.9^{4}683$	$.9^{4}696$	$.9^{4}709$	$.9^{4}721$	$.9^{4}733$	$.9^{4}744$	$.9^{4}755$	$.9^{4}765$	$.9^{4}775$	$.9^{4}784$

Source: Adapted from Gerald J. Hahn and William Q. Meeker, *Statistical Intervals: A Guide for Practitioners.* Wiley Interscience, New York (1991).

Table IV Percentiles of the Student's t Distribution with df Degrees of Freedom: $t_{(1-\alpha,\, df)}$

	$(1-\alpha)$									
df	0.750	0.800	0.850	0.900	0.950	0.975	0.980	0.990	0.995	0.999
1	1.000	1.376	1.963	3.078	6.314	12.71	15.89	31.82	63.66	318.3
2	0.816	1.061	1.386	1.886	2.920	4.303	4.849	6.965	9.925	22.33
3	0.765	0.978	1.250	1.638	2.353	3.182	3.482	4.541	5.841	10.21
4	0.741	0.941	1.190	1.533	2.132	2.776	2.999	3.747	4.604	7.173
5	0.727	0.920	1.156	1.476	2.015	2.571	2.757	3.365	4.032	5.893
6	0.718	0.906	1.134	1.440	1.943	2.447	2.612	3.143	3.707	5.208
7	0.711	0.896	1.119	1.415	1.895	2.365	2.517	2.998	3.499	4.785
8	0.706	0.889	1.108	1.397	1.860	2.306	2.449	2.896	3.355	4.501
9	0.703	0.883	1.100	1.383	1.833	2.262	2.398	2.821	3.250	4.297
10	0.700	0.879	1.093	1.372	1.812	2.228	2.359	2.764	3.169	4.144
11	0.697	0.876	1.088	1.363	1.796	2.201	2.328	2.718	3.106	4.025
12	0.695	0.873	1.083	1.356	1.782	2.179	2.303	2.681	3.055	3.930
13	0.694	0.870	1.079	1.350	1.771	2.160	2.282	2.650	3.012	3.852
14	0.692	0.868	1.076	1.345	1.761	2.145	2.264	2.624	2.977	3.787
15	0.691	0.866	1.074	1.341	1.753	2.131	2.249	2.602	2.947	3.733
16	0.690	0.865	1.071	1.337	1.746	2.120	2.235	2.583	2.921	3.686
17	0.689	0.863	1.069	1.333	1.740	2.110	2.224	2.567	2.898	3.646
18	0.688	0.862	1.067	1.330	1.734	2.101	2.214	2.552	2.878	3.610
19	0.688	0.861	1.066	1.328	1.729	2.093	2.205	2.539	2.861	3.579
20	0.687	0.860	1.064	1.325	1.725	2.086	2.197	2.528	2.845	3.552
21	0.686	0.859	1.063	1.323	1.721	2.080	2.189	2.518	2.831	3.527
22	0.686	0.858	1.061	1.321	1.717	2.074	2.183	2.508	2.819	3.505
23	0.685	0.858	1.060	1.319	1.714	2.069	2.177	2.500	2.807	3.485
24	0.685	0.857	1.059	1.318	1.711	2.064	2.172	2.492	2.797	3.467
25	0.684	0.856	1.058	1.316	1.708	2.060	2.167	2.485	2.787	3.450
26	0.684	0.856	1.058	1.315	1.706	2.056	2.162	2.479	2.779	3.435
27	0.684	0.855	1.057	1.314	1.703	2.052	2.158	2.473	2.771	3.421
28	0.683	0.855	1.056	1.313	1.701	2.048	2.154	2.467	2.763	3.408
29	0.683	0.854	1.055	1.311	1.699	2.045	2.150	2.462	2.756	3.396
30	0.683	0.854	1.055	1.310	1.697	2.042	2.147	2.457	2.750	3.385
35	0.682	0.852	1.052	1.306	1.690	2.030	2.133	2.438	2.724	3.340
40	0.681	0.851	1.050	1.303	1.684	2.021	2.123	2.423	2.704	3.307
50	0.679	0.849	1.047	1.299	1.676	2.009	2.109	2.403	2.678	3.261
60	0.679	0.848	1.045	1.296	1.671	2.000	2.099	2.390	2.660	3.232
70	0.678	0.847	1.044	1.294	1.667	1.994	2.093	2.381	2.648	3.211
80	0.678	0.846	1.043	1.292	1.664	1.990	2.088	2.374	2.639	3.195
90	0.677	0.846	1.042	1.291	1.662	1.987	2.084	2.368	2.632	3.183

Table IV (*continued*) Percentiles of the Student's t Distribution with df Degrees of Freedom: $t_{(1-\alpha,\, df)}$

	(1 – α)									
df	0.750	0.800	0.850	0.900	0.950	0.975	0.980	0.990	0.995	0.999
100	0.677	0.845	1.042	1.290	1.660	1.984	2.081	2.364	2.626	3.174
120	0.677	0.845	1.041	1.289	1.658	1.980	2.076	2.358	2.617	3.160
∞	0.675	0.842	1.036	1.282	1.645	1.960	2.054	2.327	2.576	3.091

Source: Adapted from Gerald J. Hahn and William Q. Meeker, *Statistical Intervals: A Guide for Practitioners*. Wiley Interscience, New York (1991).

TABLE V Factors $g_{(1-\alpha, p, n)}$ for Calculating Normal Distribution Two-Sided $100(1-\alpha)\%$ Tolerance Intervals

	$p = 0.500$			$p = 0.900$			$p = 0.950$			$p = 0.990$			
	$1-\alpha$												
n	0.90	0.95	0.99	0.90	0.95	0.99	0.90	0.95	0.99	0.90	0.95	0.99	n
2	6.808	13.652	68.316	15.512	31.092	155.569	18.221	36.519	182.720	23.423	46.944	234.877	2
3	2.492	3.585	8.122	5.788	8.306	18.782	6.823	9.789	22.131	8.819	12.647	28.586	3
4	1.766	2.288	4.028	4.157	5.368	9.416	4.913	6.341	11.118	6.372	8.221	14.405	4
5	1.473	1.812	2.824	3.499	4.291	6.655	4.142	5.077	7.870	5.387	6.598	10.220	5
6	1.314	1.566	2.270	3.141	3.733	5.383	3.723	4.422	6.373	4.850	5.758	8.292	6
7	1.213	1.415	1.954	2.913	3.390	4.658	3.456	4.020	5.520	4.508	5.241	7.191	7
8	1.143	1.313	1.750	2.754	3.156	4.189	3.270	3.746	4.968	4.271	4.889	6.479	8
9	1.092	1.239	1.608	2.637	2.986	3.860	3.132	3.546	4.581	4.094	4.633	5.980	9
10	1.053	1.183	1.503	2.546	2.856	3.617	3.026	3.393	4.294	3.958	4.437	5.610	10
11	1.021	1.139	1.422	2.473	2.754	3.429	2.941	3.273	4.073	3.849	4.282	5.324	11
12	0.996	1.103	1.357	2.414	2.670	3.279	2.871	3.175	3.896	3.759	4.156	5.096	12
13	0.974	1.073	1.305	2.364	2.601	3.156	2.812	3.093	3.751	3.684	4.051	4.909	13
14	0.956	1.048	1.261	2.322	2.542	3.054	2.762	3.024	3.631	3.620	3.962	4.753	14
15	0.941	1.027	1.224	2.285	2.492	2.967	2.720	2.965	3.529	3.565	3.885	4.621	15
16	0.927	1.008	1.193	2.254	2.449	2.893	2.682	2.913	3.441	3.517	3.819	4.507	16
17	0.915	0.992	1.165	2.226	2.410	2.828	2.649	2.868	3.364	3.474	3.761	4.408	17
18	0.905	0.978	1.141	2.201	2.376	2.771	2.620	2.828	3.297	3.436	3.709	4.321	18
19	0.895	0.965	1.120	2.178	2.346	2.720	2.593	2.793	3.237	3.402	3.663	4.244	19
20	0.887	0.953	1.101	2.158	3.319	2.675	2.570	2.760	3.184	3.372	3.621	4.175	20
21	0.879	0.943	1.084	2.140	2.294	2.635	2.548	2.731	3.136	3.344	3.583	4.113	21
22	0.872	0.934	1.068	2.123	2.272	2.598	2.528	2.705	3.092	3.318	3.519	4.056	22
23	0.866	0.925	1.054	2.108	2.251	2.564	2.510	2.681	3.053	3.295	3.518	4.005	23
24	0.860	0.917	1.042	2.094	2.232	2.534	2.494	2.658	3.017	3.274	3.489	3.958	24
25	0.855	0.910	1.030	2.081	2.215	2.506	2.479	2.638	2.981	3.254	3.462	3.915	25
26	0.850	0.903	1.019	2.069	2.199	2.480	2.464	2.619	2.953	3.235	3.437	3.875	26
27	0.845	0.897	1.009	2.058	2.184	2.456	2.451	2.601	2.925	3.218	3.415	3.838	27
28	0.841	0.891	1.000	2.048	2.170	2.434	2.439	2.585	2.898	3.202	3.393	3.804	28
29	0.837	0.886	0.991	2.038	2.157	2.413	2.427	2.569	2.874	3.187	3.373	3.772	29
30	0.833	0.881	0.983	2.029	2.145	2.394	2.417	2.555	2.851	3.173	3.355	3.742	30

TABLE V *(continued)* Factors $g_{(1-\alpha, p, n)}$ for Calculating Normal Distribution Two-Sided $100(1-\alpha)\%$ Tolerance Intervals

	$p = 0.500$			$p = 0.900$			$p = 0.950$			$p = 0.990$			
	$1-\alpha$												
35	0.817	0.859	0.950	1.991	2.094	2.314	2.371	2.495	2.756	3.114	3.276	3.618	35
40	0.805	0.843	0.925	1.961	2.055	2.253	2.336	2.448	2.684	3.069	3.216	3.524	40
50	0.787	0.820	0.889	1.918	1.999	2.166	2.285	2.382	2.580	3.003	3.129	3.390	50
60	0.775	0.804	0.864	1.888	1.960	2.106	2.250	2.335	2.509	2.956	3.068	3.297	60
120	0.740	0.759	0.797	1.805	1.851	1.943	2.151	2.206	2.315	2.826	2.899	3.043	120
240	0.719	0.731	0.756	1.753	1.783	1.844	2.088	2.125	2.197	2.744	2.793	2.887	240
480	0.705	0.713	0.730	1.718	1.739	1.780	2.048	2.073	2.121	2.691	2.724	2.787	480
∞	0.674	0.674	0.674	1.645	1.645	1.645	1.960	1.960	1.960	2.576	2.576	2.576	∞

Source: Adapted from Gerald J. Hahn and William Q. Meeker, *Statistical Intervals: A Guide for Practitioners*. Wiley Interscience, New York (1991).

TABLE VI Factors $g'_{(1-\alpha, p, n)}$ for Calculating Normal Distribution One-Sided $100(1-\alpha)\%$ Tolerance Bounds

	$p = 0.900$			$p = 0.950$			$p = 0.990$			
	$1-\alpha$									
n	0.900	0.950	0.990	0.900	0.950	0.990	0.900	0.950	0.990	n
2	10.253	20.581	103.029	13.090	26.260	131.426	18.500	37.094	185.617	2
3	4.258	6.155	13.955	5.311	7.656	17.370	7.340	10.553	23.896	3
4	3.188	4.162	7.380	3.957	5.144	9.083	5.438	7.042	12.387	4
5	2.742	3.407	5.362	3.400	4.203	6.578	4.666	5.741	8.939	5
6	2.494	3.006	4.411	3.092	3.708	5.406	4.243	5.062	7.335	6
7	2.333	2.755	3.859	2.894	3.399	4.728	3.972	4.642	6.412	7
8	2.219	2.582	3.497	2.754	3.187	4.285	3.783	4.354	5.812	8
9	2.133	2.454	3.240	2.650	3.031	3.972	3.641	4.143	5.389	9
10	2.066	2.355	3.048	2.568	2.911	3.738	3.532	3.981	5.074	10
11	2.011	2.275	2.898	2.503	2.815	3.556	3.443	3.852	4.829	11
12	1.966	2.210	2.777	2.448	2.736	3.410	3.371	3.747	4.633	12
13	1.928	2.155	2.677	2.402	2.671	3.290	3.309	3.659	4.472	13
14	1.895	2.109	2.593	2.363	2.614	3.189	3.257	3.585	4.337	14
15	1.867	2.068	2.521	2.329	2.566	3.102	3.212	3.520	4.222	15
16	1.842	2.033	2.459	2.299	2.524	3.028	3.172	3.464	4.123	16
17	1.819	2.002	2.405	2.272	2.486	2.963	3.137	3.414	4.037	17
18	1.800	1.974	2.357	2.249	2.453	2.905	3.105	3.370	3.960	18
19	1.782	1.949	2.314	2.227	2.423	2.854	3.077	3.331	3.892	19
20	1.765	1.926	2.276	2.208	2.396	2.808	3.052	3.295	3.832	20
21	1.750	1.905	2.241	2.190	2.371	2.766	3.028	3.263	3.777	21
22	1.737	1.886	2.209	2.174	2.349	2.729	3.007	3.233	3.727	22
23	1.724	1.869	2.180	2.159	2.328	2.694	2.987	3.206	3.681	23
24	1.712	1.853	2.154	2.145	2.309	2.662	2.969	3.181	3.640	24
25	1.702	1.838	2.129	2.132	2.292	2.633	2.952	3.158	3.601	25
26	1.691	1.824	2.106	2.120	2.275	2.606	2.937	3.136	3.566	26
27	1.682	1.811	2.085	2.109	2.260	2.581	2.922	3.116	3.533	27
28	1.673	1.799	2.065	2.099	2.246	2.558	2.909	3.098	3.502	28
29	1.665	1.788	2.047	2.089	2.232	2.536	2.896	3.080	3.473	29
30	1.657	1.777	2.030	2.080	2.220	2.515	2.884	3.064	3.447	30
35	1.624	1.732	1.957	2.041	2.167	2.430	2.833	2.995	3.334	35
40	1.598	1.697	1.902	2.010	2.125	2.364	2.793	2.941	3.249	40
50	1.559	1.646	1.821	1.965	2.065	2.269	2.735	2.862	3.125	50
60	1.532	1.609	1.764	1.933	2.022	2.202	2.694	2.807	3.038	60
120	1.452	1.503	1.604	1.841	1.899	2.015	2.574	2.649	2.797	120
240	1.399	1.434	1.501	1.780	1.819	1.896	2.497	2.547	2.645	240
480	1.363	1.387	1.433	1.738	1.766	1.818	2.444	2.479	2.545	480
∞	1.282	1.282	1.282	1.645	1.645	1.645	2.326	2.326	2.326	∞

Source: Adapted from Gerald J. Hahn and William Q. Meeker, *Statistical Intervals: A Guide for Practitioners*. Wiley Interscience, New York (1991).

C. ANSWERS TO SELECTED QUESTIONS

Chapter 2

2.1.6(b,c)	$\mathbf{x}^* = (1100, 1900)$,	$y^* = \$148{,}000$
2.1.7(c,d)	$\mathbf{x}^* = (9000, 1000)$,	$y^* = \$350{,}000$
2.1.8(b,c)	$\mathbf{x}^* = (1050, 0)$,	$y^* = \$322{,}000$
2.1.9(b,c)	$\mathbf{x}^* = (125, 375)$,	$z^* = \$196{,}000$
2.1.10(b)	$\mathbf{x}^* = (75, 25)$,	$y^* = 1{,}750$
2.2.2		
(a)	$\mathbf{x}^* = (0.6, 1.8)$,	$y^* = 1.8$
(c)	$\mathbf{x}^* = (0, 2.25)$,	$y^* = 2.25$
2.2.3		
(a)	$\mathbf{x}^* = \left(\frac{2}{3}, \frac{4}{3}\right)$,	$y^* = 6$
(c)	$\mathbf{x}^* = \left(\frac{2}{3}, \frac{4}{3}\right)$,	$y^* = 6$
(d)	$\mathbf{x}^* = \left(\frac{6}{5}, \frac{12}{5}\right)$,	$y^* = 10.8$
2.2.4 (a)		
A.	$\mathbf{x}^* = (1100, 1900)$,	$y^* = \$148{,}000$
B.	$\mathbf{x}^* = (1100, 1900)$,	$y^* = \$203{,}000$
(b)	$u_1 = \$120$ per ton	$y^* = \$345{,}000$
2.2.5		
$v = 1.$	$\mathbf{x}^* = (9000, 1000)$,	$y^* = \$350{,}000$
$v = 0.5.$	$\mathbf{x}^* = (9500, 500)$,	$y^* = \$425{,}000$
$v = 0.$	$\mathbf{x}^* = (10{,}000, 500)$,	$y^* = \$500{,}000$
2.2.6(a)		
A.	$\mathbf{x}^* = (75, 25)$,	$y^* = \$1{,}750$ ton-miles

B. $\mathbf{x}^*$ = boundary line between (75, 25) and (100, 0), y^* = \$1000 ton-miles

C. No feasible solution exists.

The following solutions in Sections 2.3 and Chapter 3 were obtained with the aid of a simplex method computer algorithm.

2.3.2

(a) $\mathbf{x}^* = (0.6, 1.8)$ $y^* = 1.8$

(b) $\mathbf{x}^* = (0, 2.25)$ $y^* = 2.25$

2.3.3

(a) $\mathbf{x}^* = (0.67, 1.33)$ $y^* = 6$

(b) $\mathbf{x}^* = (1.2, 2.4)$ $y^* = 10.8$

2.3.4

(a)	$x_1^* = 25$,	$x_2^* = 0$,	$x_3^* = 75$,	y^* = \$975
(b)	$x_1^* = 0$,	$x_2^* = 50$,	$x_3^* = 50$,	y^* = \$850
(c)	$x_1^* = 0$,	$x_2^* = 0$,	$x_3^* = 100$,	y^* = \$1100

2.3.5 (b) y^* = \$1275

2.3.6

(a) $x_1^* = 1$, $x_2^* = 2$, $x_3^* = 2$, $x_4^* = 0$,
y^* = 28 mg per liter BOD

(b) $x_1^* = 0$, $x_2^* = 3$, $x_3^* = 2$, $x_4^* = 0$,
y^* = 32 mg per liter BOD

Chapter 3

3.1.4

start → 2 → 7 → 10 → finish. y^* = 25 minutes

3.1.5

start → 2 → 9 → finish or start → 2 → 5 → 10 → finish. y^* = 9 hours

3.1.6

z^* = 8000 vehicles per hour

$x_9^* = 6000$, $x_2^* = x_5^* = x_6^* = x_{10}^* = 2000$ and $x_4^* = 4000$.

3.1.7

z^* = 11 mgd

3.1.8

$z^* = 3000$

3.1.9

excavate site A → excavate site B → pour concrete at site B → pour conc… weeks

order and delivier material for foundation → excavate foundation → pour concrete … prefabricate structural members → erect structure, $y^* = 36$ days

3.2.1

$x_{12}^* = 2$, $x_{13}^* = 5$, $x_{15}^* = 3$, $x_{45}^* = 5$, $y^* = 59$ miles

3.2.2

$x_{12}^* = 10$, $x_{23}^* = 5$, $x_{34}^* = 3$, $x_{45}^* = 8$, $y^* = 65$ miles

3.2.3

$x_{12}^* = 2$, $x_{13}^* = 3$, $x_{16}^* = 10$, $x_{43}^* = 2$, $x_{45}^* = 8$, $y^* = 43$ minutes

3.2.4

$x_{12}^* = 2$, $x_{13}^* = 3$, $x_{43}^* = 2$, $x_{45}^* = 8$, $y^* = 43$ minutes

3.2.5

$x_{14}^* = 50M$, $x_{25}^* = 60M$, $x_{32}^* = 25M$, $x_{34}^* = 25M$, $y^* = 4150$ minutes

3.2.6

$x_{26}^* = 4$, $x_{34}^* = 3$, $x_{41}^* = 6$, $x_{54}^* = 2$, $x_{56}^* = 2$, $y^* = 265$ minutes

3.2.7

$x_{13}^* = 10$, $x_{24}^* = 20$, $x_{35}^* = 10$, $x_{45}^* = 20$, $y^* = 2200$ minutes

3.2.8

Route A. $y^* = 67.5$ minutes

3.3.1

$x_{11}^* = x_{21}^* = x_{31}^* = 1820$, $y^* = \$100,100$

3.3.2

$x_1^* = 3640$, $x_{31}^* = x_{42}^* = 1820$, $y^* = \$127,400$

3.3.3

(a) $x_{11}^* = x_{21}^* = x_{31}^* = 1820$, $z^* = 1820$ tons

(b) No feasible solution exists.

3.3.4

$x_{51}^* = 300$, $x_1^* = 3040$, $x_2^* = 300$, $x_{31}^* = 1520$, $y^* = \$118{,}400$

3.3.5

$x_{21}^* = 1500$, $x_{22}^* = 6000$, $x_{31}^* = 147$, $x_{32}^* = 1218$, $y^* = \$599{,}730$

Incineration of hazardous waste is recommended.

Chapter 4

4.1.9(b) Break-even points

A. none

B. $x' = 12.7$ units

4.1.10(c) $q^* = 25000$ units

4.1.11(b) $d = 0$ miles, $m' = 1$

$d = 40$ miles, $m' = 0.72$

$d = 80$ miles, $m' = 0.42$

4.2.3 simple interest: $w(10) = \$2000$

compound interest: $w(10) = \$2593$

4.2.4(a)

A. $npw = -\$1.39$k $\rightarrow$ infeasible

B. $npw = \$0 \rightarrow$ feasible

Recommend B.

(b)

A. $anw = -\$1.5$k $\rightarrow$ infeasible

B. $anw = \$0 \rightarrow$ feasible

Recommend B.

4.2.6(b)

$npw = \$6417.66$

4.2.8

$nw(3) = \$17{,}381$

4.2.9

t = \$60.60 per ton

4.2.10

p = \$.04 per kWh

4.2.11

$k(3)$ = \$2736

4.2.12

npw = –\$22.10M → infeasible

4.2.13

t = \$786.80 per year.

4.3.6

i' = 25.7%

4.3.7(a) and (c)

I	*npw*	*anw*
3	\$1.72M	\$0.116M
6	\$0.87M	\$0.076M
9	\$0.28M	\$0.031M
12	–\$0.13M	–\$0.017M
15	–\$0.44M	–\$0.070M

(b) $i' \approx$ 11% per year

4.3.8

A. *npw* = \$0.32M

B. *npw* = \$0.40M

Recommend B.

4.3.9

. *npw* = –\$0.56M

pw = –\$0.46M

mend a do nothing alternative.

4.4.7

(a) $i' = 8\%$ per year

(b) $i' = 18.8\%$ per year

4.4.9

$g' = 16.8\%$ per year

4.4.10

	f
PPI	0.037
CPI	0.055
BCI	0.038
CCI	0.044

4.4.11(c) Recommend A.

4.5.5

(a) $\varepsilon' = \infty$ and $q = 1M$

(b) $p = 1 + \frac{q}{3}$ in \$ per unit and $p' = \$1.33$ and $q' = 1M$

(c) $w' = \$1.33M$

(d) $nw = \$0.167M$

(e) $nw = \$0.167M$

4.5.6

(a) $p' = \$15.70$ per unit and $q' = 477$ units

(b) $npw = \$2833$

4.5.7(a)

A. $npw = 2.09M$ for $n = 9$ trucks per day

B. $npw = \$2.67M$ for $n = 10$ trucks per day

Least cost alternative: A

(b) $npw < 0$, recommend A.

4.5.8 $npw = \$864{,}000 \rightarrow$ a feasible plan

4.5.9

(a) $nw^* = \$100$ where $q^* = 100$ items

(b) $nw^* = \$2256$ where $q^* = 47.5$ items

Chapter 5

5.1.6 $R(t)$ = \$1.6M per year

5.1.7 $w(t)$ = \$0.18M per year

5.2.5

(a) $\theta(d) = 2.65 \times 10^{-4}$ (child)

(b) $\theta(d) = 4.45 \times 10^{-3}$ (adult)

5.2.6

(a) MCL = 0.00175 mg/l (adult)
MCL = 0.0005 mg/l (child)

(b) MCL = 0.000175 mg/l (adult)
MCL = 0.0001 mg/l (child)

5.2.7

Model	LD_{50}	VSD at $\theta(d) = 10^{-4}$	VSD at $\theta(d) = 10^{-6}$
Linear	46 mg/kg	0.009 mg/kg	0.0008 mg/kg
One-hit	180	0.03	0.004
Multistage	190	0.064	0.008
Weibull	45	0.045	0.008

5.3.1

(a) npw = \$0.28M → feasible

(b) npw = -\$0.08M → infeasible

5.3.2 npw_A = –\$0.08M, npw_B = \$0.16M. B is recommended.

5.3.3 npw_A = -\$0.19M, npw_B = \$0.37M. B is recommended.

pter 6

6.1.2

(b) $P(S_2|S_1) = 0.09$

d) R = \$0.104M

$= p^3 + 3p^2(1-p)$

.00725

6.1.4 (b) $P(S_2) = 0.02$

6.1.5

(a) $P(S_1 \cap S_2 \cap S_3) = 0.0001$

(b) $P(S_3) = 0.0198$

6.1.6

(a) $P(S) = 0.03$

(b) $P(S) = 0.104$

6.1.7

(b) $P(S) = 0.089$

(c) $R(t) = \$8900$

6.2.2 $P(T > 10) = 0.25$

6.2.3(a) $a = 0.0444$, $b = 0.0111$

6.2.4 (a) $f(t) = \rho\gamma t^{\gamma-1} e^{-\rho t^{\gamma}}$

(b) $h(t) = \rho\gamma t^{\gamma-1}$

6.2.5 (a) $R = \$0.5\text{M}$

6.2.6 (a) $R = \$0.5\text{M}$

6.2.7 (a) $E(X) = 25$ mg/gm

6.3.1

(a) $F_V(v) = 1 - e^{-v/30}$ for $v = 0$ where $v \geq 0$

(b) $w = 6.88\, v^2$ with units of w in pounds and v in mph.

(c) $F_W(w) = 1 - e^{-0.0186\sqrt{w}}$ for $w \geq 0$

6.3.2

(a) $f_T(t) = t\rho^2 e^{-\rho t}$ for $t \geq 0$

(b) $F_T(t) = 1 - e^{-\rho t} - \rho t e^{-\rho t}$

(c) $P(T \leq 25) = 0.81$ without backup system

$P(T \leq 25) = 0.496$ with backup system

6.3.3 $P(T \leq 25) = F_T(25) = 0.30$ for a landfill built on sand.

6.3.4 $P(T \leq 25) = F_T(25) = 0.052$ for a landfill built on silt.

$$\text{(a)}\ f_T(t) = \begin{cases} \frac{1}{6}(t-2) & \text{for } 2 \leq t \leq 4 \\ \frac{1}{3} & \text{for } 4 < t \leq 5 \\ \frac{1}{6}(7-t) & \text{for } 5 < t \leq 7 \end{cases}$$

(b) $\mu = 4.5$ weeks and $\sigma^2 = 13/12$

(c) $P(T < 4) = 1/3$

6.4.4 (a) $a(1) = 0.669$, (b) $a(1) = 0.967$, (c) $a(1) = 0.934$, (d) $a(1)$ ·

6.4.5 npw_A = \$3.77M and npw_B = \$4.99M. Recommend B.

6.4.6

npw = \$3.77M for repairable system A.

npw = \$6.82M for nonrepairable system C. Recommend C.

6.4.7

npw_A = \$4.66M and npw_B = \$4.99M. Recommend B.

6.4.8

$P_0(t) = e^{-\rho t}$

$P_1(t) = \rho t e^{-\rho t}$

$P_2(t) = \frac{1}{2}(\rho t)^2 e^{-\rho t}$

Chapter 7

7.1.2

(a) $\mu_Y \pm \sigma_Y = 0 \pm 102$ ppmv

(b) $\mu_H \pm \sigma_H$ = \$0 ± \$10,200

7.1.3

(a) $\mu_X \pm \sigma_X = 0.36 \pm 1.38$

(b) $\mu_H \pm \sigma_H$ = \$4.24M ± \$29M

7.1.4

(a) $\mu_E \pm \sigma_E = 0.25 \pm 0.354$ mck

(b) $\mu_H \pm \sigma_H$ = \$25.00 ± \$35.40

7.1.5

(a) $\mu_H \pm \sigma_H$ = \$5.41M ± \$52.4M with $\rho_{12} = 0.0$

(b) $\mu_H \pm \sigma_H$ = \$5.41M ± \$54.0M with $\rho_{12} = 0.8$

7.2.2 (a) 0.683, (b) 0.955, (c) 0.997

7.2.3 (a) $\sigma = 8.33$ ppmv, (b) $P(Y \leq 21) = 0.641$

7.2.4 $P(Y \geq 175) = 0.043$

0) = 0.24

7.2.5 (a) \$24.00

E) = 0.0003

) $E(H)$ = \$0.03

7.3.4

(b) $\bar{k} \pm s_K$ = \$49.04M ± 85.13M

(c) $\bar{c} \pm s_C$ = \$66,900 ± 17,880 per ton-day

(d) $b(\bar{c} \pm s_C)$ = \$32.1M ± 8.58M for b = 480 tons per day

7.3.5

	$\bar{y} \pm s$		
	Europe	USA	Middle East
a, Ash, %	40.8 ± 40.0	26.0 ± 36.8	9.5 ± 19.6
m, Moisture content, %	47.4 ± 31.2	22.0 ± 27.7	47.4 ± 31.2
h, heat value, as collected, kcal/kg	2235 ± 1680	2670 ± 1720	2290 ± 1790

7.3.6

	$\bar{h} \pm s_H$		
	Europe	USA	Middle East
Paper and plastics removed, kcal/person-day	731 ± 535	1109 ± 916	684 ± 238
No paper and plastics removed, kcal/person-day	1062 ± 778	2099 ± 1733	993 ± 345

7.4.2

(a) $\bar{u} \pm s_U$ = \$87,710 ± 14,310

(b) U_U = \$94,870 per ton

7.4.3

$[L_\mu, U_\mu]$ = [87.60, 112.40] for known $\sigma = 20$

$[L_\mu, U_\mu]$ = [84.92, 115.08] for estimated σ of $s = 20$

7.4.4

(a) $\bar{x} + s$ = 10.05 ± 3.75 for $n = 10$

(b) $[L_X, U_X]$ = [7.37, 12.73] for $n = 10$

7.4.5

(a) $\bar{x} + s = 0.410 \pm 0.469$ for $n = 20$

(b) $\left[L_X , U_X \right] = [0.191, 0.629]$ for $n = 20$

7.4.6

(a) $\hat{\rho} = 0.13$

(b) $\left[L_\rho , U_\rho \right] = [0.11, 0.16]$

(c) $\left[P_L(T > 10) , P_U(T > 10) \right] = [0.19, 0.35]$

7.4.7

(a) $E(NW) = \$5000$

(b) $\left[L_{NW} , U_{NW} \right] = [-\$35{,}600, \$41{,}600]$

7.4.8

$n = 384$ samples

Chapter 8

8.3.3(a) $\hat{k} = 0.0915$b

8.3.4(a) $\hat{y} = 25.6 - 0.036x$

8.3.5(a) $\hat{y} = 0.0915x$

8.3.6 $\tilde{\Theta} = 1 - e^{-0.0383d}$

8.4.2

(a) $\hat{k} = -7.38 + 0.95b$

(b) $\left[L_{k_f} , U_{k_f} \right] = [-50.3, 35.5]$

$\left[L_U , U_U \right] = [0.0730, 0.116]$

8.4.3

(a) $\hat{k} = 0.0915b$

(b) $\left[L_C , U_C \right] = [0.0811, 0.1102]$

8.4.4 (b) $P(Y \leq 175) = 0.97$

8.4.5

(a) $\hat{y} = -0.000567x, + 0.00555x_2$

(b) $\left[L_{b_1}, U_{b_1}\right]$ = [–0.00565, 0.00432]

$\left[L_{b_2}, U_{b_2}\right]$ = [0.000531, 0.106]

8.4.7

(b) $\hat{y} = 106 + 9.81x_1 + 19.6x_2$

(c) $\hat{y} = 342 - 9.95x_1$

(d) $\hat{y} = 223 - 9.94x_2$

Index